Statistical Analysis Methods for Chemists

A Software-based Approach

Statistical Analysis Methods for Chemists

A Software-based Approach

William P. Gardiner
*Department of Mathematics, Glasgow Caledonian University,
Glasgow, UK*

THE ROYAL
SOCIETY OF
CHEMISTRY
Information
Services

ISBN 0-85404-549-X

Published by The Royal Society of Chemistry,
Thomas Graham House, Science Park, Milton Road, Cambridge CB4 4WF, UK

Typeset by Computape (Pickering) Ltd, Pickering, North Yorkshire, UK
Printed and bound by Athenaeum Press Ltd, Gateshead, Tyne and Wear, UK

Preface

Chemists carry out experiments to help understand chemical phenomena, to monitor and develop new analytical procedures, and to investigate how different chemical factors such as temperature, concentration of catalyst, pH, storage conditions of experimental material, and analytical procedure used affect a chemical outcome. All such forms of chemical experimentation generate data which require to be analysed and interpreted in respect of the goals of the experiment and with respect to the chemical factors which may be influencing the measured chemical outcome. To translate chemical data into meaningful chemical knowledge, a chemist must be able to employ presentational and analysis tools to enable the data collected to be assessed for the chemical information they contain.

Statistical data analysis techniques provide such tools and as such should be an integral part of the design and analysis of applied chemical experiments irrespective of complexity of experiment. A chemist should therefore be familiar with statistical techniques of both exploratory and inferential type if they are to design experiments to obtain the most relevant chemical information for their specified objectives and if they are to use the data collected to best advantage in the advancement of their knowledge of the chemical phenomena under investigation.

The purpose of this book is to develop chemists' appreciation and understanding of statistical usage and to equip them with the ability to apply statistical methods and reasoning as an integral aspect of analysis and interpretation of chemical data generated from experiments. The theme of the book is the illustration of the application of statistical techniques using real-life chemical data chosen for their interest as well as what they can illustrate with respect to the associated data analysis and interpretational concepts. Illustrations are explained from both exploratory data analysis and inferential data analysis aspects through the provision of detailed solutions. This enables the reader to develop a better understanding of how to analyse data and of the role statistics can play within both the design and interpretational aspect of chemical experimentation. I concur with the trend of including more exploratory

v

data analysis in statistics teaching to enable data to be explored visually and numerically for inherent trends or groupings. This aspect of data analysis has been incorporated in all the illustrations. Use of statistical software enables such data presentations to be produced readily allowing more attention to be paid to making sense of the collected data.

I have tried to describe the statistical tools presented in a practical way to help the reader understand the use of the techniques in context. I have de-emphasised the mathematical and calculational aspects of the techniques described as I would rather provide the reader with practical illustrations of data handling to which they can more easily relate and to show these illustrations based on using software (Excel and Minitab) to provide the presentational components. My intention, therefore, is to provide the reader with statistical skills and techniques which they can apply within practical data handling using real-life illustrations as the foundation of my approach. Each chapter also contains simple, practical, and applicable exercises for the reader to attempt to help them understand how to present and analyse data using the principles and techniques described. Summary solutions are presented to these exercises at the end of the text.

I have not attempted to cover all possible areas of statistical usage in chemical experimentation, only those areas which enable a broad initial illustration of data analysis and inference using software to be presented. Many of the techniques that will be touched on, such as Experimental Design and Multivariate Analysis (MVA), have wide ranging application to chemical problem solving, so much so that both topics contain enough material to become texts in their own right. It has therefore only been possible to provide an overview of the many statistical techniques that should be an integral and vital part of the experimental process in the chemical sciences if chemical experimental data are to be translated into understandable chemical knowledge.

Contents

Acknowledgements

I wish to express my thanks to Professor George Gettinby of the University of Strathclyde who sparked and has continually encouraged my interest in the application of statistics to practical problems and to Dr Charles Barnard of Glasgow Caledonian University who has helped to develop my interest in the chemical applications of statistics. Without these contacts and the many interesting and insightful discussions they have generated, my enthusiasm for the application of statistics would never have reached its current state, namely this book.

Special thanks also to my University colleagues, chemists Dr Ray Ansell and Dr Duncan Fortune, and statistician Dr Willie McLaren, for reviewing the manuscript. Their many constructive suggestions and helpful criticisms have improved the structure and explanations provided in the text.

I also wish to thank the many journals and publishers who graciously granted me permission to reproduce materials from their publications. Thanks are also due to the many chemistry students at Glasgow Caledonian University whose project data I have used.

Thanks are also due to my editor, Janet Freshwater, for the helpful comments made on the draft material and the questions asked concerning manuscript preparation.

Finally and most importantly, I must express my appreciation of the support and reluctant enthusiasm of my spouse, Moira, especially in respect of the long hours spent on the preparation of the manuscript and the nightly click-click of the laptop. Thanks are also due to my long suffering children, Debbie and Greg, who have had to put up with their dad constantly working and typing when they would rather I played with them or let them onto the laptop to write their stories!

Dr Bill Gardiner
Department of Mathematics
Glasgow Caledonian University
January 1997

Glossary

Absolute error The difference between the true and measured values of a chemical response.

Accuracy The level of agreement between replicate determinations of a chemical property and the known reference value.

Alternative hypothesis A statement reflecting a difference or change being tested for (denoted by H_1 or AH).

Analysis of Variance (ANOVA) The technique of separating, mathematically, the total variation within experimental measurements into sources corresponding to controlled and uncontrolled components.

Bias The level of deviation of experimental data from their accepted reference value.

Blocking The grouping of experimental units into homogeneous blocks for the purpose of experimentation.

Boxplot A data plot comprising tails and a box from lower to upper quartile separated in the middle by the median for detecting data spread and patterning together with the presence of outliers.

Chemometrics The cross-disciplinary approach of using mathematical and statistical methods to extract information from chemical data.

Cluster analysis An MVA sorting and grouping procedure for detecting well-separated clusters of objects based on measurements of many response variables.

Confidence interval An interval or range of values which contains the experimental effect being estimated.

Correspondence analysis An MVA ordination method for assessing structure and pattern in multivariate data.

Data reduction The technique of reducing a multivariate data set to uncorrelated components which explain the chemical structure of the data.

Decision rule Mechanism for using test statistic or p value for deciding whether to accept or reject the null hypothesis in inferential data analysis.

Degrees of freedom (df) Number of independent measurements that are available for parameter estimation. It generally corresponds to number of measurements minus number of parameters to estimate.

Descriptive statistics Covers data organisation, graphical presentations, and calculation of relevant summary statistics.

Distance A measure of the similarity or dissimilarity of samples or groups of samples based on shared characteristics with small values indicative of similarity.

Dotplot A data plot of recorded data where each observation is presented as a dot to display its position relative to other measurements within the data set.

Eigenvalues The measure of the importance of a 'derived variable' within MVA methods in terms of what is explains of the structure of multivariate data.

Eigenvectors The coefficient estimates of the response variables within each 'derived variable' in MVA methods.

Error Deviation of a chemical measurement from its true value.

Estimation Methods of estimating the magnitude of an experimental effect within a chemical experiment.

Experiment A planned inquiry to obtain new information on a chemical outcome or to confirm results from previous studies.

Experimental design The experimental structure used to generate chemical data.

Experimental plan   Step-by-step guide to chemical experimentation and subsequent data analysis.

Experimental unit   An experimental unit is the physical experimental material to which one application of a treatment is applied, *e.g.* chemical solution, water sample, soil sample, or food specimen.

Exploratory data analysis (EDA) Visual and numerical mechanisms for presenting and analysing experimental data to help gain an initial insight into the structure of the data.

Factor analysis (FA)   An MVA data reduction technique for detection of data structures and patterns in multivariate data.

Heteroscedastic   Data exhibiting non-constant variability as the mean changes.

Homoscedastic   Data exhibiting constant variability as the mean changes.

Inferential data analysis   Inference mechanisms for testing the statistical significance of collected data through weighing up the evidence within the data for or against a particular outcome.

Location   The centre of a data set which the recorded responses tend to cluster around, *e.g.* mean, median.

Mean   The arithmetic average of a set of experimental measurements.

Median   The middle observation of a set of experimental measurements when expressed in ascending order of magnitude.

Model   The statistical mechanism where an experimental response is explained in terms of the factors controlled in the experiment.

Multiple linear regression (MLR)   The technique of modelling a chemical response Y as a linear function of many characteristics, the X variables.

MVA   A shorthand notation for multivariate methods applied to multi-variable data sets comprising measurements on many variables over a number of samples.

Non-parametric procedures Methods of inferential data analysis, many based on ranking, which do not require the assumption of normality for the measured response.

Normal (Gaussian) The most commonly applied population distribution in statistics and is the assumed distribution for a measured response in parametric inference.

Null hypothesis A statement reflecting no difference between observations and target or between sets of observations (denoted H_0 or NH).

Observation A measured data value from an experiment.

Ordinary least squares (OLS) A parameter estimation technique used within regression modelling to determine the best fitting relationship for a response Y in terms of one or more experimental variables.

Outlier A recorded chemical measurement which differs markedly from the majority of the data collected.

Paired sampling A design principle where experimental material to be tested is split into two equal parts with each part tested on one of two possible treatments.

Parameters The terms included within a response model which require to be estimated for their statistical significance.

Parametric procedures Methods of inferential data analysis based on the assumption that the measured response data conform to a normal distribution.

Power Defines the probability of correctly rejecting an incorrect null hypothesis.

Power analysis An important part of design planning to assess design structure based on chemical differences likely to be detected by the experimentation planned.

Principal component analysis (PCA) An MVA data reduction technique for multivariate data to detect structures and patterns within the data.

Principal components (PC) Uncorrelated linear combinations of the response variables in PCA which measure aspects of the variation within the multivariate data set.

Principal components regression (PCR)   The method of modelling a chemical response on the basis of a PCA solution for measured multi-variate data.

Precision   The level of agreement between replicate measurements of the same chemical property.

p **value**   The probability that a calculated test statistic value could have occurred by chance alone.

Quality assurance (QA)   Procedures concerned with monitoring of laboratory practice and measurement reporting to ensure quality of analytical measurements.

Quality control (QC)   Mechanisms for checking that reported analytical measurements are free of error and conform to acceptable accuracy and precision.

Quantitative data   Physical measurements of a chemical characteristic.

Random error   Causes chemical measurements to fall either side of a target response and can affect data precision.

Randomisation   Reduces the risk of bias by ensuring all experimental units have equal chance of being selected for use within an experiment.

Range   A simple measure of data spread.

Ranking   Ordinal number corresponding to the position of a measurement when measurements are placed in ascending order of magnitude.

Relative standard deviation (RSD)   A magnitude independent measure of the relative precision of replicate experimental data.

Repeatability   A measure of the precision of a method expressed as the agreement attainable between independent determinations performed by a single analyst using the same apparatus and techniques in a short period of time.

Replication   The concept of repeating experimentation to produce multiple measurements of the same chemical response to enable data accuracy and precision to be estimated.

Reproducibility A measure of the precision of a method expressed as the agreement attainable between determinations performed in different laboratories.

Residuals Estimates of model error determined as the difference between the recorded observations and the model fits.

Response The chemical characteristic measured in an experiment.

Robust statistics Data summaries which are unaffected by outliers and spurious measurements.

Sample A set of representative measurements of a chemical outcome.

Significance level The probability of rejecting a true null hypothesis (default level 5%).

Similarity The commonality of characteristics shared by different samples or groups of samples.

Skewness Shape measure of data for assessing their symmetry or asymmetry.

Smoothing The technique of fitting different linked relationships across different ranges of experimental X data in regression modelling.

Sorting and grouping The technique of grouping a multivariate data set into specific groups sharing common measurement characteristics.

Standard deviation A magnitude dependent measure of the absolute precision of replicate experimental data.

Statistical discriminant analysis (SDA) An MVA sorting and grouping procedure for deriving a mechanism for discriminating known groups of samples based on measurements across many common characteristics.

Systematic error Causes chemical measurements to be in error affecting data accuracy.

Test statistic A mathematical formula numerically estimable using experimental data which provides a measure of the evidence that the experimental data provide in respect of acceptance or rejection of the null hypothesis.

Transformation   A technique of re-coding experimental data so that the non-normality and non-constant variance of reported data can be corrected.

Type I error (False positive)   Rejection of a true null hypothesis, the probability of which refers to the significance level of a test of inference.

Type II error (False negative)   Acceptance of a false null hypothesis.

Variability (Spread, Consistency)   The level of variation within collected experimental data in respect of the way they cluster around their 'centre' value.

Weights A measure of the correlation between the response variables and the PCs in PCA in terms of how much contribution the variable makes to the structure explained by the associated PC.

Weighted least squares (WLS)   The technique of least squares estimation for determining the best fitting regression model for a response Y in terms of one or more X variables when replicate data are collected

CHAPTER 1

Introduction

1 INTRODUCTION

Most analytical experiments produce measurement data which require to be presented, analysed, and interpreted in respect of the chemical phenomena being studied. For such data and related analysis to have validity, methods which can produce the interpretational information sought need to be utilised. Statistics provides such methods through the rich diversity of presentational and interpretational procedures available to aid scientists in their data collection and analysis so that information within the data can be turned into useful and meaningful scientific knowledge.

Pioneering work on statistical concepts and principles began in the eighteenth century through Bayes, Bernoulli, Gauss, and Laplace. Individuals such as Francis Galton, Karl Pearson, Ronald Fisher, Egon Pearson, and Jerzy Neyman continued the development in the first half of the twentieth century. Development of many fundamental exploratory and inferential data analysis techniques stemmed from real biological problems such as Darwin's theory of evolution, Mendel's theory of genetic inheritance, and Fisher's work on agricultural experiments. In such problems, understanding and quantification of the biological effects of intra- and inter-species variation was vital to interpretation of the findings of the research. Statistical techniques are still developing mostly in relation to practical needs with the likes of artificial neural networks (ANN), fuzzy methods, and structure–activity relationships (SAR) finding favour in the chemical sciences.

Statistics can be applied within a wide range of disciplines to aid data collection and interpretation. Two quotations neatly summarise the role statistics can play as an integral part of chemical experimentation, in particular:

'The science of Statistics may be defined as the study of chance

1

variations, and statistical methods are applicable whenever such variations affect the phenomena being studied.'[1]
'Statistics is a science concerned with the collection, classification, and
interpretation of quantitative data, and with the application of probability theory to the analysis and estimation of population parameters.'[2]

Both quotations highlight that statistics is a scientifically-based tool
appropriate to all aspects of experimentation from planning through to
data analysis to help understand the data and to provide interpretations
relevant to experimental objectives. Since all chemical measurements
are subject to inherent variation, statistical methods provide a beneficial
tool for explaining the features within the data accounting for such
inherent variation. Knowledge of statistical principles and methods
(strengths as well as weaknesses) should therefore be part of the skills of
any scientist concerned with collecting and interpreting data and should
also be an integral part of design planning. Statistics should not be
considered as an afterthought only to be brought into play after data
are collected, the 'square peg into round hole' syndrome, which is how
the application of statistical methods is often viewed within the scientific
community.

Applied chemical experimentation generally falls into one of three
categories: *monitoring*, *optimisation*, and *modelling*. Monitoring is
primarily concerned with process checking such as monitoring pollution levels, investigating how data are structured, quality assurance of
analytical laboratories, and quality control of experimental material
such as house reference materials (HRMs) and certified reference
materials (CRMs). Optimisation, often through exploratory or investi-
gative studies, comes into play when wishing to optimise a chemical
process which may influenced by a number of inter-related factors.
Instances where such experimentation may occur include optimisation
of analytical procedures, optimisation of a new chemical process, and
assessment of how different chemical factors cause changes to a
chemical outcome. Often, this type of experimentation is based on the
classical *one-factor-at-a-time* (OFAT) approach which is inefficient and
provides only partial outcome information. Through simple and
logical modification of the OFAT structure to ensure that all possible
factor combinations are tested, the experiment can be made more
efficient and provide more relevant information on factor effects, such
as factor interaction. Modelling, on the other hand, attempts to build
a model of the chemical process under investigation for predictive

---

[1] O.L. Davies and P.L. Goldsmith, 'Statistical Methods in Research and Production', 4th Edn.,
Longman, London, 1980, p. 1.
[2] 'Collins English Dictionary', Collins, London, 1979, p. 1421.

purposes. It is often also based on the results obtained from an optimisation experiment where the importance of factors has been assessed and the most important factors retained for the purpose of model building.

I will consider all of these forms of applied chemical experimentation in relation to illustrating how statistical methods can be used to provide understanding and interpretations of collected data in relation to the experimental objectives. Chapter 2 provides an introduction to exploratory data analysis (plots and summaries) and inferential data analysis (hypothesis testing and estimation) for one- and two-sample experimentation. Chapters 3 and 4 extend this introduction into more formal design structures for one-, two-, and three-factor experimentation with Chapter 4 concentrating on factorial designs, the easily implemented alternative to the classical OFAT approach. An introduction to modelling is provided in Chapter 5 through regression methods for the fitting of relationships (linear, multiple) to chemical data. Analytical applications of these techniques in the form of calibration and comparison of two linear equations will also be discussed. Chapter 6 introduces non-parametric methods as alternatives to the previously discussed parametric procedures. Experimental methods pertaining to optimisation are further developed in Chapter 7 through two-level factorial designs for multi-factor experimentation. The final chapter, Chapter 8, introduces multivariate methods appropriate to the handling of multi-response data sets. Many of the techniques and principles that will be explored are often discussed under the heading of *Chemometrics*, the name given to the cross-disciplinary approach of using mathematical and statistical methods to help extract relevant information from chemical data.

The increased power and availability of computers and software has enabled statistical methods to become more readily available for the treatment of chemical data. On this basis, all analysis concepts will be geared to using software (Excel and Minitab) to provide the data presentation on which analysis can be based. The mathematical and calculational aspects of statistics will be ignored, intentionally so, in order to be able to build up a picture of how statistics can turn chemical measurements into chemical information through interpretation of software output. Most of the methods discussed are of classical type though application methods are still developing.

2 WHY USE STATISTICS?

A question often asked by chemists is 'What use and relevance has statistics for chemistry?'. Statistics can best be described as a combination of techniques which cover the design of experiments, the collection of experimental data, the modes of presentation of data, and the ways in which data can be analysed for the information they contain. Statistical concepts, therefore, are relevant to all aspects of experimentation ranging from planning to interpretation. The latter can be subjective (exploratory data analysis, EDA) as well as objective (inferential data analysis, estimation) but the basic rule must be to understand the data as fully as possible by presenting and analysing them in a form whereby the information sought can be readily found.

Examples where statistical methods could be useful include:

- Assessing whether analytical procedures and/or laboratories differ in accuracy (systematic error) and precision (random error) of reported measurements,
- Assessing how changing experimental conditions affect a particular chemical outcome,
- Assessing the effect of many factors on the fluorescence of a chemical complex.

Such experimentation would produce numerical data which would require to be presented and analysed in order to extract the information they provide in respect of the experimental objective. Statistics, through its presentational and interpretational procedures, can provide such means of turning data into useful chemical information which explain the phenomena investigated.

Statistics can also provide tools for designing experiments ranging from simple laboratory experiments to complex experiments for analytical procedures. As assessment of chemical data is becoming more technical and demanding, this, in turn, is requiring chemists to consider more actively design structures that are efficient and to put greater emphasis on how they present and analyse their data using statistical methods. Such pressure encourages a greater awareness of the role of statistics in scientific experimentation[3] together with a greater level of usage.

Use of statistical techniques are advocated by professional bodies such as The Royal Society of Chemistry (RSC) and the Association of Official Analytical Chemists (AOAC) for the handling and assessment of analytical data to ensure their quality and reliability. Statistical procedures appropriate to this type of approach form the basis of the Valid Analytical Measurement (VAM) scheme produced by the Laboratory of the Government Chemist (LGC),[4] the National Measure

---

[3] H. Sahai, *The Statistician*, 1990, **39**, 341.
[4] B. King and G. Phillips, *Anal. Proc.*, 1991, **28**, 125.

ment and Accreditation Service (NAMAS) of the United Kingdom Accreditation Service (UKAS), and other schemes including ISO9000, BS5750, and GLP for the reporting of analytical measurements. These support initiatives and accreditation schemes highlight the importance placed on using statistical methods as integral to chemical data handling.

3 PLANNING AND DESIGN OF EXPERIMENTS

In designing an experiment, we need to have a clear understanding of the purpose of the experiment (objective), how and what response data are to be collected (measurements to be made), and how these are to be displayed and analysed (statistical analysis methods). Design and statistical analysis must be considered as one entity and not separate parts to be put together as necessary. A well planned experiment will produce useful chemical data which will be easy to analyse by the statistical methods chosen. A badly designed and planned experiment will not be easy to analyse even if statistical methods are applied.

Why is design so important? Inadequate designs provide inadequate data, so if we wish to assess experimental objectives properly, we need to design the experiment so that appropriate information for assessing the experimental objective is forthcoming. In addition to the statistical considerations of design structure, we also need to ensure that instruments are properly calibrated, experimental material is uncontaminated, the experiment is performed properly, and the data being recorded are suitable for their intended purpose. We must also ensure that there are no trends in the data through, for example, technicians operating instruments differently and batches of material being non-uniform, and that the influence of unrecognised causal factors is minimised. In comparing the measurement of two analytical procedures, for instance, it would be advisable to use comparable samples of known chemical content or else it may be impossible to know whether the procedures are efficient in their recording of the chemical response. In the chemical sciences, reduction in response variability (improved precision) by appropriate choice of factor levels may also be an important consideration. Cost, problem knowledge, and ease of experimentation also come into play when designing a chemical experiment.

It is therefore important that an experiment be carefully planned *before* implementation and data collection. If necessary, advice on structure and analysis should be sought in order to ensure that choice of, for example, number of samples to be tested, amount of replication to carry out, statistical analysis routine, and software are most appropriate for the experimentation planned. With such advice, experimentation, data collection, and data analysis can readily take place with the experimenter knowing how each part comes together to address the experimental objectives. Planning of experiments is not an easy process but by producing an *experimental plan*, or *protocol* as it is referred to in clinical trials, we can develop a useful step-by-step guide to the experimentation and subsequent data analysis. The four aspects associated with the specification of an experimental plan are as follows:

1 *Statement of the objectives of the investigation*
    This refers to a clear statement of the aims and objectives of the proposed experiment. Specification of the experimental objective(s) is the most important and fundamental aspect of scientific experimentation as it lays down the question(s) the experiment is going to try to answer. This, in turn, helps focus the subsequent planning, data collection, and data analysis towards the goal(s) of the experiment.

2 *Planning of the experiment*
    Planning entails considering how best to implement the experiment to generate relevant chemical responses. It encompasses choice of factors and ranges for experimentation, how such are to be controlled, how the experimental material is to be prepared, choice of most appropriate chemical outcome best reflective of the objective(s), the decision on how many measurements to collect, and how best to display and analyse the outcome measured (the statistical data analysis). These decisions are largely within the control of the experimenter through their knowledge of the subject area and any constraints affecting experimentation such as instrument usage and preparation of experimental material. The statistical data analysis components chosen may also influence these aspects of experimental planning.

3 *Data collection*
    This refers to the physical implementation aspect of the experiment which will produce the chemical response data. Consideration must be given to whether instrument calibration is necessary, how experimental material is to be prepared and stored, how the experiment itself is to be conducted, and how the chosen chemical response is to be recorded through either measurement or observation.

4 *Data analysis*
    Statistical methods, incorporating exploratory and inferential data analysis, should be employed in the analysis of the experimental data though choice of which technique(s) depends on the experimental objective(s), the design structure, and the type of chemical response to be measured. Inferential data analysis (significance

tests and confidence intervals) enable conclusions to be objective rather than subjective, providing an impartial basis for deciding on the chemical implications of the findings. The relevance and chemical validity of these conclusions hinge on the experimenter's ability to translate the statistical findings into useful and meaningful chemical information.

Choice of experimental design structure is important to the conduct of a good experiment. Why design choice is so important in chemical experimentation can be simply summarised through the following points:

- The experiment should have specified objective(s) to assess in respect of the chemical phenomena associated with it.
- The design should be efficient by maximising the information gained using the minimum of experimental effort (small and efficient designs).
- The design should be practical (easy to implement and analyse) and, where practicable, follow a well documented design structure (commonly used design, known structure to data analysis).

These points reinforce the need to consider a planned experiment carefully and to try to use a design structure which will provide requisite data as efficiently as possible. In addition, they show that structure should also be such that the data collected can be analysed using simple and easily understood statistical methods.

Design efficiency can be measured by the experimental error which arises from the variation between experimental units and the variation from the lack of uniformity in the execution of the experiment. The smaller the experimental error the more efficient the design. By introducing various kinds of control such as increasing number of experimental units and number of factors in the experiment, the effects of this uncontrolled variation (noise) may be reduced and the design made more efficient. Statistical methods essentially attempt to separate the signal (the response) from the noise (the error) so that the level of the signal relative to the noise can be measured, large values being indicative of significant explanatory effect and small values providing evidence of chance, and not explanatory, effect.

4 DATA ANALYSIS

Data from chemical experiments can take a variety of forms but the fundamental principle is that they require to be interpreted according to the experimental objectives. Both subjective and objective elements of analysis should be considered, the former corresponding to *exploratory data analysis* (EDA) principles and the latter to *inferential data analysis* principles.

EDA is based on using graphs and charts to present the data visually for interpretation. Graphical modes of presentation vary but the important point is to use one which helps present the data in a form relevant to the data assessment. In conjunction with data plots, it is also useful to present numerical summaries which provide succinct descriptions of the nature of the collected data. Generally, we use a summary of location (mean) and a summary of variability (standard deviation, *RSD*), the former measuring accuracy and the latter precision. For precision, low values signify closely clustered data indicative of good precision (low variability, high consistency). Use of such measures neatly summarises the two important features of most types of chemical data, accuracy and precision.

Inferential data analysis covers those formal statistical procedures (*t* tests, *F* test, confidence intervals) used to draw objective conclusions from the experimental data. They provide the means of assessing the evidence within the data in favour or against the specified experimental objective, *i.e.* the likely meaning of the results. Numerous inference procedures exist with those most appropriate dependent on the objectives of the experiment, the experimental structure, and the nature of the collected data.

When the experiment is complete and the statistical analysis has been carried out, a report can be written highlighting the conclusions reached and recommendations made. As experimentation is usually a sequential process, with one experiment answering some questions and simultaneously posing others, the conclusions reached may suggest a further round of experiments. It must always be remembered that the conclusions reached are only valid for the set of conditions used in the experiment with a wide choice of experimental conditions therefore likely to make the conclusions more applicable.

5 CONSULTING A STATISTICIAN FOR ASSISTANCE

Many experimenters believe that a statistician's role is only to help with the analysis of data once an experiment has been conducted and data collected. This is fundamentally wrong. A statistician can provide assistance with all aspects of experimentation from planning through to data analysis so that the complete experimental process can be constructed sequentially and not as a sequence of hurdles to be crossed when reached with no possibility of recourse to a previous aspect.

Through this co-operation, advice on design structure and consequent data analysis can be developed at the planning stage in association with the experimental objectives enabling the experimentation and data analysis to be better co-ordinated.

When consulting a statistician for advice, background information on the proposed experiment should be provided to help them determine, with the experimenter, the best approach to suit the experimental objectives and experimental constraints. Such information can be provided within a short résumé which should contain information on many of the following points:

1 *Purpose*
   What is the experiment being designed to investigate? What is the proposed experimental structure? Why use this structure?

2 *Previous work*
   Have any previous similar studies been carried out? How did these studies gather their information? Are any of their procedures appropriate to the planned experiment? How do these other studies relate to the planned experimentation?

3 *Response data*
   What type of response data will be collected? How do such data relate to the experimental objectives? Has size of sample been decided upon? How many experiments are planned and is replication necessary?

4 *Data analysis*
   How are the data to be presented and statistically analysed? Why use these methods and not others? What might they show as regards the experimental objectives?

5 *Use of statistical software*
   Can statistical software be used to produce the data presentations and statistical inference results (easily checked using a dummy data set)? Can the software used be tied in with word processing facilities to simplify the report writing and presentation element?

In essence, consideration must be given to as many aspects of the planned experimentation as possible before consulting a statistician for assistance. Ideally, a statistician's role should be to try to guide the experimenter through those aspects associated with data collection, display, and analysis which an experimenter is unsure of, with compromise between what is ideal and what is practical often necessary. Appropriate interpretation of the results in respect of the experimental objectives is the responsibility of the experimenter, taking account of the objectives, the statistical analysis methods employed, and the chemical implications of the results.

6 INTRODUCTION TO THE SOFTWARE

Spreadsheets, such as Excel,[5] and statistical software, such as Minitab,[6] are important tools in data handling. They provide access to an extensive provision of commonly used graphical and statistical analysis routines which are the backbone of statistical data analysis. They are simple to use and, with their coverage of routines, enable a variety of forms of data presentation to be available to the experimenter. Such software can be available across many platforms though most are now utilised within the personal computer environment under Windows.

I have chosen to base usage of software on the spreadsheet Excel and the statistical software Minitab. The latter has been included as Excel has yet to develop fully into a dedicated piece of statistical software and does not cater, by default, for many important statistical analysis tools appropriate to chemical experimentation. Procedures missing include diagnostic checking in ANOVA procedures, two-level factorial designs, 'best' regression procedures for multiple regression modelling, and multivariate methods. Other software, such as SAS,[7] S-Plus,[8] and GLIM,[9] could equally be used but I believe Minitab is best as it is simple to use and compatible in most of its operation with the operational features of Excel. The data presentation principles I will instil can be easily carried forward to other software packages.

In the statistical data analysis illustrations, I will present and explain briefly the dialog window associated with the analysis routine for the software being used to generate analysis output. In addition, in most software outputs presented, I will provide information on how the output was obtained within the software using menu command procedures. Output editing has also occurred to enable the outputs to be better presented than would initially have been the case.

---

[5] Microsoft Excel is a registered trademark of the Microsoft Corporation, One Microsoft Way, Redmond, WA 98052-6399, USA.

[6] Minitab is a registered trademark of Minitab Inc., 3081 Enterprise Drive, State College, PA 16801, USA.

[7] SAS (Statistical Analysis System) is a registered trademark of the SAS Institute Inc., SAS Campus Drive, Cary, NC 27513, USA.

[8] S-Plus is a registered trademark of StatSci Europe, Osney House, Mill Street, Oxford, OX2 0JX, UK.

[9] GLIM is a registered trademark of NAG Ltd, Wilkinson House, Jordan Hill Road, Oxford ON2 8DR, UK.

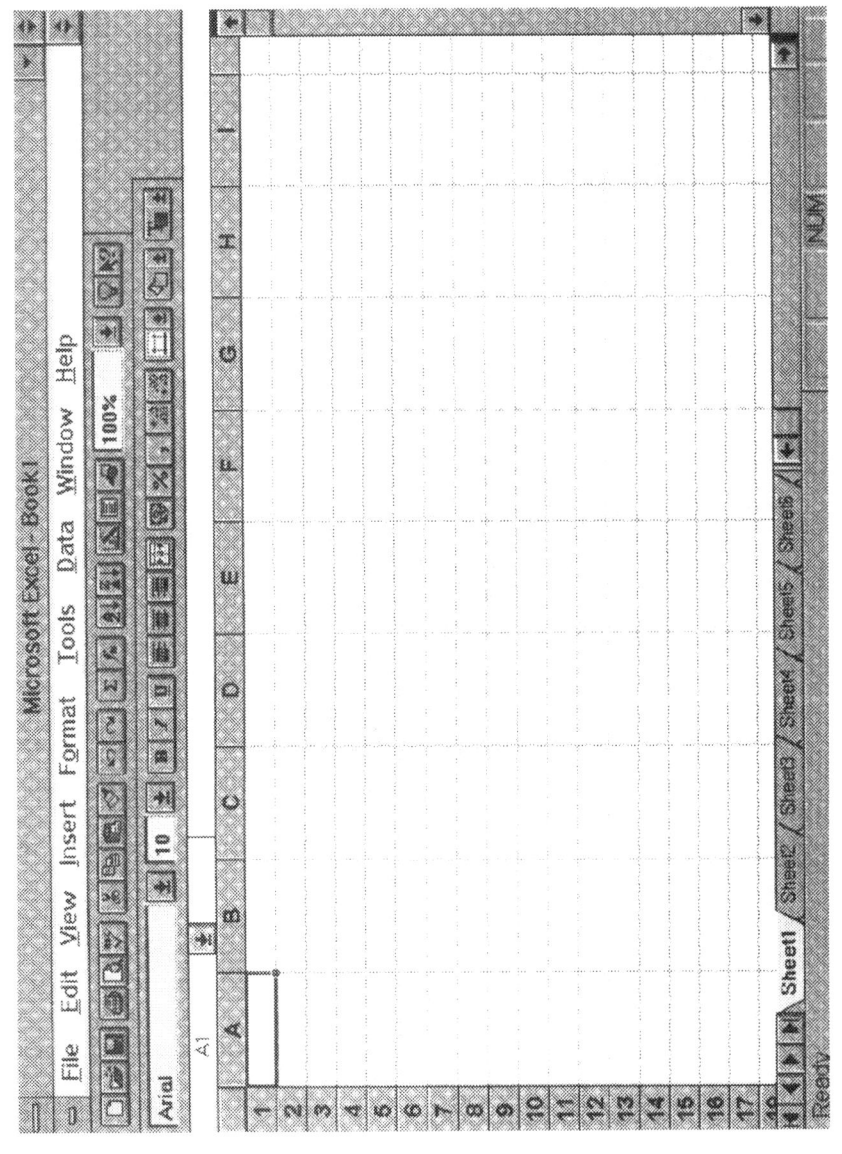

Figure 1.1 *Excel entry screen*

11

6.1 Excel

The information presented in this text refers to Excel release 5.0 where entry will result in a VDU screen similar to that of Figure 1.1 based on a new workbook start-up. The menubar at the top of the screen displays the menu procedures available in Excel. The toolbar buttons below cover most of the same procedures and by resting the mouse pointer on a button (without clicking), a short description of the procedure is displayed with the status bar at the bottom of the screen showing a fuller definition. Description of Excel operation in this text will be based on using the menubar.

The **File** menu contains access to workbook opening and saving and file printing while the **Edit** menu provides access to Excel's copy and paste facilities for copying and moving cells and data plots. **View**

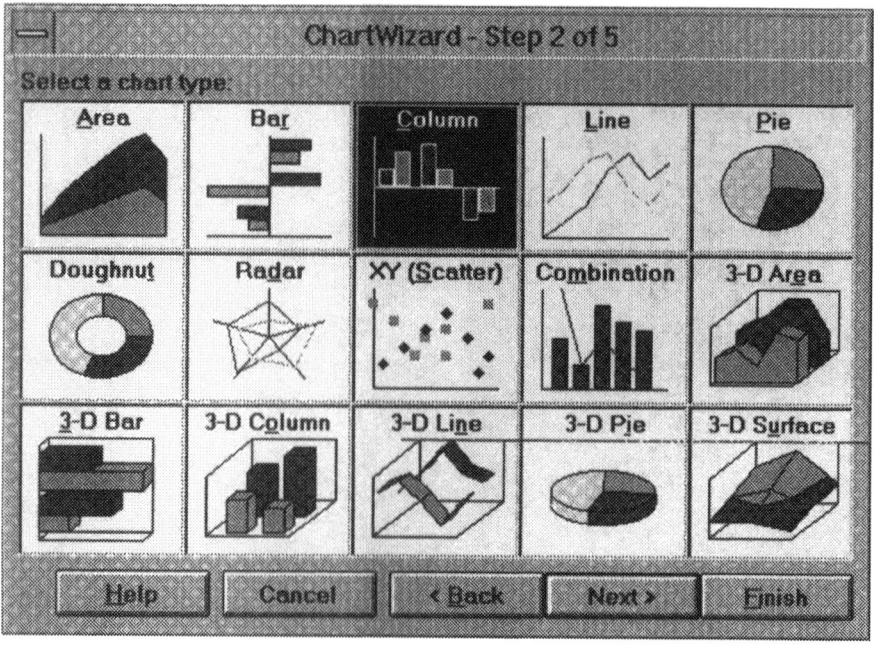

Figure 1.2 *Chart types available in Excel through Chart Wizard*

provides access to ways of providing different worksheet views with **Insert** enabling rows, columns, charts, or range of blank cells to be inserted into the worksheet. Formatting of the cells of the spreadsheet is available through the **Format** menu. The **Data** menu can be accessed for sorting and tabulating data. The **Window** menu allows for movement between workbooks while extensive on-line help and tutorial support can be accessed through the **Help** menu.

Graphical output is produced by clicking the **ChartWizard** button located immediately below the 't' in Data in Figure 1.1, the button looking like a histogram with a smoking cigarette on top. Numerous

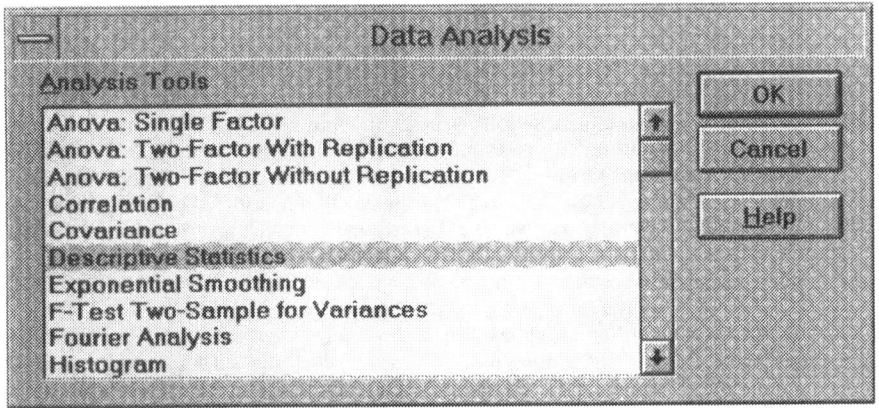

Figure 1.3 *Data analysis tool dialog window in Excel*

graphical presentations, as illustrated in Figure 1.2, are available ranging from simple *X–Y* plots [*XY* (Scatter)] to multi-sample plots (Line). Figure 1.2 corresponds to the choice available in Step 2 of ChartWizard where Step 1 is used to indicate the data to be plotted. The graph required is chosen by checking the appropriate box, checking the box referring to the form of graph of the chosen type required, and following the step-by-step instructions provided.

The default statistical data analysis features of Excel are contained within the Data Analysis commands in the **Tools** menu as shown in Figure 1.3. The tools available range from simple descriptive statistics (Descriptive Statistics) through ANOVA procedures (Anova: Two-Factor With Replication) to regression modelling methods (Regression). If Data Analysis is not available when the Tools menu is chosen, it can be added in by choosing **Tools ▷ Add-ins** and loading in the Analysis ToolPak. If this ToolPak is not available under add-ins, then it will need to be loaded into Excel using a customised installation of the Microsoft Excel Setup program.

The information presented in this text will be kept simple and will be based on using a single worksheet to display all data, charts, and numerical information. A summary of conventions adopted for explaining the menu procedures in Excel is provided in Table 1.1. Only a proportion of the operation potential of Excel for data manipulation will be described.

Workbooks created in Excel can contain data alone or data with statistical data analysis elements (graphical presentations, summaries,

Table 1.1 *Excel conventions*

Item	Convention
Menu command	The menu to be chosen is specified with the first letter in capital form, *e.g.* Tools for access to the Data Analysis tools in Excel.
Emboldened text	Bold text corresponds to either the text to be typed by the user, *e.g.* **Total Nitrogen**, the menu option to be selected, or the button to be checked within a selected option.
Menu instructions	Menu instructions are set in bold with entries separated by a pointer. For example, Select **Tools** ▷ **Data Analysis** ▷ **Descriptive Statistics** ▷ click **OK** means select the Tools menu option, open the Data Analysis sub-menu by clicking the Data Analysis heading, choose the Descriptive Statistics procedure by clicking the Descriptive Statistics heading, and click OK to activate it. This will result in the dialog window for the Descriptive Statistics analysis tool being displayed.
	When such as 'for *Input range*, click and drag across cells **A1:A14**' is presented, this means click the cell A1, hold the mouse down, and drag down the cells to cell A14. This activates the data in cells A1 to A14 for use in the routine selected.
	When such as 'select the **Chart Title** box and enter **Plot of Total Nitrogen Measurements**' is presented, this means click the box specified Chart Title and type the emboldened information in the box. This enables the entered label to be used within the Excel procedure being implemented.
	When such as 'select **Output Range**, click the empty box, and enter **C1**' is presented, this means check the label Output Range, activate the associated box, and enter the emboldened information. This specifies the location in the Excel worksheet where the numerical output to be created is to be placed.

inference elements). The analysis elements can be placed in either the same worksheet (Sheet 1) of the same workbook or in separate worksheets (Sheet 1, Sheet 2, *etc.*). Access to separate worksheets is available by selecting the Sheet tabs at the bottom of the Excel screen (see Figure 1.1). Charts created in the same worksheet as the data are called *embedded charts*.

Data entry in Excel requires that measurements be entered down each column, or along the rows, of the spreadsheet. If appropriate, an optional label can be entered at the head of the column (row). Other textual information concerning the data could also be entered if necessary. Once data have been entered, we check them for accuracy and then save them in a workbook (.xls extension) on disc. When first saving data, we choose the menu commands **File** ▷ **Save As** and fill in the resultant dialog windows accordingly. Subsequent savings, after data update or analysis generation, can be based on the menu commands **File** ▷ **Save**. Such files can be readily imported into many windows-based software systems such as Word and Minitab (data

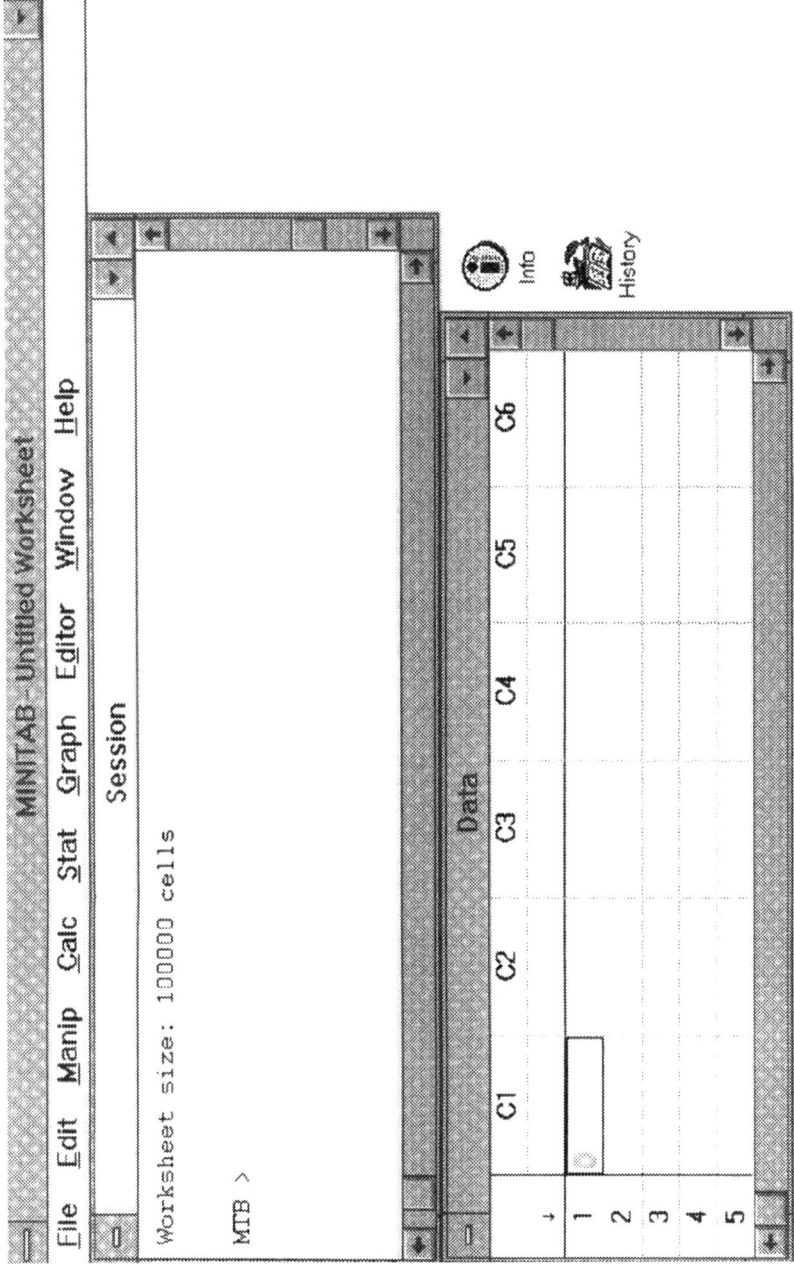

Figure 1.4 *Minitab entry screen*

15

only). Opening of previously saved workbooks is easily achieved through the **File** ▷ **Open** menu commands. The **Edit** ▷ **Copy** and **Edit** ▷ **Paste** facilities provide another means of importing data from Excel to other Windows-based software via the clipboard when operating the software packages simultaneously.

6.2 Minitab

The information presented in this text refers to Minitab release 10.5 Xtra where entry will result in the VDU screen shown in Figure 1.4. Minitab is operated using a sequence of Windows for storage of data and printing of the presentational elements of data analysis. The **Data** window, which is the active window on entry, is the spreadsheet window for data entry. The **Session** window is for entering session commands and displaying, primarily, numerical output. The Info icon provides access to the **Info** window which contains a compact overview of the data and number of observations in the worksheet displayed in the Data window. The History icon accesses the **History** window which displays all session commands produced during a Minitab session. The **Graph** window, which will only be activated when a data plot is requested, displays the professional graphs produced by Minitab. When using Minitab, the Data, Session, and Graph windows are the most commonly accessed.

The menubar at the top of the screen refers to the menu procedures available within Minitab. The **File** menu contains access to worksheet opening and saving, output file creation, file printing, and data display while the **Edit** menu provides access to Minitab's copy and paste facilities for copying output, session commands, and moving cells in the data window. The **Manip** and **Calc** menus provide access to data manipulation and calculation features. Statistical data analysis routines are accessed through the **Stat** menu and choice of appropriate sub-menu associated with the required data analysis. The routines available are comprehensive and cover most statistical data analysis procedures from basic statistics (Basic Statistics), incorporating such as descriptive statistics and two sample inference procedures, through ANOVA procedures (ANOVA), including one-factor and multi-factor designs, to multivariate methods (Multivariate), such as principal component analysis and discriminant analysis. The **Graph** menu, as the name suggests, provides access to Minitab's extensive plotting facilities incorporating simple X–Y plots (Plot), boxplots (Boxplot), dotplots (Character Graphs ▷ Dotplot), interval plots (Interval Plot), and normal plots (Normal Plot). The **Editor** menu enables session command language and fonts to be modified. Movement between open windows

Table 1.2 *Minitab conventions*

Item	Convention
Menu command	The menu to be chosen is specified with the first letter in capital form, *e.g.* Graph for access to the data plotting facilities and Stat for access to the statistical analysis facilities.
Emboldened text	Bold text corresponds to either the text to be typed by the user, *e.g.* **Absorbance**, the menu option to be selected, or the button to be checked within a dialog box window.
Menu instructions	Menu instructions are set in bold with entries separated by a pointer. For example, Select **Stat** ▷ **ANOVA** ▷ **Balanced ANOVA** means select the Stat menu, open the ANOVA sub-menu by clicking the ANOVA heading, and choose the Balanced ANOVA procedure by clicking the Balanced ANOVA heading. This will result in the Balanced ANOVA dialog window being displayed on the screen.
	When such as 'for *Classification variable*, select **Lab** and click **Select**' is presented, this means click the specified Lab label appearing in the variables list box on the top left of the sub-menu window and click the Select button. This procedure specifies that the Lab data are to be used in the routine selected.
	When such as 'select the **Label 2** box and enter **Concentration**' is presented, this means click the box specified Label 2 and type the emboldened information in the box. This enables the entered label to be used within the Minitab procedure being implemented.
	When such as 'for *Display*, select **Data**' is presented, this means check the box labelled Data to specify that the information generated by this choice is to be included in the output created.

can be achieved through use of the **Window** menu while comprehensive on-line help is provided through the **Help** menu.

Minitab can be operated using either session commands or menu commands. I will only utilise the latter for each illustration of Minitab usage. A summary of the conventions adopted for explanation of the menu procedures is shown in Table 1.2. Only those features of Minitab appropriate to the statistical techniques explored will be outlined though this represents only a small fraction of the operational potential of Minitab in terms of data handling and presentation.

In Minitab, data entry is best carried out using the spreadsheet displayed in the Data window though session commands could equally be used. Data are entered down the columns of the spreadsheet (see Figure 1.4) where C1 refers to column 1, C2 column 2, C3 column 3, and so on. Unlike Excel, it is not possible to mix data types in a column by entering a label for the data in the same column as the data as the spreadsheet in Minitab is not a true spreadsheet. Only one type of data can therefore be entered in a column with labelling (up to eight characters) achieved using the empty cell immediately below the column heading.

Once data have been entered, it is always advisable to check them for accuracy and then save them in a worksheet file (.mtw extension) on disc. When first saving data, we use the menu commands **File ▷ Save Worksheet As** and fill in the resultant dialog box accordingly. For subsequent savings, if data are edited or added to, we would use the menu commands **File ▷ Save Worksheet** which will automatically update the data file. Minitab can also save data in a Microsoft Excel format (.xls extension) if desired by changing the 'Save File as Type' entry in the 'Save Worksheet As' dialog window, enabling interchange of data files between Minitab and Excel. The **File ▷ Open Worksheet** menu command enables previously saved worksheets to be retrieved as well as the importing of Excel saved workbooks providing a further means of interchange between Excel and Minitab. The **Edit ▷ Copy** and **Edit ▷ Paste** facilities provide an alternative means of importing data or output from Minitab to other Windows-based software via the clipboard when operating the software packages simultaneously

CHAPTER 2

Simple Chemical Experiments: Parametric Inferential Data Analysis

1 INTRODUCTION

Often, practical chemical experiments are simple in nature involving comparison of, for example, two analytical procedures, two analysts, two temperature levels, or two levels of concentration in respect of a specific chemical outcome. The purpose of such experiments is to compare the two sets of data collected and to assess such for the chemical information they contain. To carry out such assessment appropriately requires the use of statistical methods comprising *graphical presentations*, *numerical summaries*, and *inferential procedures* in order to gain as much chemical information as possible from the measurements accounting for the inherent variability present in them.

The advantages of considering all of these analysis approaches lie in their ability to provide the user with a comprehensive view of the data interpretation from both subjective and objective standpoints. Graphical procedures enable the data to be presented visually while use of numerical summaries provides the means for quantitatively representing the data. Inference procedures provide the techniques for objective inference based on using the collected data to weigh up the evidence associated with the experimental objective. Both of these analysis concepts will be addressed in this chapter and are those advocated by such as the Laboratory of the Government Chemist (LGC) for assessing bias, accuracy, and precision in analytical measurements.[1]

Inference procedures are numerous, providing mechanisms for dealing with a wide variety of experimental situations. They are split into two families: *parametric procedures* and *non-parametric procedures*, the difference being in the assumptions underpinning them, the type of data to which they are most suited, the way in which the data are used, and their associated power. Parametric tests use the data as collected

---

[1] Laboratory of the Government Chemist, Vamstat Software.

(generally quantitative continuous data) and are based on the assumption of normality for the measured response. By contrast, non-parametric procedures are less restrictive in terms of assumptions and in the type of data to which they are best suited, with data re-specification into ranks, or classes, the basis of most. This results in such procedures generally having lower power than comparable parametric ones.

Experimental data refer to the measurements or observations made on experimental material for the purpose of data assessment. The variable measured is generally referred to as the *response variable*. In chemical experimentation, this could be absorbance of a compound, fluorescence intensity of a chemical complex, measurements of analyte concentration, or peak area from a chromatograph.

Such forms of chemical data represent *quantitative* data in that they correspond to physical measurements of a chemical characteristic. Such data can be placed on either the *interval* or *ratio* measurement scale, representative of measurement systems for measurement comparison on the basis of magnitude. Data of this type can be further split into *discrete*, where the response only takes whole number values, *e.g.* number of radioactive particles recorded, and *continuous* where the response can take any value within a given range, *e.g.* analyte concentration, pH of a buffer solution, and titration measurements. Chemical data can also be *qualitative* in nature corresponding to categorical data of *nominal* or *ordinal* type, *e.g.* colour change in litmus paper and level of exposure to harmful chemicals.

All of the analysis mechanisms to be described in this text will be geared to handling quantitative chemical data and will be approached from two angles: use of *descriptive statistics* and use of *inferential statistics* including *estimation*. The former covers data organisation, graphical presentations (simple data plots), and evaluation of relevant summary statistics (mean, standard deviation, *RSD*). Such components are often used as the principal aspects of *Exploratory Data Analysis* (EDA) when analysing and interpreting experimental data. The inferential statistics and estimation aspects cover those statistical procedures (*t* tests, *F* tests, confidence intervals) used to help draw objective inferences from experimental data. They enable the statistical significance of an experimental objective to be formally assessed by weighing up the evidence within the data and using this to assert whether the data agree or disagree with the objective. Emphasis throughout this text will be on illustrating how these mechanisms can be used together to interpret chemical data.

2 SUMMARISING CHEMICAL DATA

To interpret experimental data, it is necessary first to present the data in summary form to provide the base for analysis and interpretation. The EDA tools of graphical and numerical summaries represent such techniques.

2.1 Graphical Presentations

Data gathered from chemical experiments can be difficult to understand and interpret in their raw form as a series of measurements. Data plots represent simple pictorial representations which provide concise summarising of data in simple and meaningful formats. Such forms of visual presentation enable experimental findings and results to be communicated to others readily and simply. The use of graphics to assess data has been advocated for many years. In fact, Florence Nightingale, as far back as the 1850s, believed that statistical analysis was a vital tool in the understanding of data and that graphical displays "affect thro' the eyes what we may fail to convey to the brains of the public through their word-proof ears". The advent of powerful statistical and graphical software has made this aspect of data analysis more widely available.

Which graphical presentations to use, however, depend on the experimental objective(s), the experimentation to be carried out, the nature and level of measurement of the response(s), and the amount of data collected. Several forms of graphical presentation exist with histogram, boxplot, dotplot, scatter diagram (X–Y plot), standard error plot, control charts, time series plots, interaction plots, and quantile-quantile (Q–Q) plots most commonly used in chemical studies. Several of these pictorial data presentations will be used in the examples covered in this text.

One of the most commonly used and simple to interpret data plots is the *dotplot*. In such a plot, each response measurement is presented separately enabling each measurement's position to be displayed. A dotplot, illustrated in Figure 2.1, consists of a horizontal (or vertical) axis covering the range of experimental measurements with each

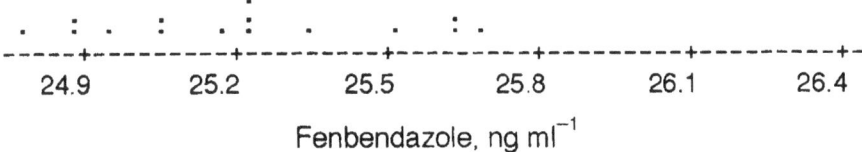

Figure 2.1 Data dotplot

measurement specified by a dot, or suitable symbol, at the requisite value on the axis. A dotplot is particularly useful when comparing measurements from two or more data sets as later examples will show.

Another useful plot within EDA is a *boxplot* shown in Figure 2.2. Boxplots illustrate the spread and patterning of data and are useful for identifying outliers (see Section 4). The plot corresponds to a box, based on the lower quartile (Q_1, 25% of data below this measure) and upper quartile (Q_3, 25% of data above this measure), where the vertical crossbar inside the box marks the position of the median (see Box 2.1). The tails (whiskers) are used to connect the box edges to adjacent values corresponding to measurements lying within the *inner fences* $Q_1 - 1.5(Q_3 - Q_1)$ and $Q_3 + 1.5(Q_3 - Q_1)$.

In respect of using boxplots to assess spread and patterning, the

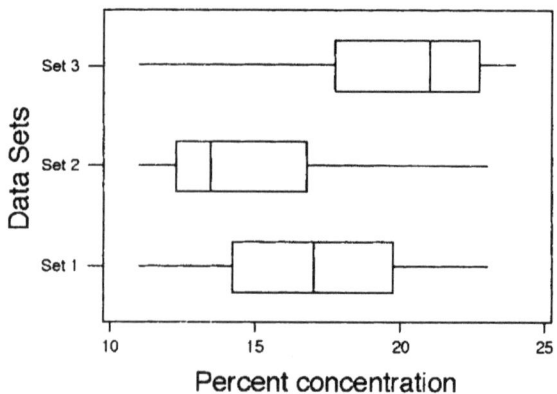

Figure 2.2 *Data boxplots*

illustrations in Figure 2.2 show three likely cases. In Set 1, the median lies near the middle of the box and the tails are roughly equal length indicative of *symmetric* data (mean ≈ median). *Right-skewed* data (mean > median) are represented by Set 2 where the median lies close to the lower quartile with the plot exhibiting a short left tail and long right tail. Set 3 corresponds to *left-skewed* data (mean < median) with median close to upper quartile and the related boxplot showing a long left tail and short right tail.

Figure 2.3 provides a further illustration of the concept of skewness. Right-skewed data are concentrated at the lower end of the scale of measurement though a few larger measurements also appear in the data. Hence, the long tail to the right and the definition right-skew. By contrast, left-skewed data are more concentrated at the top end of the associated scale of measurement though a few numerically smaller measurements also occur in the data. Such data therefore have a long left tail and so are defined as left-skew. A numerical summary of data skewness is provided by the statistic

$$\frac{n\sum(x-\bar{x})^2}{(n-1)(n-2)s^3} \tag{2.1}$$

whereby negative values hint at left-skewed data, positive right-skewed, and values near zero suggest symmetrical data.

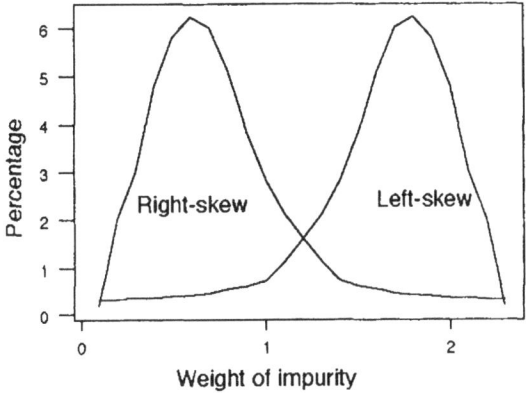

Figure 2.3 *Illustrations of skew-type data*

Example 2.1

The total Kjeldahl nitrogen (TKN) measurements in Table 2.1 were recorded from wastewater samples collected at the outlet of a treatment plant. The measurements, in mg l^{-1} N, were made using the Kjeldahl digestion method.

Exploratory data analysis − data plot

Table 2.1 *Total Kjeldahl nitrogen (TKN) measurements for Example 2.1*

11.6	39.2	4.9	7.3	50.6	9.8	11.6	6.7	42.1	14.4	5.1	48.8	15.9

Source: A. Cerda, M. T. Oms, R. Forteza and V. Cerda, *Analyst (Cambridge)*, 1996, **121**, 13.

An Excel produced dotplot of these data is presented in Output 2.1 using a vertical axis for the experimental measurements. We can see from this presentation of the data that there are two distinct groups of measurements, one around 10 mg l^{-1} N and another between 40 and 50 mg l^{-1} N. No wastewater samples appeared to provide mid-range measurements. The majority of measurements lie at the lower end of the scale indicating right-skewed data. There is also wide variability in the TKN data evidenced by the range of measurements presented. Such groupings and wide variation could be a result of the time of sampling of the discharge.

Example 2.2

Output 2.1 *Plot of the total nitrogen measurements of Example 2.1*

Total nitrogen data in cells A1:A14 (label in A1). Select **ChartWizard**. Click cell **A18**.

Chart Wizard Step 1 of 5: for *Range*, click and drag from cell **A1:A14** ▷ click **Next**.

Chart Wizard Step 2 of 5: select **Line** ▷ click **Next**.

Chart Wizard Step 3 of 5: select **Line chart 3** ▷ click **Next**.

Chart Wizard Step 4 of 5: for *Data Series in*, select **Rows** ▷ click **Next**.

Chart Wizard Step 5 of 5: for *Add a Legend?*, select **No** ▷ select the **Chart Title** box and enter **Plot of Total Nitrogen Measurements** ▷ select the **Axis Titles Value (Y)** box and enter **Total Nitrogen** ▷ click **Finish**.

Double click the plot. Double click the X axis ▷ in the *Format Axis* folder, select **Scale** and click **Value (Y) Axis Crosses between Categories** ▷ click **OK**. Points modified by double clicking the points and modifying the Format Data Series dialog box provided.

Plot of Total Nitrogen Measurements

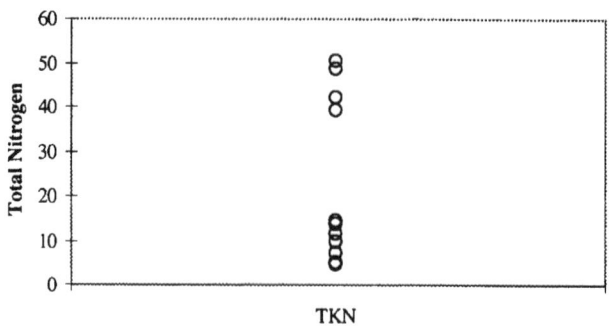

Two catalysts that may affect the concentration of one component in a three-component liquid mixture were investigated. The concentrations, in g per 100 ml, are presented in Table 2.2.

Table 2.2 *Concentration measurements for Example 2.2*

Catalyst A	58.2	57.2	58.4	55.8	54.9	56.3	58.7	56.5	55.3	57.8
Catalyst B	56.3	54.5	57.0	55.3	56.5	55.7	54.2	55.1	56.0	54.8

Source: D.C. Montgomery, 'Design and Analysis of Experiments', 4th Edn., Wiley, New York, 1997: reproduced with the permission of John Wiley & Sons, Inc.

Output 2.2 *Plot of the concentration measurements of Example 2.2*

Catalyst A data are in cells A1:A11 (label in A1) and catalyst B data in cells B1:B11 (label in B1). Select **ChartWizard**. Click cell **C1**.

Chart Wizard Step 1 of 5: for *Range*, click and drag from cell **A1:B11** ▷ click **Next**.

Chart Wizard Step 2 of 5: select **Line** ▷ click **Next**.

Chart Wizard Step 3 of 5: select **Line chart 3** ▷ click **Next**.

Chart Wizard Step 4 of 5: for *Data Series in*, select **Rows** ▷ click **Next**.

Chart Wizard Step 5 of 5: for *Add a Legend?*, select **No** ▷ select the **Chart Title** box and enter **Plot of Concentration Measurements Against Catalyst** ▷ select the **Axis Titles Category (X)** box and enter **Catalysts** ▷ select the **Axis Titles Value (Y)** box and enter **Concentration** ▷ click **Finish**.

Double click the plot. Double click the X axis ▷ in the *Format Axis* folder, select **Scale** and click **Value (Y) Axis Crosses between Categories** ▷ click **OK**. Double click the Y axis ▷ in the *Format Axis* folder, select **Scale** and change **Minimum** to **53.5** ▷ click **OK**. Points modified as per Output 2.1.

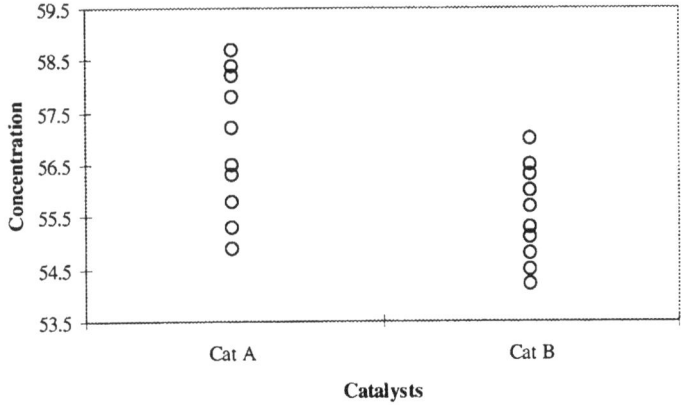

Plot of Concentration Measurements Against Catalyst

Exploratory data analysis — data plot

Output 2.2 contains a simple dotplot presentation of these two data sets
from Excel. We can see from this visual presentation that the concentrations resulting from the two catalysts differ and produce relatively
symmetric data. Catalyst A produces higher concentrations compared
with catalyst B though there is a degree of similarity in the measurements quoted. Additionally, we can see that the measurements are also
more variable for catalyst A (larger range covered) suggesting that
catalyst B appears to result in marginally more consistent component
concentrations.

2.2 Numerical Summaries

Plotting chemical data is the first step in analysis. The second step
requires determination of numerical summary measures which are
thought to be 'typical' of the measurements collected. Two basic forms
of measure are generally used: a *measure of location* and a *measure of
variability*.

The former provides a single measurement value specifying the
position of the 'centre' of the data set, corresponding to the measurement value the data tend to cluster around. The *arithmetic mean*, or
average, and *median* are two such summaries with the mean the most
commonly used for quantitative chemical data. Derivation of both is
explained in Box 2.1. Comparison of a sample mean to a reference
measurement can be used to assess data *accuracy* with similarity
implying good accuracy and dissimilarity inaccuracy, the latter often
affected by systematic errors.

Box 2.1 *Summary measures of location*

Mean refers to the sum of the experimental measurements divided by the
number (n) of measurements collected and is denoted by $\bar{x}$, *i.e.*

$$\bar{x} = \left(\sum x\right)/n \qquad (2.2)$$

Median refers to the value, denoted by Q_2, which splits the data 50:50 with
50% of the measurements below it and 50% above. Obtained by expressing
the measurements in order of magnitude from smallest to largest and
finding the value which splits the data into two equal halves. Corresponds
to the $[(n + 1)/2]$th ordered observation for a sample of n measurements.

A measure of data variability is used to provide a numerical summary of the level of variation present in the measurements, in respect of how the data cluster around their 'centre' value. Variability is often also referred to as *spread, consistency*, or *precision*. Variation measures in common usage for summarising chemical data include *range, standard deviation*, and *relative standard deviation* (RSD), the latter also known as coefficient of variation (CoV). The computational aspects of these measures are described in Box 2.2. Low values tend to signify closely clustered data exhibiting low variability, good consistency, or high precision. In contrast, high values are indicative of wide dispersion in the data, reflective of high variability or imprecise data. Random errors are generally the major cause of imprecision in chemical measurements.

Box 2.2 *Summary measures of variability*

Range specified as the difference between maximum and minimum measurements, *i.e.* (maximum − minimum). Simple to compute but ignores most of the available data.
Standard deviation provides a summary of variability which reflects how far the measurements deviate from the mean measurement. The sample standard deviation, for the case of n measurements, is expressed as

$$s = \sqrt{\frac{\sum(x - \bar{x})^2}{n - 1}} = \sqrt{\frac{\sum x^2 - \frac{(\sum x)^2}{n}}{n - 1}} \qquad (2.3)$$

where numerator refers to *corrected sum of squares* (always non-negative). *Relative standard deviation (RSD)* is simply the standard deviation expressed as a percentage of the mean, *i.e.*

$$RSD = 100(s/\bar{x})\%. \qquad (2.4)$$

The reason behind dividing by $(n-1)$ in the standard deviation calculation (2.3) lies with the concept called *degrees of freedom*. It is known that the sum of the deviations of each measurement from their mean will always be equal to zero, *i.e.* $\sum(x - \bar{x}) = 0$. If we know $(n-1)$ of these deviations, this constraint means the nth deviation must be fixed. Therefore, in a sample of n observations, there are $(n-1)$ independent pieces of information, or degrees of freedom, available for standard deviation estimation. Hence, the use of a divisor of $(n-1)$ in the calculation. Summary statistics can be generated easily in Excel

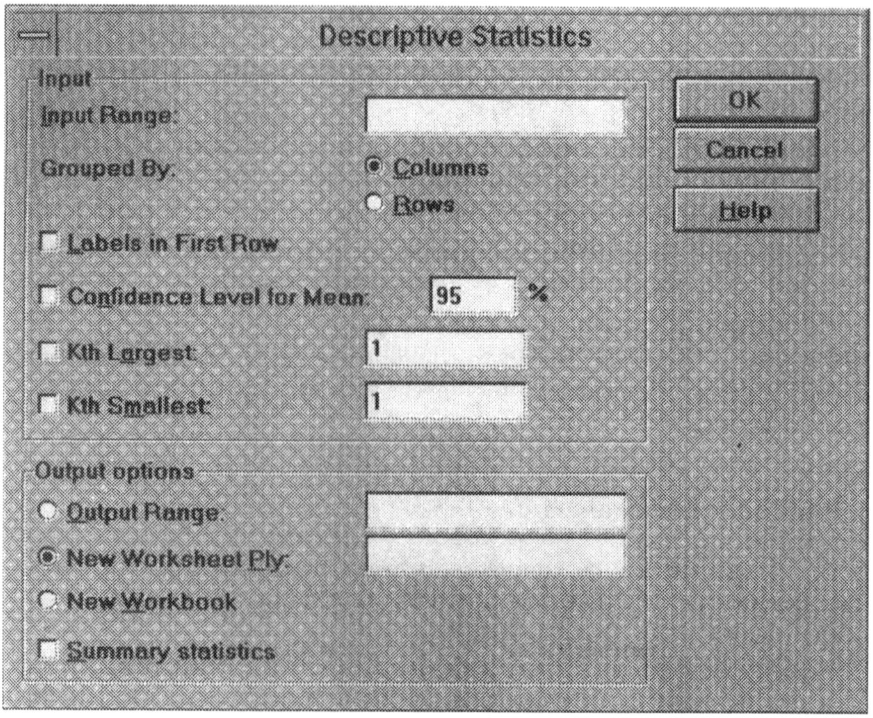

Figure 2.4 *Descriptive statistics dialog window in Excel*

through use of the Descriptive Statistics analysis tool, the dialog window for which is shown in Figure 2.4. The blank boxes shown require to be either filled in, left blank, or checked where appropriate.

Squaring the standard deviation (2.3) provides the data *variance*. The standard deviation is a measure of the *absolute precision* of data, the value of which depends on the magnitude of the data. The *RSD* provides a measure of the *relative precision* of data and is a magnitude independent measure of data variability. In this sense, the *RSD* is more useful when comparing different data sets which may differ in magnitude or scale of measurement. In laboratory instrumentation, for example, an *RSD* target of 5% is often used for the reported measurements.

Example 2.3

Using the total nitrogen measurements presented in Example 2.1, we will illustrate determination of relevant summary statistics both manually and using Excel.

Manual determination

We have $n = 13$ measurements of total nitrogen. Using these measurements, we first calculate $\sum x$ as $39.2 + 50.6 + \ldots + 15.9 = 268$ and $\sum x^2$ as $39.2^2 + 50.6^2 + \ldots + 15.9^2 = 9224.38$.

Mean: Using $\sum x = 268$ and $n = 13$, the mean (2.2) becomes $\bar{x} = 268/13 = 20.62$ mg l^{-1} N.

Median: Ordering the measurements with respect to magnitude provides the sequence 4.9 5.1 6.7 7.3 9.8 11.6 11.6 14.4 15.9 39.2 42.1 48.8 50.6. As $n = 13$, the median will be the $(13 + 1)/2 = 7$th observation which is 11.6 so $Q_2 = 11.6$ mg l^{-1} N. The median differs markedly from the mean indicating skew data, and as mean exceeds median, the data appear right-skewed as highlighted in Example 2.1.

Range: As maximum $= 50.6$ and minimum $= 4.9$, the range will be $50.6 - 4.9 = 45.7$ mg l^{-1} N.

Standard deviation: Using the summations found initially, the standard deviation (2.3) becomes

$$ s = \sqrt{\frac{9224.38 - {268^2}/{13}}{13 - 1}} = \sqrt{308.288} = 17.56 \text{ mg l}^{-1}\text{N}. $$

RSD: The *RSD* (2.4) will be $100(17.56/20.62) = 85.2\%$ which is very high indicating large variability in the TKN measurements.

Excel determination

The summaries generated by Excel's Descriptive Statistics tool for the total nitrogen measurements are presented in Output 2.3. Routine procedures in most software generate more statistical summaries than we would normally determine manually. Only a subset of these summaries is generally used to summarise a data set.

The figures presented in Output 2.3 for mean, median, range, and standard deviation agree with those determined manually providing for the same data interpretation. The skewness coefficient is specified as 0.91 ('Skewness') highlighting that the data appear right-skewed as hinted at earlier. The *RSD* is not produced by default so requires manual derivation.

Example 2.3 has been included primarily to illustrate manual derivation of summary statistics. All future illustrations of data handling will be oriented towards using software (Excel or Minitab) to provide the summary measures. Obviously, chemical experiments can often involve collection of more than one sample of data. Comparison of summaries is straightforward in these cases, with means compared on basis of magnitude differences, and variabilities compared on basis of

Output 2.3 *Summary statistics for the total nitrogen measurements of Example 2.1*

Data in cells A1:A14, label in cell A1.

Select **Tools** ▷ **Data Analysis** ▷ **Descriptive Statistics** ▷ click **OK** ▷ for *Input Range*, click and drag across cells **A1:A14** ▷ select **Labels in First Row** ▷ select **Output Range**, click the empty box, and enter **C1** ▷ select **Summary statistics** ▷ click **OK**.

	TKN
Mean	20.62
Standard Error	4.87
Median	11.60
Standard Deviation	17.56
Sample Variance	308.29
Kurtosis	−1.04
Skewness	0.91
Range	45.70
Minimum	4.90
Maximum	50.60
Sum	268.00
Count	13
Confidence Level (95%)	10.61

multiplicative difference with a ratio in excess of 2:1 generally indicative of important differences in data variability.

Example 2.4

Refer to the concentration measurements presented in Example 2.2. In this example, we will only illustrate Excel approaches to summary determination.

Excel determination

Summary statistics can be generated easily in Excel through the Descriptive Statistics analysis tool as in Example 2.3. The summaries generated by Excel for the concentration measurements of Example 2.2 are presented in Output 2.4.

Mean: The mean (2.2) is specified to be 56.91 for catalyst A and 55.54 for catalyst B. These summaries differ by about 1.4 suggesting higher concentrations, on average, for catalyst A. However, the importance of such a difference cannot be fully interpreted without additional infor-

Output 2.4 *Summary statistics for the concentration measurements of Example 2.2*

Data in cells A1:B11, labels in cells A1 and B1.
  Select **Tools** ▷ **Data Analysis** ▷ **Descriptive Statistics** ▷ click **OK** ▷ for *Input Range*, click and drag across cells **A1:B11** ▷ select **Labels in First Row** ▷ select **Output Range**, click the empty box, and enter **D1** ▷ select **Summary statistics** ▷ click **OK**.

	Cat A	*Cat B*
Mean	56.91	55.54
Standard Error	0.43	0.29
Median	56.85	55.5
Standard Deviation	1.349	0.916
Sample Variance	1.819	0.838
Kurtosis	−1.48	−1.08
Skewness	−0.12	0.09
Range	3.8	2.8
Minimum	54.9	54.2
Maximum	58.7	57
Sum	569.1	555.4
Count	10	10
Confidence Level (95%)	0.84	0.57

mation on the size of difference which would be reflective of chemical importance.

Range: For catalyst A, range is stated to be 3.8 while for catalyst B, it is 2.8.

Standard deviation: The standard deviation (2.3) is specified to be 1.349 for catalyst A and 0.916 for catalyst B.

RSD: The *RSD*s (2.4) are $100(1.349/56.91) = 2.4\%$ for catalyst A and $100(0.916/55.54) = 1.6\%$ for catalyst B. These summaries suggest that the concentration measurements for catalyst A are marginally less consistent (higher *RSD*) than those of catalyst B as the ratio of *RSD*s is approximately 1.5:1.

Skewness: The skewness coefficient (2.1) is near zero for both data sets indicating symmetric data as suggested in the plot analysis of Example 2.2.

  In conclusion, we could say that concentrations appear to differ with catalyst. Catalyst A appears to produce marginally higher and more variable concentrations compared to catalyst B.

Exercise 2.1

In the spectrophotometric determination of sulphate ions in tap water, the following data were obtained using method blank measurement. The data refer to the analytical signal in g × absorbance units. Carry out an exploratory analysis on these data to assess how length of time standing before filtration affects the analytical signal.

---

10 minutes standing before filtration

30.93 22.04 21.27 21.40 21.87 25.32 30.73 33.74 28.80 24.69 28.85 27.33 21.28 31.74 22.84

24 hour standing before filtration

11.05 11.43 11.88 11.50 11.57 11.45 10.86 11.25 11.20 10.99 10.97 11.58 11.02 11.40 11.21

Source: F. Torrades, J. Garcia and M. Castellvi, *Analyst (Cambridge)*, 1995, **120**, 2417.

In analytical experimentation, *accuracy* and *precision* are the two important components underpinning the quality of reported analytical measurements. Ideally, all such data should be both accurate (close to target, free of systematic error) and precise (little variation, free of random error) though often different combinations of these characteristics will occur. The dotplots in Figure 2.5 refer to the replicate measurements of fenbendazole reported by four analysts for milk samples spiked with 25.6 ng ml^{-1} fenbendazole. Analyst A's results are very close together and near to the known target representative of data which can be said to be both accurate and precise. By contrast, B's results are more variable though they are evenly distributed around the target indicative of accurate but imprecise data. The measurements reported by analyst C have low spread but are above the target (inaccurate and precise) while analyst D has reported results which are primarily below the target and are also wide ranging (inaccurate and imprecise).

Chemical measurement precision is often also explained in terms of *reproducibility* and *repeatability*. Repeatability is essentially a measure of *within-run precision* based on replicate measurements being collected under identical experimental conditions. Reproducibility measures *between-run precision* and would be an appropriate measure of variability if measurements were collected under different experimental

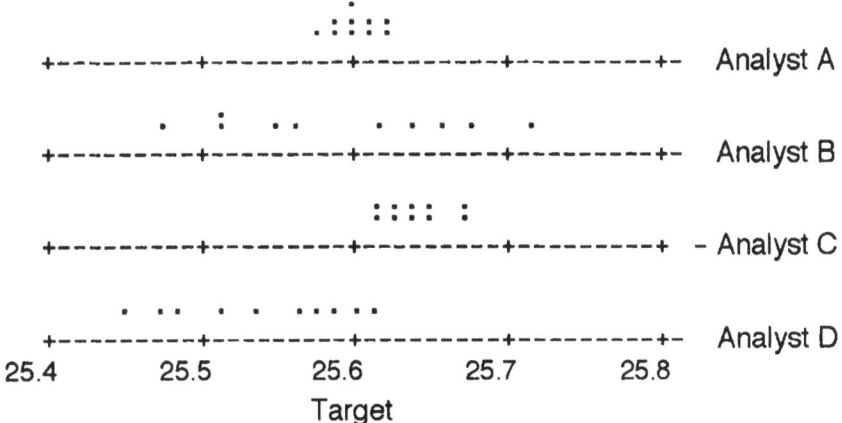

Figure 2.5 *Dotplots illustrating accuracy and precision*

conditions using different analytical procedures. In inter-laboratory comparisons, such as collaborative and co-operative trials where both of the aforementioned forms of measurement reporting occur, it is expected that reproducibility will generally exceed repeatability.

3 THE NORMAL DISTRIBUTION WITHIN DATA ANALYSIS

The *normal*, or *Gaussian*, *distribution* is the most important of all the probability distributions which underpin statistical reasoning, where distribution refers to the theoretical pattern (relative frequency histogram) expected to be exhibited by the random phenomena being measured. The normal distribution is the foundation for all parametric inference procedures through the assumption that the measured response is normally distributed. A knowledge of its shape and pattern is therefore useful as a base to the understanding of inferential data analysis. The normal distribution function is specified by the expression

$$f(x) = \frac{1}{\sqrt{2\pi}\sigma}\, e^{-\frac{(x-\mu)^2}{2\sigma^2}} \qquad (2.5)$$

where μ *(mu)* is the mean and σ *(sigma)* the standard deviation of the population of experimental measurements, and e represents the exponential constant. Graphically, this function describes a bell-shaped curve symmetrical about the mean μ with equal tails to left and right. This distribution has been used to describe, either exactly or approximately, many different chemical responses such as pH of soil, concentration of albumin in blood sera, analyte concentration, and peak areas in HPLC.

Figure 2.6 provides a diagrammatic representation of equation (2.5) for a mean $\mu = 100$ and three values of standard deviation σ (5, 10 and 15). For $\sigma = 5$, the curve is sharply peaked with short tails signifying that data with such a pattern would be representative of a highly precise experiment (low variability, good precision). Increasing variability flattens the peak and lengthens the tails corresponding to data with greater variability and less precision. Data conforming to any of the patterns exhibited in Figure 2.6 are described as *symmetrical* and generally have similar mean and median.

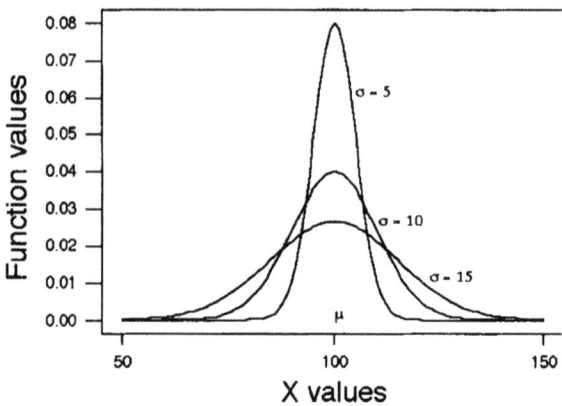

Figure 2.6 *Normal distribution: $\mu = 100$; $\sigma = 5$, 10 and 15*

Our interest in the normal distribution, in parametric inferential data analysis, lies in its use as the assumed distribution for the measured experimental response and the role this plays in the underlying theory of the parametric statistical test procedure being implemented.

4 OUTLIERS IN CHEMICAL DATA

Summary statistics like the mean and standard deviation are sensitive to *outliers*, measurement values untypical of the majority of measurements. Inclusion of outliers can affect data interpretation while exclusion may be justified on the basis of knowledge of why they differ from the majority. However, inclusion of outliers is often necessary in order to assess the collected data fully, and also because agencies monitoring experimental analyses may not accept omission of data measurements, e.g. Federal Drug Administration (FDA) policy on clinical trial data.

The point is that we should be aware of how to detect outliers and of their effect on data interpretation.

Numerical detection of outliers can be based on evaluating the z score

$$z = (\text{value} - \text{mean})/\text{standard deviation} \qquad (2.6)$$

for each untypical measurement and comparing it with ± 2 or ± 3. If a z score lies between 2 and 3 numerically, then the corresponding measurement can be considered a *possible*, or *suspect*, outlier while a measurement with score exceeding 3 suggests a *probable*, or *highly suspect*, outlier. Graphical detection of outliers can be based on use of *boxplots* (see Section 2.1) through the inclusion of *outer fences* located at a distance $3(Q_3 - Q_1)$ below the lower quartile and above the upper quartile. Observations between the inner and outer fences of the boxplot are considered *possible* outliers while values beyond the outer fences are classified as *probable* outliers. In software produced boxplots, outliers are generally highlighted by a '*' beyond the tails. Example 2.5 will be used to discuss the detection and distortive effects of outliers.

Example 2.5

Table 2.3 contains an analyst's twelve replicate determinations of the percent potash content of a sample of fertiliser. Checking these data, we see that the measurement 15.0% appears different from the rest which lie between 15.3% and 15.6%. We need, therefore, to assess whether this measurement is an outlier and to investigate the effect it has on the data summaries.

Table 2.3 *Percent potash measurements for Example 2.5*

15.5	15.0	15.6	15.4	15.5	15.4	15.5	15.3	15.6	15.4	15.3	15.5

Source: C.J. Brookes, I.G. Betteley and S.M. Loxston, 'Fundamentals of Mathematics and Statistics for Students of Chemistry and Allied Subjects', Wiley, Chichester, 1979: reproduced with the permission of John Wiley & Sons, Ltd.

Detection of outlier

The measurement of 15.0 has a z score (2.6) of $z = (15.0 - 15.417)/0.164 = -2.54$ within the range of it being classified as a possible outlier. A boxplot of the potash data presented in Figure 2.7 clearly highlights this measurement as a possible outlier (presence of * symbol). The remainder of the measurements are similar as shown by the compactness of the boxplot. Both numerical and graphical detection methods clearly highlight the 15.0% measurement as a possible outlier.

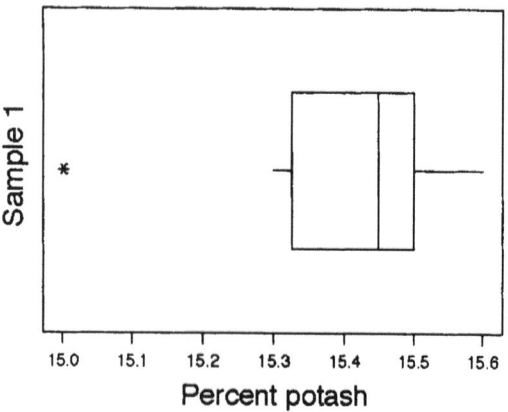

Figure 2.7 *Boxplot of percent potash measurements for Example 2.5*

Summary statistics

Summaries of the potash measurements with and without the possible outlier are presented in Table 2.4. Inclusion of the value 15.0 decreases the mean marginally due to the outlier being a low lying observation. The standard deviation is higher with the outlier as its inclusion is effectively increasing the range, and therefore decreasing the precision, of the recorded measurements. For this case, only the standard deviation is markedly affected by the outlier suggesting less precision in the analytical measurements than is actually occurring in the majority.

Table 2.4 *Data summaries for Example 2.5*

	Mean	Median	Standard deviation	Minimum	Maximum
With outlier	15.417	15.45	0.164	15.0	15.6
Without outlier	15.455	15.50	0.104	15.3	15.6

From Example 2.5, we can see that both outlier detection methods arrive at the same conclusion on the unusual observation of 15%. The differences shown in the summaries highlight the effect outliers can have and the importance of detecting such values prior to analysis. However, if there is more than one outlier in a data set, the boxplot and fences method is better for detection purposes as they are unaffected by the presence of untypical observations. An inflated standard deviation value caused by an outlier(s) can affect use of the z score method so making it not as reliable as the boxplot method for outlier detection.

In the presence of outliers, the expression $1.5*MAD$ can be used to provide an approximate standard deviation estimate for the data. MAD refers to *median absolute difference* and is the median of the absolute differences of each measurement from the group mean. For Example 2.5, MAD is 0.083, resulting in a standard deviation estimate of 0.1245 close to the result obtained when the possible outlier is excluded. This form of variability estimate is *robust* as it is unaffected by outliers and spurious data, though it only describes the 'good' part of the data. Dixon's Q test[2] and Grubb's test provide formal statistical tests for outliers. Information on how to deal with outliers within analytical data through the use of robust statistics is also available.[3,4]

5 BASIC CONCEPTS OF INFERENTIAL DATA ANALYSIS

Inferential data analysis by means of *hypothesis*, or *significance, testing* is based on the principle of attempting to model an experimental response as an additive model of the explanatory factors and an error term. The testing aspect assesses how the variation explained by the controlled components differs from that associated with the error. If controlled variation exceeds error variation markedly, the evidence within the data is pointing to a particular effect being the reason for the difference and that the effect is potentially of chemical significance. If controlled variation does not exceed error, then the differences are likely to be random and not chemical. Significance testing therefore tests whether result differences are significant or are likely to be random variations in the measurements. Parametric inference procedures, which this text primarily discusses, are based on the assumption that the error component is normally distributed. It is this assumption which provides the foundation for the underlying theory of statistical inference and parameter estimation.

Significance testing of experimental data rests on the construction of *statistical hypotheses* which describe the likely responses to the experimental objective being assessed. Two basic hypotheses exist, the *null* and the *alternative* (*experimental, research*). The null hypothesis, denoted H_0 or NH, is always used to reflect no difference either between the observations and their target or between sets of observations. The alternative, denoted H_1 or AH, describes the difference or system change being tested for. These hypotheses are often expressed in terms of the population parameter being tested (parameterised format), μ for mean and σ^2 for variability, though it is feasible to express hypotheses in simpler ways as Example 2.6 will show. It should be noted, as Fisher[5]

[2] J.C. Miller and J.N. Miller, 'Statistics for Analytical Chemistry', 3rd Edn., Ellis Horwood, Chichester, 1993, pp. 62–65.
[3] Analytical Methods Committee, *Analyst (London)*, 1989, **114**, 1693.
[4] Analytical Methods Committee, *Analyst (London)*, 1989, **114**, 1699.
[5] R.A. Fisher, 'Statistical Methods for Research Workers', 14th Edn., Hafner Publishing Company, Darien, Connecticut, 1970.

pointed out, that the null hypothesis is never proved or established but only disproved with experiments being used to give the facts a chance of disproving H_0.

A further aspect, particularly in one and two sample experimentation, is that of the *nature of the test*. This reflects whether the question to be assessed concerns a general directional difference or a specific directional difference. The former is signified by the presence of the not equal to symbol ($\neq$) in H_1 (two-sided alternative, two-tailed test) and means that both < and > are being tested. Presence of solely less than (<) or solely greater than (>) corresponds to specific directional difference (one-sided alternative, one-tailed test).

Example 2.6

A study is to be set up to assess whether the rate of diffusion of CO_2 through two different soils, S1 and S2, differs in respect of both mean rate and variability in rate. Any differences detected may be reflective of differences in chemical content. As the investigation is to assess both mean and variability, hypotheses relevant to each case need to be constructed.

Mean

The null hypothesis, as always, reflects no difference, *i.e.* H_0: no difference in the mean rate of diffusion of CO_2 ($\mu_{S1} = \mu_{S2}$). Assessment of a difference in rate between the two soils reflects a general directional difference corresponding to a two-sided alternative of the form H_1: difference in the mean rate of diffusion ($\mu_{S1} \neq \mu_{S2}$). The alternative is specified as two-sided because the difference being tested for corresponds to either $\mu_{S1} < \mu_{S2}$ or $\mu_{S1} > \mu_{S2}$ with $\mu_{S1} \neq \mu_{S2}$ summarising these two cases.

Variability

As with the mean, the associated null hypothesis corresponds to no difference in the variabilities, *i.e.* H_0: no difference in variability in the rate of diffusion of CO_2 ($\sigma^2_{S1} = \sigma^2_{S2}$). The question posed is concerned with assessing if there is a difference in the variability of the two rates of diffusion, *i.e.* whether $\sigma^2_{S1} < \sigma^2_{S2}$ or $\sigma^2_{S1} > \sigma^2_{S2}$. This again reflects a general directional difference requiring specification of a two-sided alternative of the form H_1: difference in the variability of the rate of diffusion ($\sigma^2_{S1} \neq \sigma^2_{S2}$).

Presence of the equality sign in null hypothesis specification *does not imply* exact equality. It simply means that the experimental measurements are near target or are similar enough to reflect that no detectable difference of importance can be found. Hypotheses for specific difference testing may be expressed differently as H_0: $\mu_A \leq \mu_B$ and H_1: $\mu_A >$ μ_B though practical definition of the hypotheses is unchanged. Another form of null hypothesis that could be considered is H_0: $\mu_A - \mu_B = 5$ if assessment of a specified level of difference between two groups A and B was appropriate.

Exercise 2.2

An analysis of two analytical procedures for determining copper in blood serum is proposed to examine the assertion that the two methods differ in their responses and precision. Construct statistical hypotheses for this investigation.

The second stage of statistical inference concerns the determination of a *test statistic*. Test statistics are simple formulae based on the sample data which provide a measure of the evidence that the experimental data provide in respect of acceptance or rejection of the null hypothesis. Large values are generally indicative that the observed difference is unlikely to have occurred by chance and so the less likely the null hypothesis is to be true, while small values provide the opposite interpretation. Essentially, significance testing is a decision making tool with a decision necessary on whether to accept or reject the null hypothesis based on the evidence provided by the experimental data.

Such a decision could be in error as in inferential data analysis; we can never state categorically that a decision reached is true, only that it has not been demonstrated to be false. Such errors in significance testing are referred to as *Type I error*, or *false positive*, and *Type II error*, or *false negative*, with their influence on the decision reached summarised in Table 2.5. The probability of a Type I error is generally specified as α, the *significance level of the test*. Given that significance level is a measure of error probability, all statistical testing is based on usage of low significance levels such as 10%, 5%, and 1% with 5% the generally accepted default. If we reject the null hypothesis at the 5% level, we specify that the difference tested is significant at the 5% level and that there is only a 5% probability that the evidence for this decision was likely to be due to chance alone. In other words, the evidence points to the detected difference being statistically important which will hopefully also reflect chemical importance.

Implementation of all statistical inference procedures for the objective assessment of experimental data are based on trying to minimise both Type I and Type II errors. In practice, the probabilities of Type I

Table 2.5 *Errors in inferential data analysis*

| | | Decision reached | |
		accept H_0	reject H_0
True	H_0 *true*	no error	Type I
result	H_0 *false*	Type II	no error

and Type II errors, denoted α and β respectively, are inversely related such that, for fixed sample size, decreasing α causes β to increase. Generally, we specify the significance level of the test α and select β to maximise the power of the test, specified as $(1 - \beta)$.

In scientific journals, inferential results are often reported using the *p value*, the probability that the observed difference could have occurred by chance alone. The *p* value provides a measure of the weight of evidence in favour of acceptance of the null hypothesis, large values indicative of strong evidence and small values indicative of little or no evidence. Often, it is expressed in a form such as $p < 0.05$ (*null* hypothesis rejected at 5% significance level) meaning that there is, on average, a 1 in 20 chance that a difference as large as the one detected was due to chance alone. In other words, there is little or no evidence in favour of acceptance of the null hypothesis and the evidence points to the existence of a significant and meaningful difference.

Box 2.3 *General decision rule mechanisms for inferential data analysis*

Test statistic and critical value approach. This approach involves comparing the computed test statistic with a *critical value* which depends on the inference procedure, the significance level, the type of alternative hypothesis (one- or two-sided), and, in many cases, the number of measurements. The latter is generally reflected in the *degrees of freedom* of the test statistic corresponding to the amount of sample information available for parameter estimation. Critical values for inference procedures are extensively tabulated as evidenced in the Tables in Appendix A.

For a two-sided alternative (two-tailed test), the decision rule is
   critical value 1 < test statistic < critical value 2 ⇒ accept H_0.
For a one-sided alternative (one-tailed test), we use
   test statistic > critical value ⇒ accept H_0
if H_1 contains the inequality <, *e.g.* $\mu < \mu_0$ and $\mu_1 < \mu_2$, and
   test statistic < critical value ⇒ accept H_0
if H_1 contains the inequality >, *e.g.* $\mu > \mu_0$ and $\mu_1 > \mu_2$. Exceptions to these rules do exist, however.

p value and significance level approach. An estimate of the *p* value generally appears as part of software output. The associated decision rule, which is dependent only on the significance level of the test, is
   p value > significance level of the test ⇒ accept H_0.

Since both test statistic and *p* value provide measures in respect of acceptance of the null hypothesis, we can use either as a means of constructing general rules for deciding whether to accept or reject the null hypothesis in inferential data analysis. Such rules are explained in Box 2.3.

Acceptance of H_0 indicates that the evidence within the data suggests that the experimental measurements are near target or are similar to one another. It does not mean that we have demonstrated equality of observations. Both test statistics and *p* value approaches will be demonstrated but, in all such cases, only one of these decision approaches need be used. Software generation of *p* values generally provides figures rounded to a given accuracy, *e.g.* to three decimal places. A value of *p* = 0.000, for instance, does not mean that the *p* value is exactly zero. Such a result should be interpreted as meaning *p* < 0.0005 and would specify that the null hypothesis H_0 can be rejected at the 1% significance level.

One drawback of classical hypothesis testing, as highlighted by Kay,[6] is that the procedures used provide no information about the magnitude of an experimental effect. They simply provide means of detecting evidence in support of or against the null hypothesis. By contrast, estimation through determination of a *confidence interval* (CI) provides a means of estimating the magnitude of an experimental effect enabling a more specific scientific interpretation of the results to be forthcoming.

Confidence intervals are defined as ranges of values within which we can be reasonably certain that the experimental effect being estimated will in fact lie. They are constructed in additive form

$$estimate \pm critical\ value\ {}^*measure\ of\ variability \qquad (2.7)$$

for location effects such as mean [see equation (2.10)] or difference of means [see equation (2.17)], and in multiplicative form

$$estimate\ {}^*function\ of\ critical\ value \qquad (2.8)$$

for variability measures [see equations (2.12) and (2.19)]. The additive nature of confidence intervals for location measures stems from the use of symmetric distributions for the requisite critical values (normal and *t*), resulting in the confidence interval being symmetrical about the target. Variability confidence intervals, by contrast, are based on critical values from skewed distributions (χ^2 and *F*) which result in the intervals not being symmetrical around target.

The level of a confidence interval, expressed as $100(1-\alpha)$%, provides

[6] R. Kay, *Clin. Res. Focus*, 1995, **5**, 4.

a measure of the confidence (degree of certainty) that the experimental effect being estimated lies in the constructed interval. Such levels are generally set at one of three possibilities: 90% ($\alpha = 0.1$), 95% ($\alpha = 0.05$), or 99% ($\alpha = 0.01$), comparable to the three significance levels appropriate to significance testing.

Having discussed the principles underpinning statistical inference, it is important to consider how statistical significance relates to chemical importance. Statistical analysis may result in a conclusion stating that a 'statistically significant' difference has been detected. This difference may not be large enough, however, for us to claim that an important chemically interpretable effect has been found. Significance of an effect on a statistical basis, therefore, does not necessarily relate directly to chemical importance. Such a decision rests with the investigator based on their expertise, understanding of the problem, and level of difference which they deem to be chemically appropriate.

Hence, the importance of considering both exploratory and inferential data analysis approaches to analyse chemical data to provide a firm foundation for the interpretations reached. Confidence intervals are particularly useful in this respect as they provide for estimating the magnitude of an experimental effect to provide a chemical measure of the findings.

6 INFERENCE METHODS FOR ONE SAMPLE EXPERIMENTS

One sample inference occurs when comparing a single set of experimental responses against a 'standard' or 'target' value. For example, comparing experimental measurements recorded by an analytical procedure on a standard sample with known amount of analyte and comparison of copper content within samples of copper sulphate of known quantity of copper. The former illustrates a test of procedure bias while the latter is an illustration of a test of purity of samples. Only mechanisms for small sample ($n < 30$) inference will be discussed.

6.1 Hypothesis Test for the Mean Response

This inference procedure is essentially a test of location effect and enables an assessment of experimental data against a known target to be considered. Such an approach is particularly useful for assessing whether there is evidence of systematic errors in experimental data. The operation of the associated procedure is explained in Box 2.4. It is not currently available in Excel by default. As all aspects of data analysis are to be geared to using software output, elements of EDA have been included in Box 2.4 for completeness.

Box 2.4 *One sample t test*

Assumption. Experimental response approximately normally distributed.
Hypotheses. The null hypothesis is always expressed as H_0: no difference from target ($\mu = \mu_0$) where μ_0 is the specified target for data comparison. The alternative hypothesis can be expressed in one of three ways, two-sided (1) or one-sided (2 and 3):

  1. H_1: difference from target ($\mu \neq \mu_0$) (used in most analytical procedure comparisons), includes both $\mu < \mu_0$ and $\mu > \mu_0$,
  2. H_1: data below target ($\mu < \mu_0$),
  3. H_1: data above target ($\mu > \mu_0$).

Exploratory data analysis (EDA) Assess data plots and summaries to achieve an initial assessment of the response data.
Test statistic. The associated test statistic, based on the Student's t distribution with $(n-1)$ degrees of freedom, is expressed as

$$t = \frac{\bar{x} - \mu_0}{s \big/ \sqrt{n}} \qquad (2.9)$$

where $\bar{x}$ is the sample mean, μ_0 the hypothesised target, s the sample standard deviation, and n the number of measurements in the sample.
Decision rule. Testing at the $100\alpha\%$ significance level can be operated in either of two ways if software produces both test statistic and p value.
test statistic approach: this approach depends on which of the three forms of alternative hypothesis is appropriate and uses the t distribution critical values displayed in Table A.1.

  1. two-sided alternative H_0: $\mu \neq \mu_0$ (two-tailed test), $-t_{\alpha/2,n-1} < t < t_{\alpha/2,n-1} \Rightarrow$ accept H_0
  2. one-sided alternative H_1: $\mu < \mu_0$ (one-tailed test), $t > -t_{\alpha,n-1} \Rightarrow$ accept H_0,
  3. one-sided alternative H_1: $\mu > \mu_0$ (one-tailed test), $t < t_{\alpha,n-1} \Rightarrow$ accept H_0.

p value approach: p value > significance level $\Rightarrow$ accept H_0 where significance level must be expressed in decimal not percent format.

The Student's t distribution first appeared in an article published by William S. Gossett in 1908. Gossett worked at the Guinness Brewery in Dublin but the company did not permit him to publish the work under his own name so he chose to use the pseudonym 'Student'. The form of the t distribution exhibits a comparable symmetric pattern to that displayed in Figure 2.6 for the normal. Critical values in Table A.1 are presented as $t_{\alpha,n-1}$ where α is related to the significance level and nature of the test, and $(n-1)$ is the degrees of freedom of the test statistic.

The decision rule for alternative hypothesis format 1 in Box 2.4 is often expressed as

$$|t| < t_{\alpha/2,n-1} \Rightarrow \text{accept } H_0$$

where $|t|$ refers to the absolute value of the test statistic (2.9) irrespective of sign and $\alpha/2$ is used to signify the two possibilities defined for the mean in the associated alternative. This form of approach to the decision rule for two-sided testing will be illustrated throughout. The basic premise of this test procedure is that small values of the test statistic will imply acceptance of the null hypothesis and large values rejection, the former indicative of sample mean close to target (*no significant evidence of difference*) and the latter indicative of sample mean differing from target (*significant evidence of difference*). With the *p* value, acceptance of the null hypothesis is associated with large *p* values and rejection with small *p* values corresponding to the same basic interpretation as the test statistic approach.

Example 2.7

An analytical laboratory was provided with a sample of poultry feed of nominal lasalocid sodium content 85 mg kg^{-1}. The laboratory was asked to make ten replicate determinations of the lasalocid sodium content in the supplied material. The collected data are presented in Table 2.6 as mg kg^{-1} lasalocid sodium. Are the laboratory's measurements acceptable?

Table 2.6 *Lasalocid sodium data for Example 2.7*

87	88	84	84	87	81	86	84	88	86

Source: Analytical Methods Committee, *Analyst (Cambridge)*, 1995, **120**, 2175.

We want to use the collected data to examine whether they agree or disagree with the nominal content specified. The question to be answered is 'is the mean lasalocid sodium recorded by the laboratory acceptably near the known content of 85 mg kg^{-1}?'. Unfortunately, Excel does not currently have an analysis tool for one sample inference. However, appropriate use of Excel's Function features can enable such information to be produced easily as Output 2.6 illustrates. Outputs 2.7

and 2.8 provide other statistical information relevant to this example based on the structure adopted for Output 2.6.

Assumptions
Lasalocid sodium measurements approximately normally distributed.

Hypotheses
The null hypothesis is H_0: no difference in mean lasalocid sodium measurement from target ($\mu = 85$) and since the test is one of general directional difference, the alternative will be two-sided and expressed in the form H_1: mean lasalocid sodium different from target ($\mu \neq 85$).

Exploratory data analysis
An Excel produced dotplot of the collected data, with response measurements as the vertical axis, is presented in Output 2.5. As some measurements are repeated, the plot contains only five points though ten are plotted. The reported results are spread reasonably evenly around the target of 85 mg kg^{-1} with a range of 81 to 88 mg kg^{-1}. No obvious clustering or patterning appears to be occurring.

Output 2.5 *Data plot for the lasalocid sodium measurements of Example 2.7*

Lasalocid sodium data entered as in cells A1:A11 (label in A1). Menu commands as Output 2.1.

Plot of Lasalocid Sodium Measurements

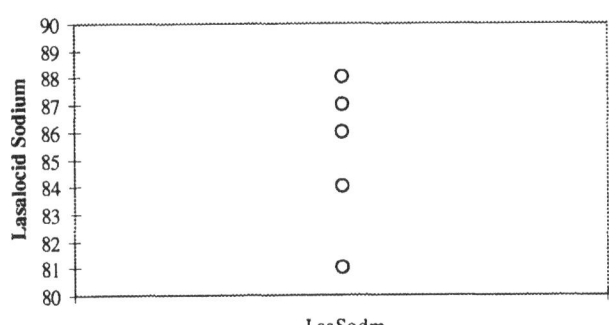

The summary statistics in Output 2.6 indicate that the mean content reported is 85.5 mg kg^{-1}, marginally higher than target. This difference is not substantial as it is less than the standard error of 0.703. The *RSD* is 2.6% which is relatively small and indicative of good consistency in

Output 2.6 *t Test information for Example 2.7*

Labels are entered in column C from cell 1 and numerical summaries in column D from cell 2.

In cell **D2**, enter **=COUNT(A2:A11)**. In cell **D3**, enter **=AVERA-GE(A2:A11)**. In cell **D4**, enter **=STDEV(A2:A11)**. In cell **D5**, enter **=G4/SQRT(G2)**. In cell **D6**, enter **=100(D4/D3)**.

In cell **D9**, enter **85**. In cell **D10**, enter **=(D3 − D9)/D5** to generate the *t* test statistic (2.9). In cell **D11**, enter **= TDIST(ABS(D10),D2 − 1,2)** to generate the *p* value of the two-tail test (2) based on $(n-1)$ degrees of freedom.

Summaries	
Count	10
Mean	85.5
St Dev	2.22
St Err	0.703
RSD	2.6
t Test	
Target	85
Test Stat	0.711
p Value	0.4951

the measurements. Both aspects of EDA point to there being no detectable difference in the laboratory's measurements compared to the proposed target.

Test statistic
The second part of Output 2.6 contains the *t* test results generated by Excel. Both test statistic and *p* value are presented enabling both decision procedures to be illustrated (accept H_0 if $|t| < t_{\alpha/2,n-1}$ as H_1 is two-sided or accept H_0 if *p* > significance level).

Test statistic approach The test statistic (2.9) is specified as $t = 0.711$ ('Test Stat'). The critical value for test statistic comparison, based on 5% ($\alpha = 0.05$) significance level and $n = 10$ ('Count'), is given as $t_{0.025,9} = 2.262$ (Table A.1). Since $|t|$ is less than the critical value, the null hypothesis is accepted and we would conclude that there appears to be sufficient evidence to suggest that the laboratory's measurements differ little from target and are therefore acceptable ($p < 0.05$).

p Value approach The *p* value in Output 2.6 is specified as 0.4951 ('p Value'). Based on testing at the 5% ($\alpha = 0.05$) significance level, we can see that the *p* value exceeds the significance level, specified as 0.05, leading to acceptance of H_0 as for the test statistic approach.

6.2 Confidence Interval for the Mean Response

Hypothesis testing provides a mechanism for weighing up the evidence for or against the null hypothesis. Often, estimation of the effect being assessed would be beneficial to understand more about the chemical data collected. Construction of a confidence interval for the mean provides such a mechanism, the determination of which is explained in Box 2.5. This estimation approach is not currently available in Excel by default. It can also be used as an alternative to two-tailed testing through the mechanism that if the interval contains the target value μ_0 then we can accept the null hypothesis H_0.

Box 2.5 *Confidence interval for one sample mean*

Assumptions. As for one sample *t* test (see Box 2.4).
100(1−α)% confidence interval. The $100(1-\alpha)$% confidence interval for the population mean based on a small sample ($n < 30$) is expressed as

$$\bar{x} \pm t_{\alpha/2,n-1}(s/\sqrt{n}) \qquad (2.10)$$

where $\bar{x}$, $t_{\alpha/2,n-1}$, s, and n are as defined in Box 2.4 for the one sample *t* test statistic, and $s/\sqrt{n}$ refers to the standard error of the mean, a measure of the precision of the sample mean estimate.

Example 2.8

Referring to the lasalocid sodium data of Example 2.7, we will now illustrate construction and use of the 95% confidence interval for the mean content.

The *assumption* for the response is as stated in Example 2.7.

Confidence interval
Output 2.7 specifies the limits for a 95% confidence interval (2.10) for mean lasalocid sodium measurement producing the confidence interval of (83.91, 87.09) mg kg^{-1}. This indicates that measurements are

acceptably accurate as the target of 85 mg kg^{-1} lies within the interval. The uneven split of the interval around the target of 85 suggests that there are more values above 85 than would be ideal suggesting that accuracy of reported measurements may not be as good as initial inference implies.

Output 2.7 *Confidence interval information for Example 2.8*

Follows on from Output 2.6.
   In cell **D14**, enter **0.95** for the required 95% confidence level. In cell **D15**, enter **=D3−TINV((1−D14),D2−1)*D5** to generate the lower limit of the confidence interval (2.10). In cell **D16**, enter **=D3+TINV((1−D14),D2−1)* D5** to generate the upper limit of the confidence interval (2.10).

Conf Int	
Level	0.95
Low Lim	83.91
Upp Lim	87.09

6.3 Hypothesis Test for the Variability in a Measured Response

In chemical experimentation, in particular, precision of measurement is an important facet of ensuring reliability of reported results. As such, statistical checking of the variability of recorded data can be a useful aid to assessing measurement precision. For single samples, the χ^2 *test of variability* can be used, the steps of which are summarised in Box 2.6. As with the t test and confidence interval procedures, Excel does not carry out this test by default but use of Excel commands can produce a requisite test statistic.
   The chi-square (χ^2) distribution is another statistical distribution for describing random phenomena. Unlike the normal and t distributions, this distribution is not symmetrical but right-skewed comparable to the data pattern illustrated in Figure 2.3. As with the critical values for the t distribution, the critical values of the χ^2 distribution depend also on the significance level of test and the degrees of freedom $(n-1)$.

Box 2.6 χ^2 *Test of variability*

Assumption. As for one sample t test (see Box 2.4).

Hypotheses. The null hypothesis is always expressed as H_0: no difference in variability from target ($\sigma^2 = \sigma_0^2$) where σ_0 is the nominal suggested target for variability, expressed as a standard deviation measurement. Three forms of alternative hypothesis are again possible:

1. H_1: variability different from target ($\sigma^2 \neq \sigma_0^2$) (used in most analytical procedure precision comparisons),
2. H_1: variability lower than target ($\sigma^2 < \sigma_0^2$),
3. H_1: variability greater than target ($\sigma^2 > \sigma_0^2$).

Test statistic. The test statistic, based on the χ^2 (chi-square) distribution with $(n-1)$ degrees of freedom, is given by

$$\chi^2 = \frac{(n-1)s^2}{\sigma_0^2} \tag{2.11}$$

where n is the number of experimental measurements, s the sample standard deviation, and σ_0 the target variability.

Decision rule. Both test statistic and p value approaches can be considered if software can produce information for both mechanisms using the $100\alpha\%$ significance level.

test statistic approach: This approach can be operated along comparable lines to those for the one sample t test using the χ^2 critical values displayed in Table A.2. For the three forms of alternative hypothesis, we have:

1. two-sided alternative H_1: $\sigma^2 \neq \sigma_0^2$ (two-tailed test), $\chi^2_{1-(\alpha/2),n-1} < \chi^2 < \chi^2_{\alpha/2,n-1} >$ accept H_0,
2. one-sided alternative H_1: $\sigma^2 < \sigma_0^2$ (one-tailed test), $\chi^2 > \chi^2_{1-\alpha,n-1}$ $\Rightarrow$ accept H_0,
3. one-sided alternative H_1: $\sigma^2 > \sigma_0^2$ (one-tailed test), $\chi^2 < \chi^2_{\alpha,n-1} \Rightarrow$ accept H_0.

p value approach: As for one sample t test (see Box 2.4).

Example 2.9

Referring to the laboratory study of Example 2.7, suppose it had been stipulated that the variability (standard deviation) in lasalocid sodium measurement should not exceed 2 mg kg^{-1}. If variability in response exceeds this target, the precision of the reported data may be in doubt. We want to carry out an appropriate test to assess this effect.

The *assumption* for the response is as stated in Example 2.7.

Hypotheses
The null hypothesis is H_0: no difference in the variability of the lasalocid sodium measurements from target ($\sigma^2 = 2^2$). Given the purposes of the assessment, the alternative hypothesis will require to be one-sided and is expressed as H_1: variability exceeds target ($\sigma^2 > 2^2$).

Statistical test
Using Excel's commands, we can generate appropriate test information as Output 2.8 illustrates. Test statistic (2.11) is specified to be 11.09 ('Test Stat'). This enables the decision approach accept H_0 if $\chi^2 < \chi^2_{\alpha,n-1}$ to be used.

Output 2.8 *Variability test information for Example 2.9*

Follows on from Outputs 2.6 and 2.7.
   In cell **D19**, enter 2. In cell **D20**, enter **=(D2 – 1)*D4^2/D19^2** to generate the χ^2 test statistic (2.11).

Var Test	
Target	2
Test Stat	11.09
p Value	0.2696

For the *test statistic approach*, the critical value from Table A.2 for the 5% significance level is $\chi^2_{0.05,9} = 16.92$ (one-sided alternative). As test statistic is less than critical value, we accept H_0 and conclude that measurement precision appears sufficiently acceptable.

Pulling together all the exploratory and inferential analyses conducted on the lasalocid sodium data of Example 2.7, we can conclude that measurements reported by the laboratory appear statistically acceptable in respect of both accuracy ($p > 0.05$) and precision ($p > 0.05$). However, the 95% confidence interval result of 83.91 to 87.09 mg kg^{-1} does suggest that the measurements are perhaps on the high side.
   One drawback of the experiment illustrated in Example 2.7 is that only one laboratory was assessed. By including other laboratories or by including comparison of analytical procedures, the information provided by such a quality assurance study could be improved markedly to be of more practical benefit. Experiments of this type will be discussed later in this chapter and developed further in other chapters.

6.4 Confidence Interval for Response Variability

As with consideration of the mean, it is also often useful to construct a confidence interval for response variability to provide an estimate of the magnitude of the related effect. Such a confidence interval is multiplicative in nature due to the skewed nature of the χ^2 distribution providing the appropriate critical values and will, therefore, not be evenly spread around the target. The $100(1-\alpha)\%$ confidence interval for the population variance σ^2 is given by

$$\frac{(n-1)s^2}{\chi_{\alpha/2,n-1}} < \sigma^2 < \frac{(n-1)s^2}{\chi_{1-\alpha/2,n-1}} \qquad (2.12)$$

Exercise 2.3

The following titration values (titre cm^{-3}) were obtained by a single analyst, applying a standard technique to separate aliquot samples of a homogeneous starting material. The procedure involves a number of pre-treatment stages before the final titration. The nominal titration measurements should be around 27.75 titre cm^{-3} with variability 0.175 titre cm^{-3}. Are these data conforming to the specified targets?

27.86 27.68 28.02 28.05 27.60 27.55 27.03 27.76 27.73 27.75 27.77 27.53

Source: D. McCormick and A. Roach, 'Measurement, Statistics and Computation', ACOL Series, Wiley, Chichester, 1987: reproduced with the permission of HMSO.

2.7 INFERENCE METHODS FOR TWO SAMPLE EXPERIMENTS

In many experimental situations, it may be preferable to compare two homogeneous groups under comparable experimental conditions rather than one sample against a nominal target. This modification of approach to two sample experimentation enables more information on the experimental objective to be obtained and also enables confounding factors which could influence the outcome of the measurement assessment to be accounted for. Two sample experiments are particularly useful when wishing to compare a new analytical procedure with a reference procedure or comparing whether changing experimental conditions affects a chemical outcome. Such a structure is often referred to as a 'treatment' versus 'control' design as its basis lies in subjecting one group to a 'treatment' and using the other to act as the 'control'. As with one sample inference, there are inference and estimation procedures for means and variability. Only small sample tests (sample size < 30) will be considered.

Box 2.7 *Two sample t test*

Assumptions. Two independent random samples and response variable approximately normally distributed.

Hypotheses. As previously, the null hypothesis reflects no difference, *i.e.* H_0: no difference between the two populations ($\mu_1 = \mu_2$) and three forms of alternative are possible:

    1. H_1: difference between the two populations ($\mu_1 \neq \mu_2$),
    2. H_1: population 1 lower ($\mu_1 < \mu_2$),
    3. H_1: population 1 higher ($\mu_1 > \mu_2$).

Exploratory data analysis (EDA). Assess data plots and summaries to achieve an initial assessment of the response data.

Test statistic. The form of two sample *t* test to use depends on whether the *ratio of larger sample variance to smaller sample variance* is below or above 3.

ratio less than 3: for this case, the test statistic, referred to as the *pooled t test*, is expressed as

$$t = \frac{\bar{x}_1 - \bar{x}_2}{s_p\sqrt{\dfrac{1}{n_1} + \dfrac{1}{n_2}}} \tag{2.13}$$

based on ($n_1 + n_2 - 2$) degrees of freedom where $\bar{x}_1$ and $\bar{x}_2$ are the sample means, and n_1 and n_2 the number of observations in the samples. The term s_p defines the pooled estimate of standard deviation

$$s_p = \sqrt{\frac{(n_1 - 1)s_1^2 + (n_2 - 1)s_2^2}{n_1 + n_2 - 2}} \tag{2.14}$$

based on the sample standard deviations s_1 and s_2.

ratio exceeds 3: for this case, the form of the test statistic, referred to as the *separate-variance t* test or *Welch's test*, is given by

$$t = \frac{\bar{x}_1 - \bar{x}_2}{\sqrt{\dfrac{s_1^2}{n_1} + \dfrac{s_2^2}{n_2}}} \tag{2.15}$$

with degrees of freedom approximated by (truncated to nearest integer)

$$df = \frac{\left(s_1^2/n_1 + s_2^2/n_2\right)^2}{\dfrac{\left(s_1^2/n_1\right)^2}{n_1 - 1} + \dfrac{\left(s_2^2/n_2\right)^2}{n_2 - 1}} \tag{2.16}$$

Decision rule. As one sample *t* test (see Box 2.4) with requisite change of degrees of freedom for test statistic approach. Again, a small test statistic or a large *p* value will correspond to acceptance of H_0.

7.1 Hypothesis Test for Difference in Mean Responses

Difference of means testing represents a simple extension of the one sample t test. There is obviously twice the amount of experimental information available with the test based on assessing whether the difference between the population means differs from zero. Box 2.7 provides a summary of the components of this inference procedure.

The check of ratio of variances enables a decision to be made on whether or not to assume equal population variances. The figure 3 stems from the fact that the variances can differ by as much as a factor of three and the equality of variability assumption is still valid for equal sample sizes. This simple check on data variability ensures its effect is properly accounted for in test statistic derivation. The numerator expressions in equations (2.13) and (2.15) correspond to the standard error of the difference between the mean measurements providing a measure of the precision of this difference estimate.

It is possible to modify the two sample t tests (2.13) and (2.15) to assess a specified difference D_0 between the population means. This requires the numerator of these test statistics to be re-expressed as $(\bar{x}_1 - \bar{x}_2 - D_0)$ to enable the specified difference to be tested.

In Excel, the test statistic (2.13) can be derived using the t-Test: Two-Sample Assuming Equal Variances analysis tool, the dialog window for which is shown Figure 2.8. The entry under 'Alpha' enables the significance level for output information to be specified with 0.05

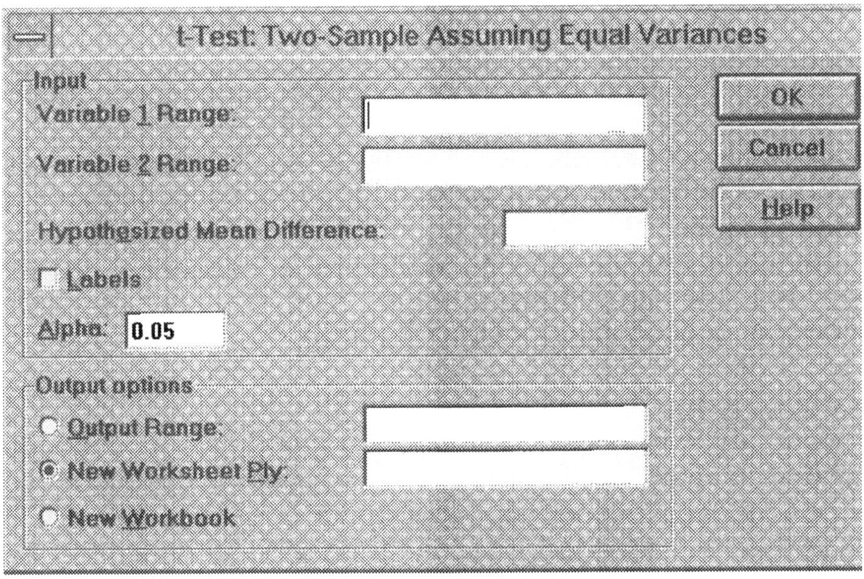

Figure 2.8 *Pooled t test dialog window in Excel*

corresponding to the default 5% significance level. A comparable dialog window is available for the t-Test: Two-Sample Assuming Unequal Variances analysis tool for cases where it is necessary to produce the test statistic (2.15) though the degrees of freedom presented are rounded and not truncated.

Example 2.10

As part of an inter-laboratory study, two analytical laboratories were asked to carry out a standard method of determining the total phosphate concentration, as $\mu g \ l^{-1}$, of a sample of river water. Each laboratory carried out eight replicate determinations of phosphate concentration with Table 2.7 showing the reported results. Do these data indicate that laboratory A's results tend to be higher and less variable than those of laboratory B? The variability aspect will be considered in Example 2.12.

Table 2.7 *Total phosphate measurements for Example 2.10*

Laboratory A	20.7	27.5	30.4	23.9	21.7	24.1	24.8	28.9
Laboratory B	20.9	21.4	24.9	20.5	19.7	26.3	22.4	20.2

Source: D. McCormick and A. Roach, 'Measurement, Statistics and Computation', ACOL Series, Wiley, Chichester, 1987: reproduced with the permission of HMSO.

Assumptions
The two sets of phosphate results are independent and normally distributed.

Hypotheses
Construction of the hypotheses is relatively straightforward. The null hypothesis for comparison of the means is simply H_0: no difference in mean phosphate measurements ($\mu_A = \mu_B$). As the supposition to be tested asserts that laboratory A's results tend to be higher, we specify this as a one-sided alternative of the form H_1: mean phosphate measurement of laboratory A higher ($\mu_A > \mu_B$).

Exploratory data analysis
Output 2.9 provides a plot of the phosphate data by laboratory. We can see from this presentation that laboratory A appears to have higher and more variable results given the wider spread of points associated with it. In addition, it appears that laboratory B's results tend to cluster at the lower end of the scale of measurements reported hinting at a skew-like pattern.

Output 2.9 *Data plots for phosphate measurements of Example 2.10*

Laboratory A data are in cells A1:A9 (label in A1) and laboratory B data in cells B1:B9 (label in B1). Menu commands as Output 2.2.

Plot of Phosphate Measurements by Laboratory

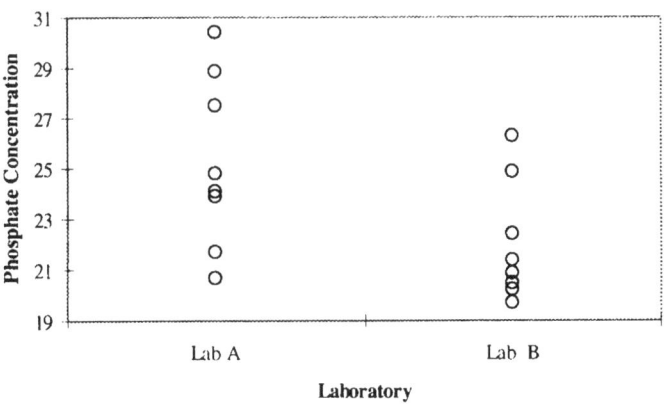

Output 2.10 *Summary statistics for the phosphate measurements of Example 2.10*

Laboratory A data are in cells A1:A9 (label in A1) and laboratory B data in cells B1:B9 (label in B1). Menu commands as Output 2.4.

	Lab A	Lab B
Mean	25.25	22.038
Standard Error	1.207	0.839
Median	24.45	21.15
Standard Deviation	3.413	2.372
Sample Variance	11.651	5.628
Kurtosis	−1.11	−0.07
Skewness	0.26	1.09
Range	9.7	6.6
Minimum	20.7	19.7
Maximum	30.4	26.3
Sum	202	176.3
Count	8	8
Confidence Level (95.0%)	2.852	1.983

The summary statistics in Output 2.10 provide further evidence of laboratory difference as the mean for laboratory A is larger than that for laboratory B by over 3 μg l^{-1}. The *RSD*s for laboratory A and B are 13.5% and 10.8% respectively suggesting a marginal difference in the consistency of the reported measurements. The skewness statistic for laboratory B suggests right-skewed data as hinted at in the data plot.

Test of difference of means

Using the Sample Variances presented in Output 2.10, the ratio of variances is 11.651/5.628 = 2.07. As this is less than 3, we therefore require to use the pooled *t* test (2.13) to determine the necessary test statistic. The output produced by Excel using the Equal Variances tool illustrated in Figure 2.8 is given in Output 2.11. Both test statistic ('t Stat') and *p* value ['P(T<=t) one-tail'] are presented.

Output 2.11 *Two sample t test information for Example 2.10*

Laboratory A data are in cells A1:A9 (label in A1) and laboratory B data in cells B1:B9 (label in B1).
Select **Tools** ▷ **Data Analysis** ▷ t-Test: Two-Sample Assuming Equal Variances ▷ click **OK** ▷ for *Variable 1 Range*, enter **A1:A9** ▷ click the **Variable 2 Range** box and enter **B1:B9** ▷ select **Labels** ▷ select **Output Range**, click the empty box, and enter **F17** ▷ click **OK**.

t-Test: Two-Sample Assuming Equal Variances

	Lab A	Lab B
Mean	25.25	22.038
Variance	11.651	5.628
Observations	8	8
Pooled Variance	8.64	
Hypothesized Mean Difference	0	
df	14	
t Stat	2.186	
P(T< = t) one-tail	0.0232	
t Critical one-tail	1.761	
P(T< = t) two-tail	0.0463	
t Critical two-tail	2.145	

For the *test statistic approach*, (2.13) is specified to be 2.186. The associated critical value is $t_{0.05,14} = 1.761$ based on 14 degrees of freedom (df), obtained either from Table A.1 or from the entry 't

Critical one-tail' in Output 2.11. As test statistic exceeds critical, we reject H_0 and conclude that there appears sufficient evidence to suggest that laboratory A is reporting higher measurements ($p < 0.05$). The p value is quoted as 0.0232 ['P(T<=t) one-tail'] and as this is less than 0.05, the p *value approach* also leads to the same statistical conclusion.

7.2 Confidence Interval for Difference in Mean Responses

As with one-sample inference, use of a confidence interval for the difference in means can be a useful part of two sample data analysis by providing a process for estimating the magnitude of this effect. Box 2.8 summarises this mechanism for two sample investigations.

Box 2.8 *Confidence interval for difference in means*

Assumptions. As two sample t test (see Box 2.7).
100(1 − α)% confidence interval. A $100(1 - \alpha)\%$ confidence interval for the difference between two-sample means is expressed as

$$\bar{x}_1 - \bar{x}_2 \pm t_{\alpha/2,\mathrm{df}} \, \mathrm{se}(\bar{x}_1 - \bar{x}_2) \qquad (2.17)$$

where df refers to degrees of freedom and $\mathrm{se}(\bar{x}_2 - \bar{x}_2)$ is the standard error of the difference between the means, *i.e.* numerator of test statistics (2.13) or (2.15) as appropriate.

Example 2.11

Referring to the phosphate data of Example 2.10, we will now construct and interpret a 95% confidence interval for the difference in phosphate means.

The *assumptions* for the response are as stated in Example 2.10.

Confidence interval
A pooled t test was used in Example 2.10 to generate the required test statistic so the necessary 95% confidence interval (2.17) will be based on df $= 14$ and $\mathrm{se}(\bar{x}_1 - \bar{x}_2) = s_p\sqrt{(1/n_1 + 1/n_2)}$ where s_p can be found using (2.14). The information and numerical summaries presented in Output 2.12 illustrate how Excel commands can be used to generate the associated summaries and confidence interval limits ('Low Lim' and 'Upp Lim').
Based on the limits shown, the 95% confidence interval can be specified

Output 2.12 *Difference of means confidence interval for Example 2.11*

Laboratory A data are in cells A1:A9 (label in A1) and laboratory B data in cells B1:B9 (label in B1). Labels are entered into column J from cell 1 and numerical summaries in column K from cell 2.

In cell **K2**, enter =AVERAGE(A2:A9). In cell **K3**, enter =STDEV(A2:A9). In cell **K4**, enter =AVERAGE(B2:B9). In cell **K5**, enter =STDEV(B2:B9). In cell **K6**, enter =COUNT(A2:A9). In cell **K7**, enter =COUNT(B2:B9).

In cell **K10**, enter **0.95** for the required 95% confidence level. In cell **K11**, enter =**K6** + **K7** − **2** to generate the degrees of freedom. In cell **K12**, enter = SQRT(((K6 − 1)*K3^2 + (K7 − 1)*K5^2)/K11) to generate s_p, the pooled estimate of standard deviation (2.14). In cell **K13**, enter =K12*SQRT(1/K6 + 1/K7) to generate the requisite standard error of the difference in means. In cell **K14**, enter =**K2** − **K4** − TINV((1 − **K10**),**K11**)***K13** to generate the lower limit of the confidence interval (2.17). In cell **K15**, enter = **K2** − **K4** + TINV((1 − **K10**),**K11**)***K13** to generate the upper limit of the confidence interval (2.17).

Summaries	
Mean1	25.25
St Dev 1	3.413
Mean 2	22.038
St Dev 2	2.372
n1	8
n2	8
Conf Int	
Level	0.95
df	14
Sp	2.939
St Error	1.4695
Low Lim	0.06
Upp Lim	6.364

as (0.06, 6.364) μg l^{-1}. As the difference is calculated as (A − B), these limits imply that phosphate measurements are generally larger in laboratory A by up to 6.36 μg l^{-1}, on average. This represents a potentially wide level of difference which casts doubt on the reliability of laboratory A's reported measurements compared to those of laboratory B.

7.3 Hypothesis Test for Variability

In addition to testing and estimation of location effects in two sample experimentation, assessment of the precision (variability) of the measurements can also be appropriate. This can be achieved through a test of the equality of the sample variances called a *variance ratio test* as summarised in Box 2.9. The test can be used to formalise the decision on which form of t test, (2.13) or (2.15), to use for comparison of the means.

The F distribution is a skewed statistical distribution similar to the right-skew pattern illustrated in Figure 2.3. The test statistic (2.18) can also be expressed in the form $F =$ smaller/larger but this leads to a more complex decision rule process. By evaluating the test statistic (2.18), the

---

Box 2.9 *Variance ratio F test*

Assumptions. As two sample t test (see Box 2.7).
Hypotheses. The null hypothesis again reflects no difference, *i.e.* H_0: no difference in variability between the two samples ($\sigma_1^2 = \sigma_2^2$). The alternative can again be specified in one of three forms:

1. H_1: difference in variability ($\sigma_1^2 \neq \sigma_2^2$),
2. H_1: variability lower in population 1 ($\sigma_1^2 < \sigma_2^2$),
3. H_1: variability higher in population 2 ($\sigma_1^2 > \sigma_2^2$).

Test statistic. The associated test statistic is simply constructed as

$$F = \frac{larger\ sample\ variance(lsv)}{smaller\ sample\ variance(ssv)} \qquad (2.18)$$

based on an F distribution with (df_1, df_2) degrees of freedom where $df_1 = n_{lsv} - 1$, $df_2 = n_{ssv} - 1$, and n_{lsv} and n_{ssv} are the respective sample sizes for the larger and smaller sample variance estimates.
Decision rule. Decision rule approaches, for testing at the $100\alpha\%$ significance level, conform to similar principles as previously outlined.
test statistic approach: again depends on which form of alternative hypothesis is appropriate and uses the critical values displayed in Table A.3.

1. $F < F_{\alpha/2, df_1, df_2} \Rightarrow$ accept H_0,
2. $F < F_{\alpha, df_1, df_2} \Rightarrow$ accept H_0,
3. As 2.

p value approach: as for one sample t test (see Box 2.4).

numerical value generated will never be less than 1 and it is this result that enables the decision rule for alternative hypotheses 2 and 3 to be expressed similarly.

In Excel, *F* test derivation can be achieved through the use of the F-Test Two Sample for Variances analysis tool with dialog window shown in Figure 2.9. In order to produce the correct estimate of test statistic (2.18), 'Variable 1 Range' must correspond to the sample data giving rise to the larger sample variance and 'Variable 2 Range' to the sample with smaller sample variance. Specification of 'Alpha' again sets the significance level for the statistical information Excel produces, 0.05 corresponding to the 5% default.

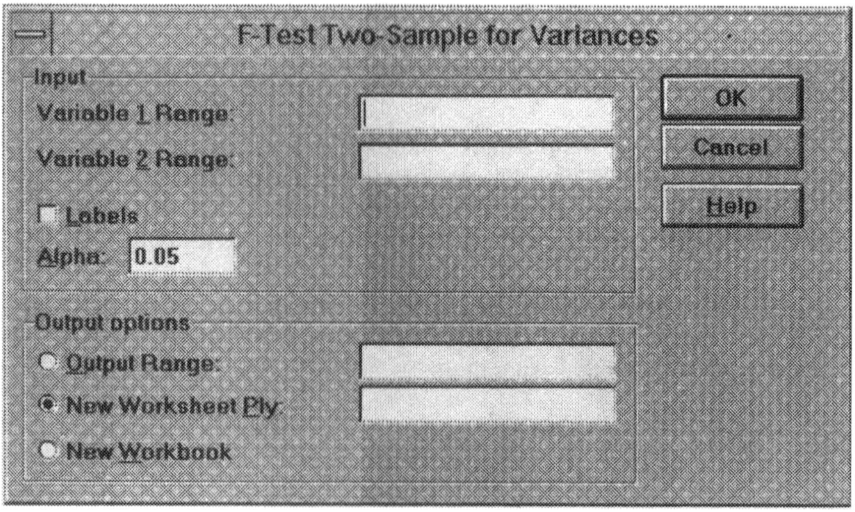

Figure 2.9 *F test dialog window in Excel*

Example 2.12

Referring to the phosphate study of Example 2.10, a part of the experimental objective was to assess whether the variability in reported results was greater for laboratory A.

The *assumptions* for the response are as stated in Example 2.10.

Hypotheses
The null hypothesis is, as usual, reflective of no difference, *i.e.* H$_0$: no difference in the variability of the reported phosphate measurements ($\sigma^2_A = \sigma^2_B$). Since the test is to assess for greater variability in laboratory

A's results, the alternative will be one-sided and specified as H_1: variability of phosphate measurements higher in laboratory A ($\sigma^2_A > \sigma^2_B$).

Variances test statistic

The F test analysis tool in Excel (see Figure 2.9) produces the output shown in Output 2.13 for the phosphate data. The test statistic (2.18) is stated as 2.07 ('F'). The critical value for the 5% significance level using degrees of freedom (7, 7), $F_{0.05,7,7}$, is 3.79. This can be read from Table A.3 or from the 'F Critical one-tail' entry in Output 2.13. As test statistic is less than critical value, we accept H_0 and conclude that there appears no statistical evidence to suggest that the variability of laboratory A's measurements exceed those reported by laboratory B ($p < 0.05$). The p value of 0.1790 ['P(F<=f) one-tail'] is greater than the significance level of 0.05 and so would provide for the same conclusion.

Output 2.13 *F test information for Example 2.12*

Laboratory A data are in cells A1:A9 (label in A1) and laboratory B data in cells B1:B9 (label in B1).
Select **Tools** ▷ **Data Analysis** ▷ **F-Test: Two-Sample for Variances** ▷ click **OK** ▷ for *Variable 1 Range*, enter **A1:A9** ▷ click the **Variable 2 Range** box and enter **B1:B9** ▷ select **Labels** ▷ select **Output Range**, click the empty box, and enter **F33** ▷ click **OK**.

F-Test Two-Sample for Variances

	Lab A	Lab B
Mean	25.25	22.038
Variance	11.651	5.628
Observations	8	8
df	7	7
F	2.07	
P(F< = f) one-tail	0.179	
F Critical one-tail	3.787	

Pulling together all the analyses carried out, exploratory and inferential, on the phosphate data of Example 2.10, we can conclude that there is support for the supposition that laboratory A's measurements appear higher ($p < 0.05$) but no support for higher variability ($p > 0.05$). The confidence interval provides further back-up of the higher nature of the results for laboratory A with a difference of up to 6.36 $\mu g\,l^{-1}$, on average, reported. We would therefore conclude that laboratory A should re-evaluate their measurement processes, though knowledge of the known phosphate levels of the sample provided would aid this considerably.

In Example 2.12, we illustrated both decision rule approaches for deciding on acceptance or rejection of the null hypothesis in a two sample variability test. The information presented by Excel, in respect of both critical and p value, assume one-tailed testing (one-sided alternative). If the test being carried out is two-tailed, the presented values are inappropriate for the conclusion phase of inferential analysis. In such a case, we would need to use the tables of critical values (Table A.3) for critical value determination, or double the presented p value to produce the correct p value for the required two-tailed test.

7.4 Confidence Interval for the Ratio of Two Variances

In addition to statistically testing variability differences, it is useful to consider construction of a confidence interval for the ratio of two variances to provide an estimate of the variability differences between the samples. As with the confidence interval for response variability for a one sample experiment, the confidence interval is multiplicative in nature and not evenly spread around the target ratio. The $100(1-\alpha)\%$ confidence interval for the ratio of two variances is expressed as

$$\frac{s_1^2}{s_2^2}\frac{1}{F_{\alpha/2,df_1,df_2}} < \frac{\sigma_1^2}{\sigma_2^2} < \frac{s_1^2}{s_2^2}F_{\alpha/2,df_2,df_1} \qquad (2.19)$$

where, in this instance, df_1 and df_2 refer to the degrees of freedom for sample 1 and sample 2 respectively.

Exercise 2.4

A comparative test was run to compare determination of the trace levels of zinc in an aqueous solution as measured by XRF and FAAS. Ten replicate determinations were performed by each procedure on a single solution resulting in the data below (in $\mu g\,l^{-1}$ Zn). Assess whether these procedures differ in terms of accuracy and precision.

XRF	138	117	108	127	116	138	120	117	112	110
FAAS	130	140	119	125	135	132	138	131	127	122

Source: D. McCormick and A. Roach, 'Measurement, Statistics and Computation', ACOL Series, Wiley, Chichester, 1987: reproduced with the permission of HMSO.

2.8 INFERENCE METHODS FOR PAIRED SAMPLE EXPERIMENTS

In two independent samples experimentation, results may be influenced by variation in the experimental material, For example, asking two analysts to measure the level of phenylalanine in two sets of blood sera samples may detect differences in the analysts' results. This difference may be due to analyst differences but could equally be as a result of variations between the sets of samples each analyst assessed. In other words, a detected difference may be due to analysts or samples tested or a mixture of both. In such a case, we should try to account for sample differences by splitting test samples and allocating a separate half of each sample to each analyst. By so doing, sample variation can be accounted for enabling any differences detected to be most likely attributable to analyst differences.

This form of experimentation is generally referred to as *paired comparison testing*, where the objective is to try to eliminate a source of extraneous variation by making the pairings chosen as similar as possible with respect to confounding variable. Such experiments are special cases of two sample experiments and involve observations on two treatments being collected in pairs under as near homogeneous conditions as possible. Tests of mean difference and difference in variability will be outlined.

8.1 Hypothesis Test for Mean Difference in Responses

This parametric procedure is based on collecting the data in pairs from homogeneous material and using the difference between the recorded measurements corresponding to each matched pair as the basis of the procedure. No detectable difference between the pairings would be considered reflective of little obvious difference between the treatments. To test this difference statistically, we use the paired comparison *t* test of the differences. This is essentially a one sample *t* test of observation differences as summarised in Box 2.10.

As with two independent samples, this test procedure can also be used to assess a specific level of difference D_0 rather than simply a mean difference of zero, as in Example 2.13. To do this, we would need to alter the numerator of the test statistic (2.20) to read $(\bar{x} - D_0)$ with appropriate modification to enable relevant assessment to occur.

Box 2.10 *Paired sample t test*

Difference. As paired comparison experimentation is based on measurement of a response from paired specimens, we require to first specify the requisite difference D. This is most often expressed as D = treatment 1 − treatment 2 though order of determination is unimportant.
Assumptions. Differences approximately normally distributed.
Hypotheses. Again, the null hypothesis will correspond to no difference between the treatments, *i.e.* H_0: no difference between the two treatments (mean difference $\mu_D = 0$). As with previous testing, three standard forms of alternative hypothesis can be considered:

 1. H_1: difference between treatments (mean difference $\mu_D \neq 0$),
 2. H_1: treatment 1 lower (mean difference $\mu_D < 0$),
 3. H_1: treatment 1 higher (mean difference $\mu_D > 0$).

Exploratory data analysis (EDA). Use data plots and difference summaries to obtain an initial picture of the data.
Test statistic. The required test statistic is essentially a re-expression of the one sample t test statistic (2.9) with $\bar{x}$ replaced by $\bar{D}$ (mean difference), μ_0 by 0, and s by s_D (standard deviation of the differences) to produce

$$t = \frac{\bar{D}}{s_D/\sqrt{n}} \tag{2.20}$$

where n is the number of data pairings and $(n-1)$ is the associated degrees of freedom.
Decision rule. As one sample t test (see Box 2.4).

In Excel, generation of the test statistic (2.20) and related p value is available through use of the t-Test: Paired Two Sample for Means analysis tool. The dialog window for this is presented in Figure 2.10 with the blank boxes requiring to be filled in, left blank, or checked where appropriate. Again, the 'Alpha' entry of 0.05 specifies that all information produced will be based on use of the 5% significance level. Output resulting from this analysis tool will be discussed in Example 2.13.

Example 2.13

A study was conducted to compare a multicomponent flow injection (FI) analysis system with AAS in respect of the determination of copper (μg ml^{-1}) in blood serum. The latter is frequently used for the

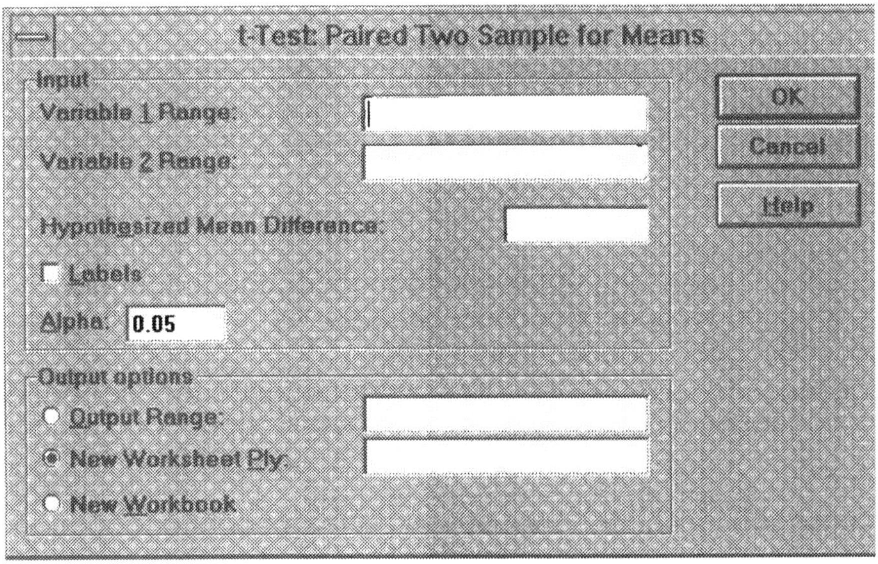

Figure 2.10 *Paired sample t est dialog window in Excel*

determination of Cu in this type of sample. Several blood serum samples were prepared with each sample split into two parts and one part randomly assigned to each analytical procedure in order to help remove the effect of sample variation from the measurements made. The results collected are presented in Table 2.8. Does the FI system provide comparable or different results from AAS in respect of accuracy and precision? The precision aspect will be discussed more fully in Example 2.15.

Table 2.8 *Cu measurements for Example 2.13*

Sample	1	2	3	4	5	6	7	8	9	10
FI	0.608	0.712	0.589	0.562	0.770	0.548	0.662	0.625	0.558	0.652
AAS	0.592	0.708	0.601	0.564	0.755	0.564	0.655	0.624	0.555	0.655

Source: O. Hernandez, F. Jimenez, A.I. Jimenez and J.J. Arias, *Analyst (Cambridge)*, 1996, **121**, 169.

Difference
For this example, the difference D will be determined as $D = \text{FI} - \text{AAS}$ based simply on the order of the data presentation.

Assumption
Differences in Cu measurements approximately normally distributed.

Hypotheses

The accuracy aspect requires assessment of difference. This can be based on testing a null hypothesis of no difference, *i.e.* H_0: no difference in the mean Cu level ($\mu_D = 0$), against a two-sided alternative of difference expressed as H_1: mean Cu level differs with analytical method ($\mu_D \neq 0$).

Exploratory data analysis

Based on the plot in Output 2.14, there appears to be little difference between the procedure results. On some occasions, FI exceeds AAS and on others, AAS exceeds FI though the differences are minor in all instances. The trend in both sets of results is comparable so it would appear the two procedures are producing comparable Cu measurements.

Output 2.14 *Data plot for Cu measurements of Example 2.13*

Sample numbers in cells A1:A11 (label in A1), Cu data for FI in cells B1:B11 (label in B1), and Cu data for FAAS in cells C1:C11 (label in cell C1). Select **ChartWizard**. Click cell **E1**.

Chart Wizard Step 1 of 5: for *Range*, click and drag from across cells **A1:C11** ▷ click **Next**.

Chart Wizard Step 2 of 5: select **Line** ▷ click **Next**.

Chart Wizard Step 3 of 5: select **Line chart 1** ▷ click **Next**.

Chart Wizard Step 4 of 5: for *Use First Column(s)*, select **1** ▷ click **Next**.

Chart Wizard Step 5 of 5: select the **Chart Title** box and enter **Plot of Cu Measurements Against Sample Number** ▷ select the **Axis Titles Category (X)** box and enter **Sample Number** ▷ select the **Axis Titles Value (Y)** box and enter **Copper Meas** ▷ click **Finish**.

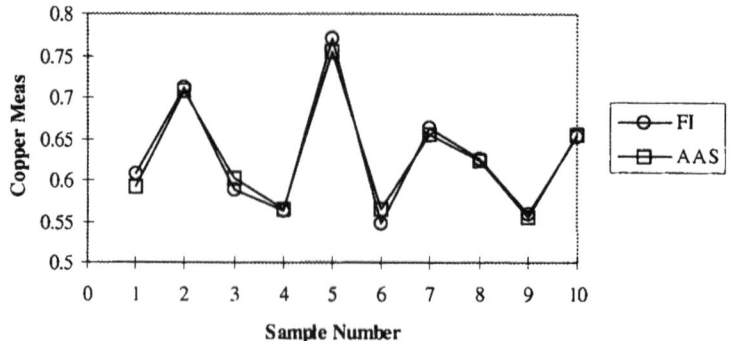

Calculation of the necessary differences is straightforward in Excel as Output 2.15 demonstrates. The differences range from -0.016 to 0.016 with no obvious trend therein. The mean difference is marginally above zero but not sufficiently to suggest procedure differences as the standard error is two-and-a-half times this difference in magnitude.

Output 2.15 *Summary statistics for difference in Cu measurements for Example 2.13*

For determination of differences, click cell **D1** and enter **FI − AAS**. In cell **D2**, enter **= B2 − C2**. Copy and paste this formula into cells **D3:D11**. Menu commands then as Output 2.3 for differences.

Sample	FI	AAS	FI − AAS		FI − AAS	
1	0.608	0.592	0.016		Mean	0.0013
2	0.712	0.708	0.004		Standard Error	0.00325
3	0.589	0.601	−0.012		Median	0.002
4	0.562	0.564	−0.002		Standard Deviation	0.0103
5	0.77	0.755	0.015		Sample Variance	0.00011
6	0.548	0.564	−0.016		Kurtosis	−0.35
7	0.662	0.655	0.007		Skewness	−0.22
8	0.625	0.624	0.001		Range	0.032
9	0.558	0.555	0.003		Minimum	−0.016
10	0.652	0.655	−0.003		Maximum	0.016
					Sum	0.013
					Count	10
					Confidence Level (95%)	0.0074

A dotplot of the procedure differences is shown in Output 2.16. The plot amply illustrates that there are six positive differences and four negative differences. Such similarity in the nature of the differences provides further evidence that there appears to be little obvious difference in the measurements reported by each procedure.

t Test for mean difference
Output 2.17 contains the Excel output for the paired sample *t* test analysis tool (see Figure 2.10) in such a form as to enable either decision rule to be considered again. Since the alternative hypothesis is two-sided, this means use of either $|t| < t_{\alpha/2,\mathrm{df}}$ or $p >$ significance level.

The paired sample test statistic (2.20) is quoted as 0.4 ('t Stat') with

Output 2.16 *Data plot for differences in Cu measurements for Example 2.13*

Differences data are in cells D1:D11 (label in D1). Menu commands as Output 2.1.

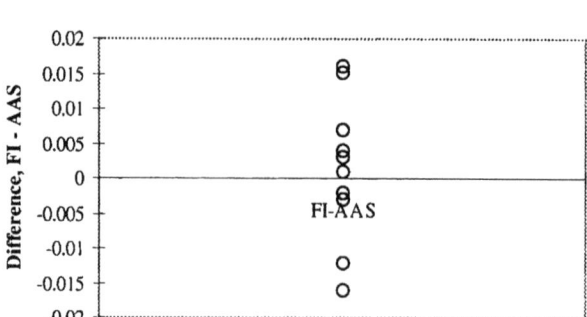

Plot of Cu Differences

Output 2.17 *t test information for Example 2.13*

Cu data for FI in cells B1:B11 (label in B1) and Cu data for FAAS in cells C1:C11 (label in cell C1).

Select **Tools** ▷ **Data Analysis** ▷ **t-Test: Paired Two Sample for Means** ▷ click **OK** ▷ for *Variable 1 Range*, enter **B1:B11** ▷ click the **Variable 2 Range** box and enter **C1:C11** ▷ select **Labels** ▷ select **Output Range**, click the empty box, and enter **C15** ▷ click **OK**.

t-Test: Paired Two Sample for Means

	FI	AAS
Mean	0.6286	0.6273
Variance	0.0052	0.0044
Observations	10	10
Pearson Correlation	0.9922	
Hypothesized Mean Difference	0	
df	9	
t Stat	0.4	
P(T<=t) one-tail	0.3494	
t Critical one-tail	1.833	
P(T<=t) two-tail	0.6987	
t Critical two-tail	2.262	

p value 0.6987 ['P(T<=t) two-tail']. Based on the 5% critical value of 2.262 ('t Critical two-tail'), we have $|t|$ less than critical value indicating acceptance of H_0. As the *p* value exceeds the significance level of 0.05, we also accept H_0. Both mechanisms again specify the same conclusion that there appears insufficient evidence to suggest that the accuracy of the reported measurement differs with procedure.

8.2 Confidence Interval for Mean Difference in Responses

For paired experimentation, a confidence interval for the mean difference can be used to provide an estimate of the magnitude of the experimental difference being tested for. The confidence interval is based on that used for the mean in one sample inference, equation (2.10), through replacing $\bar{x}$ by $\bar{D}$ (mean difference) and s by s_D (standard deviation of the differences). The resultant $100(1-\alpha)\%$ confidence interval is expressed as

$$\bar{D} \pm t_{\alpha/2.n-1}(s_D/\sqrt{n}) \tag{2.21}$$

where *n* represents the number of data pairings.

Example 2.14

Referring to the copper data of Example 2.13, we will determine and interpret a 95% confidence interval for the mean difference in Cu recording.

The *difference* calculated and the related *assumption* are as stated in Example 2.13.

Confidence interval
Derivation of the confidence interval (2.21) can be carried out manually using summary information provided in Excel. In Output 2.15, we have $\bar{D} = 0.0013$ μg ml^{-1} ('Mean') and a term 'Confidence Level (95%)' of 0.0074. The latter is simply the $\pm$ part of equation (2.21) for a 95% interval, so the full 95% confidence interval for the mean difference in Cu measurements can be expressed as $0.0013 \pm 0.0074 = (-0.0061, 0.0087)$ μg ml^{-1}. As the interval straddles zero, we have further evidence for the conclusion of measurement similarity between the procedures. The interval is relatively evenly distributed either side of zero indicating that there is a similar number of negative and positive

differences reflecting no trend in the procedure differences and therefore little difference between the procedures.

8.3 Hypothesis Test for Variability

In paired experiments, interest may also lie in comparing the variability of the treatment measurements. For example, in a study comparing a new analytical procedure with a reference procedure using paired specimens, we may wish to assess whether the precision (variability) of the reported measurements from each procedure is comparable irrespective of any differences in accuracy of the reported measurements.

The variance ratio F test (2.18) for two independent samples is inappropriate in this case as the sample variance estimates s_1^2 and s_2^2 may be correlated because data recordings are based on a natural relationship between the specimens tested. In such a case, a special form of t test, summarised in Box 2.11, can be used for testing variability in paired sample experiments. The test is a modification of the t test for correlation and is referenced, in a modified form, in Snedecor and Cochran.[7] It is currently not available in Excel by default.

Box 2.11 *Variance test for paired sample experiments*

Hypotheses. As for variance ratio F test (see Box 2.9).
Test statistic. The t test statistic for testing paired sample variability is expressed as

$$t = \frac{\left(s_1^2 - s_2^2\right)\sqrt{n-2}}{2s_2 s_2 \sqrt{1-r^2}} \qquad (2.22)$$

based on the t distribution with $(n-2)$ degrees of freedom. s_1 and s_2 are the standard deviations of each treatment group, and r defines the correlation (degree of linear association) between the two data sets.
Decision rule. As one sample t test (see Box 2.4).

The correlation measure r is included to provide a measure the similarity in the trend of the measurements recorded for each pairing. If each pair of observations are of comparable magnitude or follow a similar trend, it will be close to $+1$, while a value near 0 would be indicative of little similarity in the measurements. An alternative way of computing statistic (2.22) is provided by

---

[7] G.W. Snedecor and W.G. Cochran, 'Statistical Methods', 7th Edn., The Iowa State University Press, Ames, Iowa, 1980, pp 190–191.

$$t = \frac{[(s_1^2/s_2^2) - 1]\sqrt{n-2}}{2\sqrt{(s_1^2/s_2^2)(1 - r^2)}} \qquad (2.23)$$

where s_1 and s_2 are the standard deviations for samples 1 and 2, respectively.

Example 2.15

Referring to the copper data of Example 2.13, part of the experimental objective was to assess whether the precision (variability) of the Cu measurements differed with analytical procedure. This will need the variances of each method to be tested using the test statistic (2.22).

Hypotheses
The precision aspect of the paired Cu measurements is based on assessing a null hypothesis of H_0: no difference in variability of Cu measurements between the two procedures ($\sigma^2_{FI} = \sigma^2_{AAS}$) against a two-sided alternative of H_1: variability in Cu differs with procedure ($\sigma^2_{FI} \neq \sigma^2_{AAS}$).

t Test for variances
Though the test statistic (2.22) is not available in Excel, it is possible to use Excel commands to derive it. Output 2.18 shows how to achieve this for the Cu measurement data and provides both test statistic of 1.84 ('Test Stat') and p value of 0.103 ('p Value'). The critical value for test statistic comparison is $t_{0.025,8} = 2.306$ (5% significance level, two-sided alternative, eight degrees of freedom). Since $|t|$ is less than this critical value, we accept H_0 and conclude, on the basis of the reported measurements, that it appears there is no difference in Cu variability with analytical procedure ($p > 0.05$). The p value of 0.103 would provide the same conclusion.

Summarising the Cu measurement analysis of Examples 2.13 to 2.15, we can conclude that there appears no difference between the two procedures in respect of both accuracy ($p > 0.05$) and precision ($p > 0.05$) of Cu measurement. The 95% confidence interval for the mean difference of $(-0.0061, 0.0087)$ μg ml^{-1} shows neither procedure differs markedly from the other. The results attained suggest that the FI system is acceptable and comparable to the standard AAS procedure.

Output 2.18 *Paired sample variability test for Example 2.15*

Cu measurements for FI in cells B1:B11 (label in B1) and for FAAS in cells C1:C11 (label in C1). Labels information entered into column I from cell 1 and numerical summaries in column J from cell 2.

In cell **J2**, enter **=STDEV(B2:B11)**. In cell **J3**, enter **=STDEV(C2:C11)**. In cell **J4**, enter **=PEARSON(B2:B11,C2:C11)** to generate the correlation coefficient *r*. In cell **J5**, enter **=COUNT(B2:B11)** to provide the number of pairings *n*.

In cell **J8**, enter **=((J2^2-J3^2)*SQRT(J5-2))/(2*J2*J3*SQRT(1-J4^2))** to generate the paired sample variability test statistic (2.22). In cell **J9**, enter **=TDIST(ABS(J8),J5-2,2)** to generate the *p* value of the two-tailed (2) variability test based on $(n-2)$ degrees of freedom.

Summaries	
s1	0.0719
s2	0.0663
r	0.9922
n	10
t Test	
Test Stat	1.84
p Value	0.103

Exercise 2.5

To compare the solubility of the steroid *hydrocortisone* at two temperatures, steroid specimens were randomly assigned to a pressure setting then split into two equal parts with each part randomly assigned to each temperature regime. The solubility data, recorded as mole fraction $\times 10^{-7}$, are given below. Do the collected data provide evidence to suggest that temperature affects steroid solubility in terms of both average and variability?

Pressure kg cm^{-2}	132	181	101	231	149	200	164	215	116
Temperature 35 °C	12.40	51.60	4.58	90.10	14.80	57.50	25.40	87.30	9.81
Temperature 55 °C	16.30	56.80	0.43	87.80	23.20	69.30	39.30	86.50	9.79

Source: J.R. Dean, M. Kane, S. Khundker, C. Dowle, R.L. Tranter and P. Jones, *Analyst (Cambridge)*, 1995, **120**, 2153.

2.9 SAMPLE SIZE ESTIMATION IN DESIGN PLANNING

In chemical experimentation, knowledge of how many measurements to make is important in order to ensure that sufficient data are collected to satisfy the required assessment of the experimental objectives. Sample size estimation is part of *power analysis*[8] and should be integral to design planning but is often neglected in experimentation. Power analysis enables a proposed design structure and related inference procedure to be assessed in respect of their ability to detect specified chemical differences if such are likely to be present within the data. This ability is specified as the probability of the inference procedure rejecting the null hypothesis if this is the result pointed to by the data, *i.e.* $1 - P$(Type II error). Experiments should be based on a power of at least 80%. Experiments with low power make little sense as it is unlikely they will be capable of properly answering the experimental objectives. If power analysis suggests inadequate experimental structure, there is an opportunity to re-design the structure and assess how such changes may improve power and consequently the usefulness of the planned experiment.

Estimation of sample size in simple experimentation requires that the experimenter considers likely outcomes to the experiment and sets conditions for the operation of the inference procedure. The former corresponds to specifying a difference in response means (d, practical difference of interest) which quantifies the level of treatment difference that is considered significant from a chemical perspective, *i.e.* a numerical response difference reflective of attainment of the experimental objective(s). The conditions to be set for operation of the statistical test include setting the $100\alpha\%$ significance level and the power for the planned statistical test. Simple formulae exist to help estimate sample size in simple chemical experiments. They are by no means exact but are sufficient for most practical purposes.

9.1 Sample Size Estimation for Two Sample Experimentation

In experimentation involving two independent samples, we can consider sample size estimation for either equal-sized samples or unequal-sized samples, the latter if inequality of sample size is necessary for the conduct of the planned experiment. Both are based on similar concepts and methods which we will now explain.

For a planned two-tailed test based on *equal sample sizes*, the

---

[8] M.W. Lipsey, 'Design Sensitivity: Statistical Power for Experimental Research', Sage Publications, Newbury Park, London, 1990.

experimenter must first specify the likely difference (d) between the two sample means which would be reflective of a chemical or scientific difference between the two groups. Using this difference, sample size (n) for each group can be estimated by the expression

$$n = 2(z_\beta + z_{\alpha/2})^2 / ES^2 \qquad (2.24)$$

where $\beta = 1 - (\text{Power}/100)$, z_β is such that $P(z > z_\beta) = \beta = P(\text{Type II error})$, $z_{\alpha/2}$ refers to the $100\alpha\%$ two-tailed critical z value for the proposed inference test (see Table A.4), and z corresponds to the standard normal $N(0, 1)$ distribution, *i.e.* equation (2.5) with $\mu = 0$ and $\sigma = 1$. The ES term is specified as d/σ and refers to effect size where σ represents the likely level of variability in the chemical data to be recorded based on past knowledge of the response planned to be measured.

For equation (2.24) to operate, the experimenter must pre-specify all of its components based on their requirements for the operation of the planned experiment and related statistical test. Using these constraints, we would estimate n from equation (2.24) and assess if such an estimate is acceptable. The approximate sample sizes generated by this procedure may be too large compared with what would be feasible within an experiment. Modification of the pre-specified constraints could reduce this to more appropriate levels, so consideration of the practicality of the estimate provided is important before using it as the basis of an experiment. For one-tailed testing, z_α replaces $z_{\alpha/2}$ in equation (2.24).

Example 2.16

Prior to use in laboratory experiments, an analytical laboratory carries out a check on the pH of stored distilled water. Two methods of storing the distilled water are to be compared, with each method requiring an independent random sample of specimens in equal numbers. However, the investigator is unsure of how many samples to use. A difference in mean pH of 0.5 would, in the investigator's opinion, be indicative of a significant difference between the storage regimes. Previous data suggest a variability in pH (standard deviation measure) of 0.39 which could be assumed for both storage methods. The associated two sample t test is planned to be carried out at the 5% significance level with power of at least 85%. How many specimens are necessary for each storage regime?

Experiment information
The planned experiment is designed to compare two storage methods for distilled water using a two sample structure (two-tailed test). Equal

sample sizes are required for each method so we will use equation (2.24) to estimate sample size.

Parameter determination
From the information provided, we know $d = 0.5$, $\sigma = 0.39$, and ES = $0.5/0.39 = 1.2821$. Power specification of at least 85% implies $\beta = 1 - (85/100) = 0.15$. From $\beta = P(z > z_\beta)$, Table A.4 provides $z_\beta = z_{0.15} = 1.0364$. The associated $z_{\alpha/2}$ term (two-tailed test, 5% significance level) is $z_{0.025}$ which, from Table A.4, is specified as 1.96.

Evaluation of number of specimens
With all the parameters known, equation (2.24) becomes

$$n = 2(1.0364 + 1.96)^2/1.2821^2 = 10.92 \approx 11,$$

rounding upwards to nearest integer. Estimation suggests that 11 specimens per storage regime be used. This should provide sufficient pH data from which to make suitable comparison of the storage methods.

Exercise 2.6

An experiment is planned into the determination of nitrogen in two soils by Kjeldahl digestion at a temperature of 260 °C. The experimenter believes that a mean difference of 0.4% N_2 would be sufficient to provide evidence of a nitrogen difference between the soils. Past such studies suggest a standard deviation estimate of 0.3% N_2 would be appropriate. A statistical test for comparison of mean N_2 content is planned to be carried out at the 5% significance level with power of at least 90%. Estimate the number of samples of each soil type necessary to satisfy these constraints.

For *unequal sample sizes*, estimation is based on assuming that one sample is a proportion (m) of the other, *i.e.* assume $n_2 = mn_1$. Such a form of experimentation may be necessary if ease of implementation and data collection differs with procedures to be compared. In this situation, equation (2.24) is modified to

$$n_1 = (1 + 1/m)(z_\beta + z_{\alpha/2})^2/\text{ES}^2 \tag{2.25}$$

to provide an estimate of the size of sample 1. Using $n_2 = mn_1$ provides the estimate of sample size for the second sample. For $m > 1$, power is only minimally affected by choice of unequal sample sizes but for $m < 1$, power can diminish markedly as m decreases.

9.2 Sample Size Estimation for Paired Sample Experimentation

Similar estimation and power analysis concepts can be applied if paired sample experimentation is planned. Their basis is again the experimenter's feel for a likely important difference (*d*) which will adequately reflect the chemical effect being investigated. For paired sample experimentation, sample size estimation can be achieved through use of the expression

$$n = (z_\beta + z_{\alpha/2})^2 / ES^2 \qquad (2.26)$$

where ES = d/σ_D (σ_D being an estimate of difference variability), β = $1 - (\text{Power}/100)$, and z_β and $z_{\alpha/2}$ are as defined earlier with z_α replacing $z_{\alpha/2}$ if the planned inference is to be one-tailed. Again, the estimate provided is only a guiding figure with consideration of its feasibility required before the planned experimentation occurs.

Example 2.17

A paired sample experiment is to be conducted to compare the recovery of fenbendazole from milk samples using two organic solvents, dichloromethane and ethyl acetate. It is planned to select a number of milk samples and split each into two with one half assigned randomly to each solvent. A mean difference in recovery of 3.1% is thought to be indicative of solvent difference. From past information, a standard deviation estimate of 4.7% can be assumed for the solvent recovery differences. The associated paired sample inference is to be carried out at the 5% significance level with power at least 80%. How many milk samples are necessary for this experiment?

Experiment information
The planned experiment is to compare two solvents in respect of their recovery levels of fenbendazole in milk samples using a paired experimental set-up. We will use equation (2.26) to estimate the number of milk samples (*n*) necessary to satisfy the experimental objective of a general directional difference (two-tailed test).

Parameter determination
From the information provided, we know $d = 3.1$, $\sigma_D = 4.7$, and ES = $3.1/4.7 = 0.6596$. The power constraint of at least 80% provides $\beta = 1 - (80/100) = 0.2$ and $z_\beta = z_{0.2} = 1.2816$ (Table A.4). The proposed two-tailed test at 5% significance level means $z_{\alpha/2} = z_{0.025} = 1.96$ (Table A.4).

Estimation of number of samples

With all the parameters known, (2.26) provides the sample size estimate of

$$n = (0.8416 + 1.96)^2/0.6596^2 = 18.04 \approx 19$$

rounding up to nearest integer, suggesting that 19 milk samples would be necessary. This, however, appears excessive not only from the perspective of number of samples required but also from the amount of experimentation it would entail. Increasing the level of difference deemed important can help to reduce the estimate. Using $d = 3.6$, for example, produces $n = 13$ whereas $d = 4.1$ results in $n = 11$, both of which may be more realistic. This highlights not only the importance of correctly specifying the level of difference reflecting the experimental objective but also of the need to modify expectations to enable acceptable sample size estimates to be provided.

Exercise 2.7

In a study of ethanol determination in beer, uncorrected first- and second-order FTIR spectrometry are to be compared. Beer samples are to be split into two equal parts with one part assigned to each method randomly. The experimenter is unsure of how many beer samples to assess. A mean difference of 0.15% v/v between the paired observations would, in the experimenter's opinion, provide sufficient evidence to back-up the assertion that the two methods differ. It is proposed to carry out the associated paired sample t test at the 5% significance level with approximate power of at least 80%. Assuming an estimate of standard deviation of the differences in ethanol of 0.19% v/v, how many beer samples would need to be assessed to satisfy the stated experimental objective?

Guidelines for power estimation in tests relating to means and variances are also produced by the British Standards Institution (BSI)[9] and the International Standards Organisation (ISO). Zar[10] describes an alternative trial and error approach to sample size estimation while Lipsey[8] discusses graphical approaches to power and sample size estimation. In some forms of chemical experimentation, consideration may also be necessary on the amount of material needed to achieve

---

[9] British Standards Institution, 'Guide to Statistical Interpretation of Data', BS2846, Part 5, 1977.
[10] J.H. Zar, 'Biostatistical Analysis', 3rd Edn., Prentice Hall, Upper Saddle River, New Jersey, 1996, pp 136–137.

certain limits of detection for the chemical outcome being measured.[11] This involves establishing how much test material is necessary to ensure that amount tested does not affect recording and measurement precision.

2.10 QUALITY ASSURANCE AND QUALITY CONTROL

When working in a laboratory, accuracy and reliability of measurements are clearly essential to good working practice and to the generation of confidence in the analytical results reported.[12,13] Minor errors in solutions, CRMs, procedures used, or equipment calibration can lead to inaccurate results. Periodic checking of laboratory practice and measurement recording can highlight these inaccuracies but cannot pinpoint their cause nor suggest a remedy. Errors can be reduced by regular testing of laboratories, solutions, reference materials, procedures, instruments, and intermediate results to determine if they are within the desired (tolerance or safety) limits.

Such checks of analytical measurements should be part of a Quality System[14] incorporating both *Quality Assurance* (QA) and *Quality Control* (QC). QA procedures are concerned with the continuous monitoring of laboratory practice and measurement reporting within and between laboratories to ensure the quality of information provided. QC procedures, on the other hand, are more concerned with ensuring that a laboratory's reported analytical measurements are free of error and are of acceptable accuracy for their intended purpose. Implementation of QC is based on statistical concepts such as mean, variability, and confidence intervals through the mechanism of control charts, a graphical presentation of analytical measurements.[15] QC principles are also used extensively in all types of industry including biotechnology companies, pharmaceutical companies, and food production companies as part of product monitoring.

In chemical analysis, the application of QA and QC plays an important role in areas such as establishment of reference materials, validation of analytical methods by collaborative trials, proficiency testing of analytical laboratories, and quality control of analytical data. These activities form the basis of analytical laboratory accreditation with the Laboratory of the Government Chemist (LGC) through the Valid Analytical Measurement (VAM) programme, the United

[11] M.J. Gardner, *Anal. Proc. Including Anal. Commun.*, 1995, **32**, 115.
[12] Analytical Methods Committee, *Analyst (London)*, 1989, **114**, 1497.
[13] B. King, *Anal. Proc.*, 1992, **29**, 134.
[14] E.J. Newman, *Anal. Proc. Including Anal. Commun.*, 1995, **32**, 275.
[15] E. Mullins, *Analyst (Cambridge)*, 1994, **119**, 369.

Kingdom Accreditation Service (UKAS) through the National Measurement and Accreditation Service (NAMAS), the American Society for Testing and Materials (ASTM), the BSI, and the ISO all providing relevant quality guidelines for data handling and analytical measurement.

CHAPTER 3

One Factor Experimental Designs for Chemical Experimentation

1 INTRODUCTION

In Chapter 2, methods for exploratory and inferential data analysis for simple chemical experiments were introduced. Such forms of experimentation can be easily extended to multi-sample experiments, commonly referred to as *One Factor Designs*, where the effect of different levels of a controllable factor or treatment on a measured chemical response can be investigated. Such design structures are simple to implement and represent a more efficient mode of experimentation, enabling more response information to be forthcoming without increasing the amount of experimentation markedly.

Illustrations of one factor chemical experiments include assessing how four chemists' replicate determinations of percent methyl alcohol in a chemical compound differ, comparing the replicate cadmium measurements for toxic samples recorded by three different laboratories, and comparing the yield of a reaction product for five different catalysts accounting for five different sets of experimental conditions. The factors to be tested in such experimental structures may be classified as either *qualitative* (laboratories, catalysts) or *quantitative* (temperature, quantity of catalyst, concentration of chemical). As the illustrations demonstrate, the design structure is based on testing different levels of treatment to ascertain if they vary in their effect on a measured chemical outcome, and if so, how.

The experimental designs to be outlined in this chapter, together with those of Chapters 4 and 7, are commonly used as the basis of chemical and industrial experiments in order to investigate particular experimental objectives. Examples of such objectives could include ascertaining best catalyst for a chemical reaction, best combination of chemicals for most active fertiliser, and best combination of factors to improve yield of a reaction product. Data from such experiments are generally analysed using the principle of *Analysis of Variance (ANOVA, AoV)*.

The pioneering work on the application of ANOVA to design structures was carried out by Ronald Fisher at the Rothamsted Experimental Station in the 1920s and 1930s based on agricultural experimentation for the assessment of different fertilisers on crop yield. Hence, the use of terminology such as treatment and block when discussing experimental designs. This pioneering work brought to the fore the three basic principles of experimentation: *randomisation, replication*, and *blocking*, as well as providing the base for the derivation of many fundamental statistical analysis techniques. The diversity of design structures available is extensive with usage occurring in a wide diversity of areas such as the analytical laboratory, the chemical process industry, the pharmaceutical industry, the electronics industry, the biotechnology industry, and clinical trials.

Randomisation is fundamental to ensuring data collection is free of experimentally accountable trends and patterns (unwanted causal variation). It represents the procedure undertaken to ensure that each experimental unit (sample of material) has an equal chance of being assigned to different treatments, or treatment combinations, together with the order in which independent runs of an experiment are to be performed. Randomisation can reduce the risk of results bias and can remove errors correlation, the latter important in the validation of the conditions (assumptions) required to undertake inferential data analysis.

Replication refers to the number of times experimentation is to be repeated to provide the chemical outcome data for analysis. It is tied to the principle of size of difference to be detected and the principle of power analysis (see Chapter 2, Section 9, Chapter 3, Section 7, and Chapter 4, Section 4). Essentially, it corresponds to the number of times each treatment or treatment combination is tested. Replication is particularly useful when assessing data accuracy and precision, an important aspect of chemical data handling in analytical laboratories. Provision of replication enables experimental error to be estimated, improves the precision of estimated parameters, and can reduce the level of noise (unexplained variation) in the chemical response data. Single replication (one observation per treatment combination) is acceptable if experimental constraints such as time to carry out a chemical analysis and accessibility to experimental material dictate that few experiments can be carried out. Experiments based on single replication will be introduced in Chapter 7.

Blocking represents a second measure which, in conjunction with randomisation, can minimise the effect of accountable trends and patterns on the measured response. Batches of material, or samples, to be tested are typical of blocking factors as the chemical response may vary with batch or sample tested. Blocking is therefore an extension of the pairing principle (see Chapter 2, Section 8) to account for an identifiable source of response variation, which is thought to be important but which we do not wish to investigate fully. Inclusion of a blocking factor can increase the likelihood of detecting a significant treatment effect if such an effect exists.

To illustrate the concept of blocking, consider an experiment to compare product yield from five catalysts. It could be that yield may vary depending on the experimental conditions under which the reaction occurs. In order to assess catalyst variation properly, it would be beneficial to test each catalyst under each set of conditions by using each set of conditions as a blocking element. The main factor of interest (the treatment) is the catalyst factor with the experimental conditions factor simply included to account for an effect which could be influencing the resultant response and which, if unaccounted for, could result in misleading conclusions on the catalyst effects. Through such a design, we can examine whether the different catalysts affect product yield and, if they do, how they do so.

ANOVA methods are in effect extensions of two sample procedures. They essentially attempt to separate, mathematically, the total variation within the experimental measurements into sources corresponding to the elements controlled within the experiment (the factors studied) and the elements not controlled (unexplained variation, experimental error). From this, the variation due to controlled factors can be compared against that of the unexplained variation, large ratios signifying a significant explanatory effect. These variance ratios, known as *F* ratios in honour of Ronald Fisher, provide the statistical test element of experimental designs.

Classical experimental designs are based on the specifying a *model for the measured response* and splitting the model into two parts: an element due to the factors controlled in the experiment and an element due to error or uncontrolled variation. In other words, the model attempts to describe the variation in response data in terms of the variation corresponding to controlled components (treatment variation) and the variation corresponding to uncontrolled components (unexplained variation, error variation). It is hoped that the effect of the controllable components is far greater than that of the uncontrolled components, enabling conclusions on how the factors affect the response to be formulated.

Statistical methods obviously play a key role in the interpretation of data collected from structured experiments. They help explain the features of the chemical data collected, accounting for the variation (noise) within the data. The data analysis associated with experimental designs attempts to separate the signal (response) from the noise (errors) with signals that are large relative to the noise indicating potentially important factors. It is therefore important that the problem being investigated is correctly formulated and the most appropriate data are being collected (best measure of experimental objectives).

The design structures considered in this chapter are comparative and are designed to compare treatments or the effects of different experimental conditions. They are structured in a deliberate manner in order to obtain relevant response measurements for the associated experimental objective and to conform to the associated statistical analysis procedures. Discussion will centre on how to analyse data from such designs using EDA principles, ANOVA techniques, and diagnostic checks in order to gauge the level of influence of the factors studied. Only balanced designs, where all observations are present, will be discussed though the concepts introduced can be easily extended to cater for unbalanced designs where observations may be missing from some of the treatments tested. The emphasis throughout will be on the use of Minitab output as the basis of the data presentation, analysis and interpretation, using similar principles to those of Chapter 2. Excel can handle these designs but Minitab provides a more comprehensive range of default analysis routines.

Often, however, an experiment may need to be conducted to investigate the influence, either independently or in combination, of many factors on a chemical response. This concept of *multi-factor experimentation*, which comes within the domain of factorial designs, will be discussed more fully in Chapters 4 and 7.

2 COMPLETELY RANDOMISED DESIGN (CRD)

This is one of the simplest experimental design structures on which practical chemical experiments can be based. It is so called because experimental units are randomly assigned to treatment groups (different levels of one *factor*) and because interest lies in examining whether any difference exists in the chemical outcomes associated with each treatment group. No other influencing elements beyond the treatment effect are accounted for, making this design a simple extension of two sample experimentation described in Chapter 2, Section 7. Discussion of this design structure will centre only on the balanced structure (same number of measurements for each treatment). Experimental constraints, however, may dictate that unequal replication occurs. The concepts and principles underpinning the data analysis do not differ if the design is unbalanced. The only effect is a modification of some of the associated formulae (see Chapter 3, Section 4).

Illustrations of where such a design would be appropriate are numerous. In a study into the purity of a chemical compound obtained using three different analytical procedures, we may be interested in assessing whether there is a difference between the procedures and if so, which appears best for measuring compound purity. A simple project into cadmium levels in soil may be concerned with comparing the levels in four areas of a particular site in order to see whether the areas sampled differ with respect to cadmium level and, if they do, which areas contain most or least cadmium. In the removal of toxic wastes in water by a polymer, interest may lie in comparing how temperature affects the removal of the toxic waste to ascertain which temperature setting appears best suited to the task.

2.1 Response Data

The experimental layout for a completely randomised design (CRD) comprising k treatments is shown in Table 3.1. It can be seen from this structure that each treatment is experimented on the same number of times providing n replicate measurements for each treatment. Unequal replication would be represented by columns of observations of differing length. The CRD structure is such that no factors other than the treatments are accounted for.

Table 3.1 *Completely randomised design (CRD) structure*

Treatments	1	2	.	.	k
	X_{11}	X_{12}	.	.	X_{1k}
	X_{21}	X_{22}	.	.	X_{2k}
	.	.	.	.	.
	.	.	.	.	.
	X_{n1}	X_{n2}	.	.	X_{nk}

X_{ij}, *measurement made on experimental unit i receiving treatment j*

2.2 Model for the Measured Response

Data collected from a CRD structure can be modelled by a *response model* in terms of the treatment tested and the unexplained variation (error, noise), the latter referring to such as mis-recording, instrument error, calibration error, environmental differences between samples, and contaminated solutions. For a recorded observation X_{ij} in the CRD structure, the *model* structure is

$$X_{ij} = \mu + \tau_j + \varepsilon_{ij}$$

assuming additivity of effects where $i = 1, 2, \ldots, n$ and $j = 1, 2, \ldots,$ k. The term μ (mu) refers to the grand mean and is assumed constant for all k treatments. The effect of treatment j, denoted τ_j (tau), is assumed constant for all units receiving treatment j but may differ for units receiving a different treatment. The experimental error, denoted ε_{ij} (epsilon), explains the level of response variation due to uncontrolled random sources of variation.

In general, τ_j is defined as a fixed component meaning that we assume we are experimenting on all possible treatments for such an investigation. Under such an assumption, we say that the model is a *fixed effects*, or *Model I experiment*. Treatment effects may also be classified as random when the treatments tested represent a sub-sample of all possible treatments. We will only consider fixed effect treatments in the illustrations of design structures within this chapter and Chapter 4. Experimental error is generally assumed normally distributed with the variance of this error requiring to remain constant within and across all treatments. Model specification, therefore, indicates how we think the measured response is constructed, in this case through additivity of effects, with the hope being that the treatment effect explains considerably more of the response variation than error.

To illustrate model specification, consider an experiment into three different extraction procedures and their effect on pesticide residue levels in lettuce. Three groups of six randomly selected lettuce samples are to be assigned on a random basis, to each extraction procedure, and pesticide residue levels in each specimen measured. In this study, only the extraction procedure is being controlled and as we wish to compare three procedures, a CRD would be a suitable design basis. Since residue level is to be the measured response, then the model we would specify for such an experiment would be pesticide residue level $= \mu + \tau_j + \varepsilon_{ij}$, $i = 1, 2, 3, 4, 5, 6$ (six samples per method) and $j = 1, 2, 3$ (three extraction methods), the τ_j term defining the contribution of the jth extraction procedure to measured residue level. Analysis of data from such an experiment would be based on measuring the magnitude of the procedure effect relative to error effect, a large value being indicative of procedure differences.

2.3 Assumptions

For a response model for a CRD, it is assumed that the error ε_{ij} is a normal identically distributed (NID) random variable with mean 0 and variance σ^2, *i.e.* $\varepsilon_{ij} \sim \text{NID}(0, \sigma^2)$. This means we assume the experimental response to be normally distributed and that each treatment has comparable levels of response variability. Departures from these

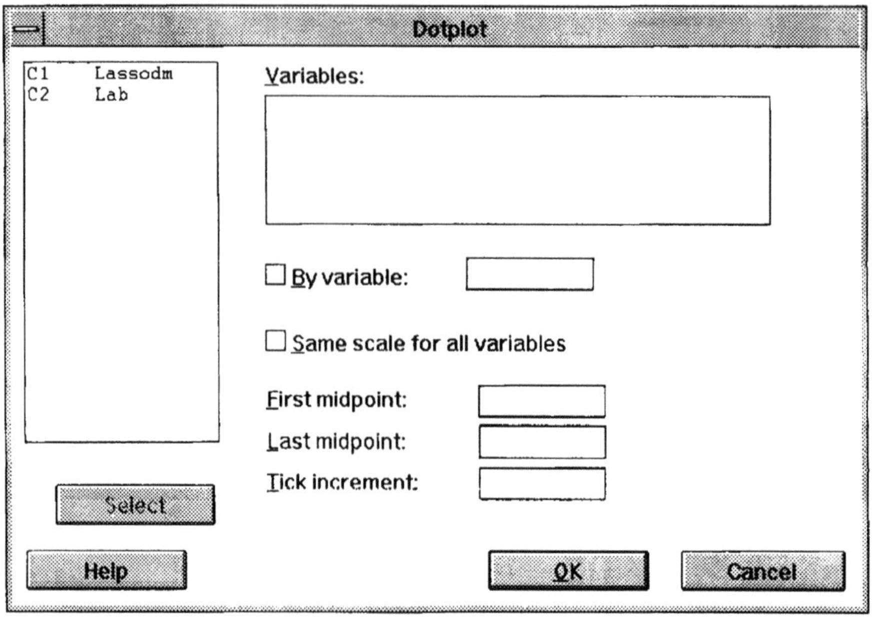

Figure 3.1 *Dotplot dialog window in Minitab*

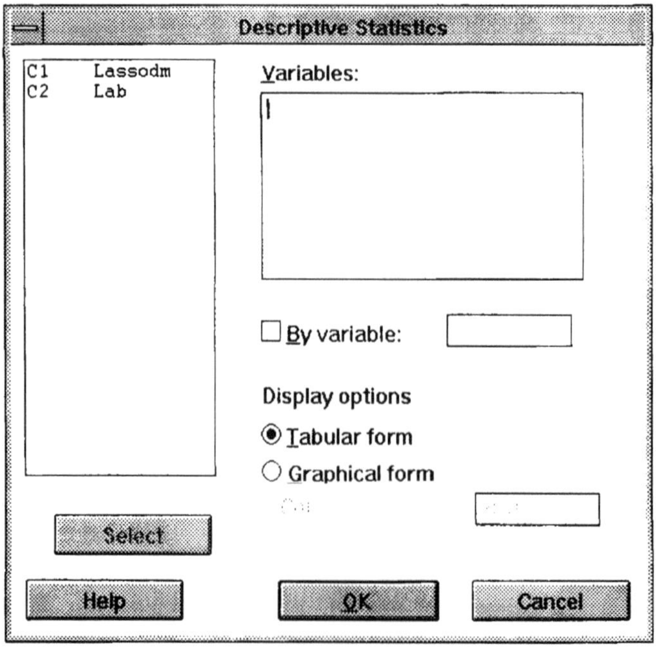

Figure 3.2 *Descriptive statistics dialog window in Minitab*

assumptions do not seriously affect the analysis results in this design structure (robust test procedure).

2.4 Exploratory Data Analysis (EDA)

As with previous illustrations, it is always advisable to examine data plots and summaries to gain insight into what they may be specifying about the experimental objectives. The associated dialog windows in Minitab for production of these illustrations of response data are shown in Figures 3.1 and 3.2. As with implementation of Excel's analysis tools, we require to fill in, leave blank, or check the relevant entries in the dialog windows for the way we wish to use the menu procedure.

Example 3.1

A reproducibility study was carried out to investigate how three laboratories performed in respect of lasalocid sodium determination in poultry feed. A portion of feed containing nominally 85 mg kg^{-1} of lasalocid sodium was sent to each laboratory who were requested to carry out 10 replicate determinations of the supplied material. The resultant lasalocid sodium measurements (mg kg^{-1}) are shown in Table 3.2. We have three laboratories with ten observations per laboratory. The experimental structure is that of a CRD with $k = 3$ and $n = 10$ with a response of lasalocid sodium in mg kg^{-1}.

Table 3.2 *Lasalocid sodium recovery measurements for Example 3.1*

Laboratory	A	B	C
	87	88	85
	88	93	84
	84	88	79
	84	89	86
	87	85	81
	81	87	86
	86	86	88
	84	89	83
	88	88	83
	86	93	83

Source: Analytical Methods Committee, *Analyst (Cambridge)*, 1995, **120**, 2175.

Minitab data entry
Data entry in Minitab utilises the Data window spreadsheet (see Figure 1.4). For this type of design, two columns of data are entered: one for

the lasalocid sodium measurements (C1) and one for the numerical codes corresponding to each laboratory (C2). Data labels are inserted in the blank row immediately below the column heading.

Entry of response data into column C1 is by laboratory, *i.e.* in three blocks of 10 measurements starting with laboratory A. The codes to be used to distinguish the laboratories are 1 for A, 2 for B, and 3 for C. This form of codes entry can be achieved as follows: Select **Calc** ▷ **Set Patterned Data** ▷ for *Store result in column*, enter **Lab** in the box ▷ for *Patterned sequence*, select the **start at** box and enter **1** ▷ for *Patterned sequence*, select the **end at** box and enter **3** ▷ select the **Repeat each value** box and enter **10** ▷ click **OK**.

Data check
Data checking can be carried out using the menu commands **Stat** ▷ **Tables** ▷ **Cross Tabulation** which can produce a data table commensurate with that displayed in Table 3.2.

Response model
As the experimental structure is that of a CRD, the response model will be,

$$\text{lasalocid sodium} = \mu + \tau_j + \varepsilon_{ij}$$

$i = 1, 2, \ldots, 10$ (10 measurement per laboratory) and $j = 1, 2, 3$ (3 laboratories). The term τ_j represents the contribution of laboratory to lasalocid sodium levels while ε_{ij} refers to the error or elements not properly considered which could affect the recorded lasalocid sodium measurements.

Assumptions
Lasalocid sodium level normally distributed and variability in lasalocid sodium similar for all laboratories.

Exploratory data analysis
Output 3.1 contains Minitab generated dotplots for each laboratory. The output highlights that lasalocid sodium levels do vary between the laboratories with B (Lab 2) appearing higher than the other two. Laboratories A (Lab 1) and C (Lab 3) provide similar results though with differing patterns. Consistency, approximated by length of the sequence of points, shows some difference with A least and B and C higher though similar in themselves.

Output 3.2 provides the numerical summaries of the lasalocid sodium

Output 3.1 *Data plot of lasalocid sodium measurements for Example 3.1*

Select **Graph** ▷ **Character Graphs** ▷ **Dotplot** ▷ for *Variables*, select **Lassodm** and click **Select** ▷ select **By variable**, click the empty box, select **Lab**, and click **Select** ▷ click **OK**.

```
Character Dotplot

 Lab                         .
  1                 .           :        :    :    :
       ---+---------+---------+---------+---------+---------+---Lassodm
 Lab                                     .
  2                      .    .    .   :    :              :
       ---+---------+---------+---------+---------+---------+---Lassodm
 Lab                 .
  3    .      .    :    .    .    :         .
       ---+---------+---------+---------+---------+---------+---Lassodm
       80.0      82.5      85.0      87.5      90.0      92.5
```

Output 3.2 *Numerical summaries of lasalocid sodium measurements for Example 3.1*

Select **Stat** ▷ **Basic Statistics** ▷ **Descriptive Statistics** ▷ for *Variables*, select **Lassodm** and click **Select** ▷ select **By variable**, click the empty box, select **Lab**, and click **Select** ▷ for *Display options*, select **Tabular form** ▷ click **OK**.

Descriptive Statistics

Variable	Lab	N	Mean	Median	TrMean	StDev	SEMean
Lassodm	1	10	85.500	86.000	85.750	2.224	0.703
	2	10	88.600	88.000	88.500	2.633	0.833
	3	10	83.800	83.500	83.875	2.616	0.827

Variable	Lab	Min	Max	Q1	Q3
Lassodm	1	81.000	88.000	84.000	87.250
	2	85.000	93.000	86.750	90.000
	3	79.000	88.000	82.500	86.000

data by laboratory where TRMean refers to the mean of the data after the bottom 5% and top 5% of observations are removed. The means (Mean) show laboratory differences with B (Lab 2) highest. Laboratories A (1) and C (3) show some difference in mean measurement but not as large as that between B and the others. All reported means differ numerically from the known nominal content of 85 mg kg^{-1} with only laboratory A providing a result that can be considered reasonably close to target. Therefore, data accuracy of the reported measurements may be called into question.

As means differ, we are best to use the *RSD* [equation (2.4)] to compare variability, the respective *RSD*s being 2.6%, 3%, and 3.1%. These figures suggest comparable precision of measurements returned by each of the laboratories though the *RSD*s do differ numerically.

2.5 ANOVA Principle and Test Statistic

The next phase of data analysis in one factor designs involves determining whether treatment differences, possibly detected in exploratory data analysis (EDA), can be shown to be statistically significant. This is where the ANOVA principle comes into play by providing the mechanism for test statistic construction and derivation based on the response model specification and statistical theory. The general procedure involves specification of hypotheses, derivation of an ANOVA table and treatment test statistic, and decision on which hypothesis best reflects the evidence the data are providing.

Hypotheses associated with the CRD experimental structure define a test of no treatment difference versus treatment difference resulting in a null hypothesis specification of H_0: no difference between treatments tested ($\mu_1 = \mu_2 = \ldots = \mu_k$, $\tau_1 = \tau_2 = \ldots = \tau_k = 0$) and alternative hypothesis specification of H_1: difference between treatments tested (at least one μ_j different, at least one $\tau_j \neq 0$). These hypotheses are general hypotheses and do not provide any information as to how treatment differences, if detected, are reflected within the response data.

The *ANOVA principle* within a CRD structure involves splitting response variation into two parts, treatment and residual, or within treatment variation, corresponding to the sources of variation within the defined response model. This variation splitting is achieved through the determination of the sum of squares (SS) terms reflecting the level of variation associated with a particular component. The main SS term is the total SS where,

$$SSTotal\ (SST) = SSTreatment\ (SSTr) + SSResidual\ (SSRes)$$

The calculational aspects and general form of ANOVA table for a CRD based experiment are summarised in Table 3.3 where the MS terms represent variance components and are part of the statistical theory associated with experimental designs.

Based on the statistical theory underpinning experimental designs, the *treatment effect test statistic* is specified as the variance ratio

$$F = MSTr/MSRes \tag{3.1}$$

Table 3.3 *General ANOVA table for a CRD experiment*

Source	df	SS		MS
Treatments	$k - 1$	$\text{SSTr} = \dfrac{\sum_{j=1}^{k} T_j^2}{n} - \dfrac{\left(\sum_{i=1}^{n}\sum_{j=1}^{k} X_{ij}\right)^2}{kn}$		$\text{MSTr} = \text{SSTr}/(k - 1)$
Residual	$k(n - 1)$	SSRes (by subtraction)		$\text{MSRes} = \text{SSRes}/[k(n - 1)]$
Total	$kn - 1$	$\text{SST} = \displaystyle\sum_{i=1}^{n}\sum_{j=1}^{k} X_{ij}^2 - \dfrac{\left(\sum_{i=1}^{n}\sum_{j=1}^{k} X_{ij}\right)^2}{kn}$		

k, number of treatments; n, number of observations per treatment; T_j, sum of the responses for treatment j; X_{ij}, the experimental response for unit i within treatment j; df, degrees of freedom; SS, sum of squares; MS, mean square

with degrees of freedom $\text{df}_1 = k-1$ and $\text{df}_2 = k(n-1)$, *i.e.* treatment degrees of freedom and residual degrees of freedom. In Minitab, generation of ANOVA information for structured experimental designs can be obtained through the menu commands **Stat** ▷ **ANOVA** ▷ **Balanced ANOVA**, the dialog window for which is illustrated in Figure 3.3. The usual procedures of selecting, filling in, and checking of the presented boxes apply.

To decide whether the evidence within the data points to acceptance or rejection of the null hypothesis, we can apply either of two *decision*

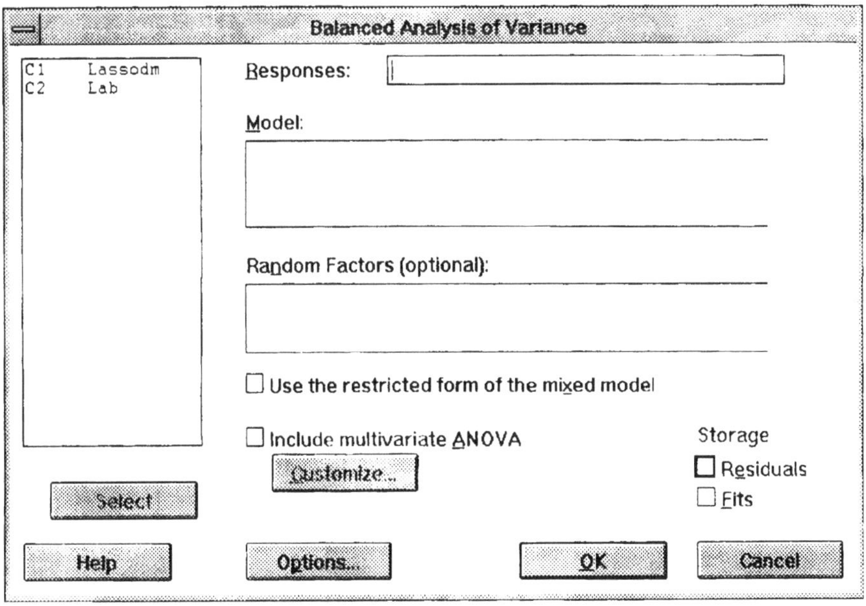

Figure 3.3 *Balanced ANOVA dialog window in Minitab*

rules provided software producing p values has been used as the basis of ANOVA table determination. The *test statistic approach* is based on the same form of decision rule as that for the variance ratio F test in two independent samples (see Chapter 2, Section 7.3) except that the degrees of freedom for critical value determination become $[k-1, k(n-1)]$, *i.e.* (treatment degrees of freedom, residual degrees of freedom). If the test statistic is less than the critical value (see Table A.3) at the $100\alpha\%$ significance level, then we accept the null hypothesis of no difference between the treatments, *i.e.* test statistic < critical value $\Rightarrow$ accept H_0 (no significant treatment differences detected). The *p value approach* depends on software usage and operates as in previous illustrations, *i.e.* p value > significance level $\Rightarrow$ accept H_0.

Example 3.2

In Example 3.1, we carried out some exploratory analysis of the lasalocid sodium data and found some evidence of laboratory differences. As the trial basis was a CRD, we now need to carry out the associated main statistical test to assess whether the detected differences are statistically significant.

Hypotheses
The basic hypotheses we are going to test are H_0: no difference in lasalocid sodium measurements between the laboratories (laboratories provide similar measurements) against H_1: difference in lasalocid sodium measurements between the laboratories (at least one laboratory returns different measurements).

ANOVA table
Output 3.3 presents the ANOVA table for the lasalocid sodium data using Minitab's ANOVA procedure. The storage variables selected are used to enable the residuals (error estimates) and predicted values (fits) to be stored in new columns for later use in diagnostic checking (see Example 3.6). The ANOVA table presented conforms to the general one shown in Table 3.3 whereby the variation in lasalocid sodium measurements is assumed to be affected by the laboratory and error (residual) only. The table shows that SSTotal = SSLabs + SSResiduals with the MS terms derived as the ratio SS/DF for both laboratory and error (residual) effects as per the theory.

Test statistic
Equation (3.1) specifies that the treatment F test statistic is the ratio of the MSTr term to the MSRes term. The figure of 9.49 in the F column

Output 3.3 *ANOVA table for Example 3.2*

Select **Stat** $\triangleright$ **ANOVA** $\triangleright$ **Balanced ANOVA** $\triangleright$ for *Responses*, select **Lassodm** and click **Select** $\triangleright$ select the **Model** box, select **Lab**, and click **Select** $\triangleright$ for *Storage*, select **Residuals** and select **Fits** $\triangleright$ click **OK**.

Analysis of Variance (Balanced Designs)

Factor	Type	Levels	Values		
Lab	fixed	3	1	2	3

Analysis of Variance for Lassodm

Source	DF	SS	MS	F	P
Lab	2	118.467	59.233	9.49	0.001
Error	27	168.500	6.241		
Total	29	286.967			

of the ANOVA table in Output 3.3 is the numerical estimate of this test statistic for the lasalocid sodium data, *i.e.* 59.233/6.241. As we have both a test statistic and p value, we can use *either* approach to assess the specified hypotheses.

For the *test statistic approach*, we know that $F = 9.49$. As $k = 3$ and $n = 10$, the degrees of freedom of the test statistic are $df_1 = k - 1 = 3 - 1 = 2$ and $df_2 = k(n-1) = 3(10-1) = 27$. The critical value, from Table A.3, for the 5% significance level, is $F_{0.05,2,27} \approx 3.32$ using the entry at (2, 30) as the nearest equivalent critical value. Since test statistic exceeds critical value, we must reject the null hypothesis and conclude that there appears sufficient evidence to suggest that the laboratories differ statistically in their lasalocid sodium measurements ($p < 0.05$).

For the *p value approach*, the column P in the ANOVA table in Output 3.3 provides the p value for the laboratories' F test as 0.001. This low value also indicates rejection of the null hypothesis and therefore implies the same conclusion of laboratory difference as reached above ($p < 0.05$).

3 FOLLOW-UP ANALYSIS PROCEDURES FOR ONE FACTOR DESIGNS

The treatment effect F test (3.1) only checks for no difference versus difference. It provides no information as to how treatment differences, if detected, are occurring. Further analysis, therefore, is necessary to enhance the conclusion and to pinpoint how treatments are differing in their level of response. Techniques for doing this include graphical approaches (*standard error plot, main effects plot*) and inferential approaches (*multiple comparisons, linear contrasts, orthogonal polyno-mials*). Generally, a mixture should be considered to enable appropriate and relevant data interpretations to be forthcoming.

3.1 Standard Error Plot

A simple way to examine the treatments for difference is to produce a standard error plot of the data by treatment, in the form of mean $\pm$ two standard errors, and examine the plot for trends and patterns. This is essentially a graphical presentation of approximate 95% confidence intervals for each of the treatment group means and is only appropriate provided most of the response variation is due to treatment differences. Overlap of intervals would suggest no statistical difference between the compared treatments with non-overlap, indicative of statistically significant differences. In Minitab, such a follow-up plot can be generated through use of the menu commands **Graph** $\triangleright$ **Interval Plot**, the dialog window for which is shown in Figure 3.4.

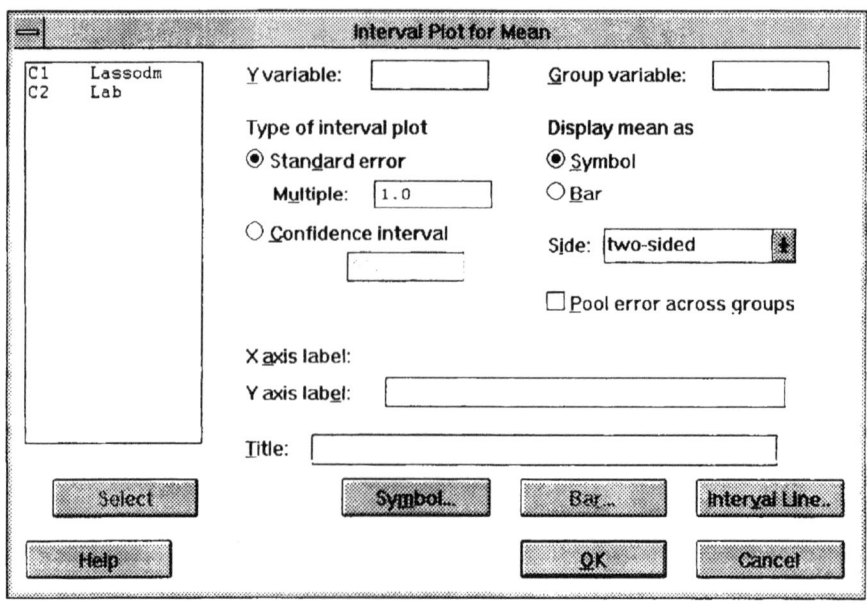

Figure 3.4 *Interval plot dialog window in Minitab*

Example 3.3

Example 3.2 indicated that mean lasalocid sodium measurement differed statistically with laboratory. However, we do not know how this difference is occurring so we need to pinpoint how the laboratories

Output 3.4 *Two standard error plot Example 3.3*

Select **Graph** ▷ **Interval Plot** ▷ for *Y variable*, select **Lassodm** and click **Select** ▷ for *Group variable*, select **Lab** and click **Select** ▷ select the **Standard error Multiple** box and enter **2.0** ▷ select **Pool error across groups** ▷ select the **X axis label** box and enter **Laboratory (1 - A, 2 - B, 3 - C)** ▷ select **Y axis label** box and enter **Lasalocid sodium** ▷ click **OK**.

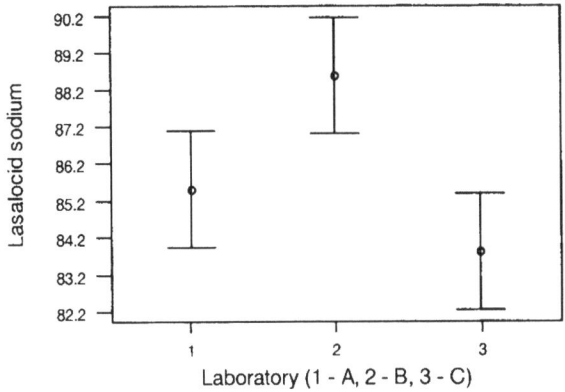

differ through production of a simple two standard error plot as shown in Output 3.4.

The plot highlights the significant difference between laboratory B (Lab 1) and C (Lab 3) as there is no interval overlap. Laboratories B and A show possible difference as there is only a partial overlap of the intervals. The statistical similarity between A (1) and C (3) is clearly seen as the intervals show a strong degree of overlap. Based on this, we can state that there appears some evidence that reported measurements laboratory B differ from those of the other two laboratories.

An alternative graphical approach to this form of data plot is to produce a *main effects plot*, a simple linear plot of treatment means. This can be achieved within Minitab through the use of the **Stat** ▷ **ANOVA** ▷ **Main Effects Plots** menu commands.

3.2 Multiple Comparisons

Often after significant treatment effect is suggested as a result of the treatment F test, we should compare each treatment against every other treatment to assess how the treatments differ. Such comparisons are called *post hoc*, or *a posteriori*, comparisons which for k treatments, result in $c = k(k-1)/2$ possible pairwise comparisons. Using a $100\alpha\%$ significance level for the c comparisons means that the overall likelihood of at least one pairwise comparison incorrectly indicating a significant difference, referred to as the *experimentwise error rate*, can grow to approximately $100[1-(1-\alpha)^p]\%$. For $\alpha = 0.05$ (5% significance level) and $k = 3$ treatments, there are three possible pairwise comparisons resulting in an experimentwise error rate of as much as 14.3%.

To protect against making this type of error, a *Multiple Comparison* should be used to help identify treatment differences. Such pairwise contrasts, which include *Fisher's Least Significant Difference (LSD)*, *Tukey's Honestly Significant Difference (HSD)*, *Duncan's Multiple Range Test*, and the *Student–Newman–Keuls (SNK)* procedure,[1-3] use the experimental error to carry out all possible comparisons in a fashion which restrains the overall significance level of all such comparisons to $100\alpha\%$. They all work on the same principle though some are more conservative whereas others are more liberal. Conservative procedures can make attaining significant differences more difficult and often result in certain treatment differences being quoted as insignificant when they really are significant (false negative). Liberal procedures provide the opposite effect and can result in significant differences being quoted for treatment differences which are not significant (false positive).

We will consider only the *Student–Newman–Keuls (SNK) procedure*. This procedure is based on the use of the studentised range statistic and focuses upon a series of ranges rather than a collection of differences using the position of each treatment mean, in an ordered list, as a base. The steps involved in the implementation of the SNK procedure are outlined in Box 3.1. It is currently not available in Minitab so requires manual derivation.

The results from this comparison are often summarised in a simple graphical form where pairs of treatment means that do not differ significantly are underlined and treatments that do are not underlined. For example, A B C would say that treatments B and C are not significantly different from each other (underlined) and that treatment A differs statistically from B and C (no underlining).

Example 3.4

For the reproducibility study of Example 3.1, we know that the laboratories differ statistically. In Example 3.3, we graphically exam-

---

[1] R.O. Kuehl, 'Statistical Principles of Research Design and Analysis', Duxbury Press, Belmont, California, 1994, pp. 84–103.

[2] D.C. Montgomery, 'Design and Analysis of Experiments', 4th Edn., Wiley, New York, 1997, pp. 93–108.

[3] J.H. Zar, 'Biostatistical Analysis', 3rd Edn., Prentice Hall, Upper Saddle River, New Jersey, 1996, pp. 211–225.

Box 3.1 *SNK multiple comparison procedure*

Hypotheses. The null hypothesis to be tested in this form of multiple comparison is H_0: no difference between compared treatments (means similar) with alternative specifying treatment difference.

Test statistics. The numerical difference between each pair of treatment means become the test statistics for the pairwise comparisons.

Critical values. Obtain, for the $100\alpha\%$ significance level, the least significant range

$$W_r = q_{\alpha,r,resdf}\sqrt{\frac{MS\ Res}{n}} \tag{3.2}$$

where $r = j + 1 - i$ is the number of steps the treatments in positions i and j are apart in the ordered list of means, resdf is the residual degrees of freedom, MSRes is the residual mean square in the ANOVA table, $q_{\alpha,r,resdf}$ is the $100\alpha\%$ critical value from the tables of the Studentised Range Statistic (Table A.5), and n is the number of observations per treatment.

Decision rule. If the numerical difference in treatment means for treatments in positions i and j ($|\bar{T}_{(i)} - \bar{T}_{(j)}|$) is less than the associated critical value W_r, we accept H_0 and declare that the compared treatments appear not to differ statistically.

ined the laboratory differences and showed laboratory B appeared to differ from the other two. We will now carry out the SNK multiple comparison procedure at the $\alpha = 0.05$ (5%) significance level following the guidelines provided in Box 3.1. The design structure provides us with $n = 10$ and $k = 3$, and the ANOVA table in Output 3.3 provides resdf = 27 and MSRes = 6.241.

Hypotheses

The null hypothesis we will test is H_0: no difference in lasalocid sodium measurements between the laboratories being compared.

Test statistics

The laboratory means and table of differences in means, the test statistics, are presented in Table 3.4. Laboratory order is shown to be C, A, and B.

Critical values

The corresponding 5% critical values of W_r, from evaluation of equation (3.2), are shown in Table 3.5 with the q values obtained

Table 3.4 *SNK test statistics for Example 3.4*

	Mean	Lab C 83.8	Lab A 85.5	Lab B 88.6
Lab C	83.8	–	1.7	4.8 *
Lab A	85.5		–	3.1 *
Lab B	88.6			–

*, difference statistically significant at the 5% significance level

Table 3.5 *Critical values for W_r for SNK multiple comparison of Example 3.4*

r	2	3	
$q_{0.05,r,27}$	2.89	3.49	(using df = 30 as nearest approximation)
W_r	2.28	2.76	

from Table A.5 for the combination of terms (df = 30, $r = 2$, 3). To illustrate the interpretation of results, we will look at all laboratory comparisons.

C versus A: the test statistic for this comparison, from Table 3.4, is 1.7. These laboratories are in positions $i = 1$ (1st) and $j = 2$ (2nd) in the ordered list so the critical for this comparison corresponds to $r = 2 + 1 - 1 = 2$ (adjacent), *i.e.* use $W_2 = 2.28$ from Table 3.5. As test statistic is less than critical value, we accept H_0 and indicate that there appears no evidence of a statistical difference in lasalocid sodium measurements between laboratories C and A ($p > 0.05$).

C versus B: the comparison test statistic from Table 3.4 is 4.8. These laboratories are in positions $i = 1$ (1st) and $j = 3$ (3rd) in the ordered list so the critical required corresponds to $r = 3 + 1 - 1 = 3$ (one step apart), *i.e.* use $W_3 = 2.76$ from Table 3.5. As test statistic exceeds critical value, we reject H_0 and conclude that there appears sufficient evidence to imply a statistical difference in lasalocid sodium measurements of these two laboratories ($p < 0.05$). A symbol '*' (significant at the 5% level) is inserted in Table 3.4 at the corresponding test statistic to indicate this result.

A versus B: the necessary test statistic from Table 3.4 is 3.1. These laboratories are in positions $i = 2$ and $j = 3$ in the ordered list so the critical value to use for this comparison corresponds to $r = 3 + 1 - 2 = 2$ (adjacent), *i.e.* use $W_2 = 2.28$ from Table 3.5. As test statistic exceeds critical value, we reject H_0 and conclude that there appears sufficient evidence to suggest a statistical difference in lasalocid sodium measurements between laboratories A and B ($p < 0.05$). Again, a symbol '*' is inserted in Table 3.4 to indicate a statistical difference has been detected.

From these comparisons, as summarised in Table 3.4, we can deduce that laboratory B differs from the others on a statistical basis and that A and C cannot be distinguished between. In summary form, we would set out these results as C̲ A̲ B.

3.3 Linear Contrasts

A contrast represents a linear combination of treatment means (or totals) that is of interest to the experimenter to explore. It is useful for specific comparisons between treatments or groups of treatments where such *planned* or *a priori* comparisons are decided upon prior to experimentation,[4] *e.g.* control versus modifications of control, or HPLC based procedures versus colorimetric procedures. The general form of a linear contrast L is

$$L = c_1\mu_1 + c_2\mu_2 + \ldots + c_k\mu_k = \sum c_j\mu_j$$

where μ_j is the mean of treatment j and the c_j terms represent constants such that $\sum c_j = 0$.

In a four treatment CRD, the general form of a linear contrast is given by $L = c_1\mu_1 + c_2\mu_2 + c_3\mu_3 + c_4\mu_4$ with specific contrasts based on this general form. For example, the contrast $L = \mu_2 - \mu_4$ compares average performance of treatments 2 and 4 ($c_1 = 0$, $c_2 = 1$, $c_3 = 0$, $c_4 = -1$, $\sum c_j = 0$), while the contrast $L = (\mu_1 + \mu_2)/2 - (\mu_3 + \mu_4)/2$ compares average performance of treatments 1 and 2 against average of treatments 3 and 4 ($c_1 = \frac{1}{2} = c_2$, $c_3 = -\frac{1}{2} = c_4$, $\sum c_j = 0$). Multiple comparisons, described in Section 3.2, are simple applications of pairwise linear contrasts. The steps associated with analysis of linear contrasts, which are currently not available in Minitab by default, are summarised in Box 3.2.

If the confidence interval contains 0, the contrast in question is said to be not statistically significant resulting in the compared treatments being declared not statistically different. If the confidence interval does not contain 0, we conclude that a statistically significant difference between the compared treatments appears to exist and interpret the result accordingly. There also exist formal t and F test statistic equivalents of equation (3.3) representing simple re-statements of the confidence interval in the form of test statistics.[1,2,3]

---

[4] Reference 1, pp. 67–75.

Box 3.2 *Linear contrasts*

Hypotheses. The null hypothesis for a contrast is expressed as H_0: $L = 0$, *i.e.* the effect being tested by the specified contrast is not significant.
Estimate of L. To estimate L, the mean response for the specified treatments are substituted into the appropriate contrast expression and a numerical estimate for the contrast calculated.
Confidence interval for contrast. The $100(1 - \alpha)\%$ confidence interval for a contrast L is given by

$$\text{estimate of } L \pm t_{\alpha/2,resdf} \sqrt{\sum_j c_j^2 \left(\frac{MS\ Res}{n} \right)} \qquad (3.3)$$

where $t_{\alpha/2,resdf}$ is the $100\alpha/2\%$ value for residual degrees of freedom (resdf) from Table A.1, n is the number of observations for each treatment, and MSRes is the residual mean square. Generally, a 95% confidence interval is used.

Example 3.5

For the reproducibility study of Example 3.1, suppose laboratory A is the laboratory used routinely for the described analysis and that it would be of interest to compare A against the other two laboratories. This comparison can be expressed in contrast form as $L = A - (B + C)/2$.

Hypotheses
The null hypothesis associated with this contrast is specified as H_0: $L = 0$, *i.e.* mean lasalocid sodium measurements of laboratory A are not statistically different from other laboratories.

Estimate of L
The contrast $L = A - (B + C)/2$ can be estimated as $85.5 - (88.6 + 83.8)/2 = -0.7$ using the laboratory means from Output 3.2. The negative value of this estimate implies that A appears to have a marginally lower mean lasalocid sodium compared to the other two laboratories, as also highlighted in the exploratory data analysis of Example 3.1.

Confidence interval
We know that resdf $= 27$, $n = 10$, and MSRes $= 6.241$ (from Output 3.3). The 95% confidence interval (3.3), using a critical value of $t_{0.025,27} = 2.052$ from Table A.1, is given by

$$-0.7 \pm (2.052)\sqrt{(1 + (-1/2)^2 + (-1/2)^2)\frac{6.241}{10}} = -0.7 \pm 1.99$$

which results in an interval $(-2.69, 1.29)$ mg kg^{-1}. As this interval contains 0, we conclude that the mean lasalocid sodium levels for A does not differ significantly from that of the other two laboratories ($p > 0.05$). However, the interval is not equally spread either side of 0 and given the negative tendency of the interval, we can conclude that A appears lower but not statistically so.

3.4 Orthogonal Polynomials

A further follow-up procedure for assessing treatment differences concerns *orthogonal polynomials* which are special types of contrasts used when a quantitative experimental factor has been set at evenly spaced intervals over a specified scale, *e.g.* three temperature levels set at 5°C apart. They enable investigation of whether the treatment effect is linear or polynomial (quadratic, cubic) in its trend, *i.e.* split effect into linear/quadratic/cubic elements where appropriate.[5]

For k such treatments, it is possible to extract from the treatment sum of squares polynomial effects up to order $(k-1)$. For example, for $k = 3$ temperature levels (treatments), a significant treatment effect may contain linear and/or quadratic trends. Orthogonal polynomials can be constructed, as with linear contrasts, using treatment means (or totals). The technique involves calculating an average effect estimate as $\sum c_j \bar{X}_j$ based on each treatment mean where c_j represents the coefficients of the appropriate orthogonal polynomials (linear or quadratic function). From this, a test statistic of the form

$$F = \frac{n(\sum c_j \bar{X}_j)^2 / \sum c_j^2}{MS\ Res} \tag{3.4}$$

with (1, resdf) degrees of freedom can be constructed to assess the statistical significance of the specific trend effect. n is the number of measurements per treatment, values for c_j and $\sum c_j^2$ are tabulated for combinations of trend type and number of treatments, and MSRes is the residual mean square.

3.5 Model Fit

This can be checked, approximately, by looking at the error ratio of residual to total sum of squares, *i.e.* (SSRes/SST)%. If this is large, say

---

[5] Reference 1, pp. 76–84.

greater than 10–20%, it implies that the suggested model for the experimental response may not be explaining all the variability in the response data (high error), suggesting there may be missing factors not included in the experiment which may be affecting the response. If this ratio is small (error low), it may suggest model fit to be good depending on the nature of the experiment.

For the reproducibility study of Example 3.1, this ratio is estimated as 58.7%. This percent value is very high indicating that, though statistical difference between the laboratories was detected, it is likely that other factors, such as technician doing the analysis, instruments used, and possible storage conditions, could be affecting the lasalocid sodium measurement reported by the laboratories.

3.6 Checks of Model Assumptions (Diagnostic Checking)

Model assumption checks are an important diagnostic tool within the data analysis process and are carried out by means of *residual (error) analysis*. The model assumptions in question are responses equally variable for each treatment and response data normally distributed. Such analysis can, in addition, assess model applicability and may show that insignificant factors in terms of mean response may be important in their effect on response variability. Invalidity of assumptions can affect the sensitivity of the statistical tests and may increase the true significance level of the tests carried out, resulting in potentially incorrect parameter estimates and invalid conclusions. Using the difference between the observed data and the corresponding predicted X values (the *residuals*) as estimates for the errors, these assumptions can be explored graphically as summarised in Box 3.3. Chemical data generally satisfy these assumptions and so conclusions drawn from ANOVA analysis are usually acceptable.

A further diagnostic check is to plot the collected response data against model predictions. Such a plot should exhibit a pattern of points lying on a 45° line passing through the origin. Deviations from this line signify data variability problems.

Illustrations of likely treatment residual plots are shown in Figure 3.5. Figure 3.5A illustrates a plot with columns of equal length and no discernible pattern, indicative that equal variability across all treatments appears acceptable. Figure 3.5B, on the other hand, shows differing column lengths implying that, for the related response data, equal variability for all treatments may be in doubt. A more formal statistical test for assessing the equal variance assumption is *Bartlett's test for homogeneity*.[6]

---

[6] Reference 3, pp. 204–206.

Box 3.3 *Mechanisms of diagnostic checking*

Equal variability. Plot residuals against each factor. The plot should generate a horizontal band of points equally distributed around 0. Patterns in the plot, *e.g.* different lengths of columns, reflect non-constant variance (heteroscedasticity), indicative that a variance stabilising transformation of the data may be required, while bunching will indicate inappropriate response model. Outliers, corresponding to untypical residuals, appear as values lying beyond $\pm 3\sqrt{\text{MSRes}}$. Their presence can affect the data interpretations and so it is necessary to check the source of such values carefully to determine why the associated measurements have resulted in an unusual residual value.

Normality. Normality of the measured response can be checked by examining a normal probability plot of the residuals (residuals against normal scores) where the normal scores are determined using the position value of each residual and the standard normal distribution (expected value of order statistics, see Chapter 6, Section 9.1). Departure from linearity may indicate response non-normality and extreme points suggest outliers.

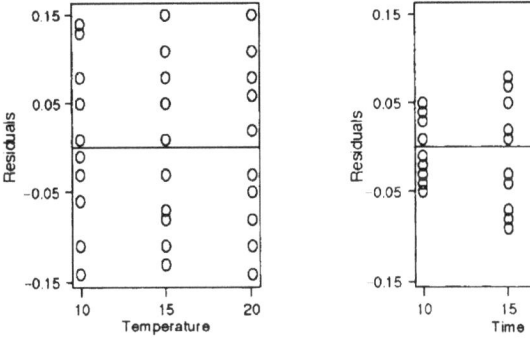

Figure 3.5 *Illustration of treatment residual plots*

Figure 3.6 provides two illustrations of likely normal plots of residuals where normal scores refer to standard normal values (normal with mean 0 and variance 1) derived from the position ordering of the residuals. Figure 3.6A illustrates a reasonable linear trend suggestive of response normality, while Figure 3.6B has more of an S-shape pattern indicative of non-normal response data. The 'straightness' of this plot can be measured by the correlation coefficient of the points in the plot and can be tested using a special form of correlation test, the Ryan–Joiner test,[7] where high correlation near 1 is consistent with assuming normality for the response variable.

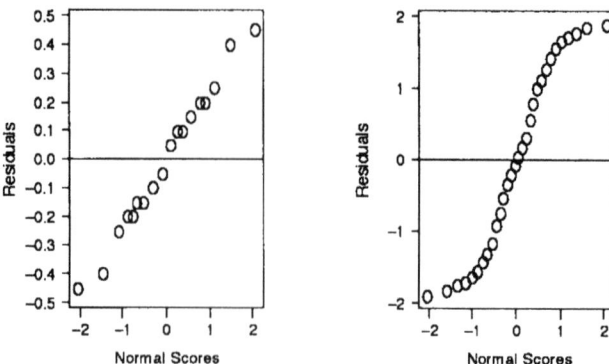

Figure 3.6 *Illustration of normal plots of residuals*

If either variability or normality is inappropriate, we can apply a data transformation to the experimental data, corresponding to a simple re-scaling of the data, to enable variability and normality assumptions to be more acceptable (see Section 8) or we can use alternative non-parametric procedures to analyse the measured response data (see Chapter 6).

Example 3.6

The final part of the analysis of the lasalocid sodium data of Example 3.1 is the diagnostic checking aspect based on the residuals (measurements minus predicted values) and fits generated by the fitted model (see Example 3.2).

Diagnostic checking
Plotting the residuals (error estimates) against the laboratories is the first step in the diagnostic checking of the reported data. The plot, shown in Output 3.5, is columnar because of the code values used to specify the laboratory factor. All columns are reasonably similar,

---

[7] T.A. Ryan and B.L. Joiner, Technical Report Minitab Inc., 1990, 1–1.

Output 3.5 *Diagnostic checking: laboratory plot for Example 3.6*

Select **Graph** $\triangleright$ **Plot** $\triangleright$ for *Graph 1 Y*, select **RESI1** and click **Select** $\triangleright$ for *Graph 1 X*, select **Lab** and click **Select**.

Select **Frame** $\triangleright$ **Axis** $\triangleright$ select the **Label 1** box and enter **Laboratory** $\triangleright$ select the **Label 2** box and enter **Residuals** $\triangleright$ click **OK**.

Select **Annotation** $\triangleright$ **Footnote** $\triangleright$ for *Footnote 1*, enter **1 - Lab A, 2 - Lab B, 3 - Lab C** in box $\triangleright$ click **OK**.

Select **Frame** $\triangleright$ **Reference** $\triangleright$ for *Direction 1*, enter **Y** in box $\triangleright$ select the **Positions 1** box and enter **0.0** $\triangleright$ click **OK** $\triangleright$ click **OK**.

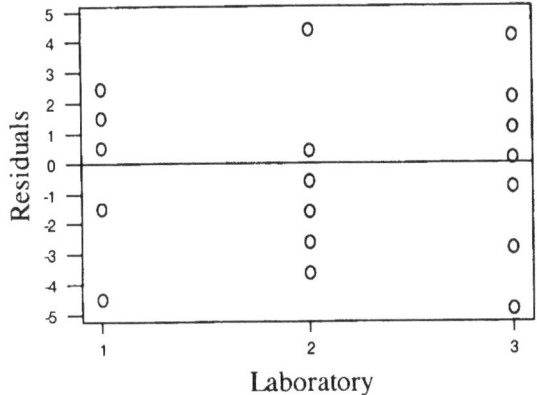

1 - Lab A, 2 - Lab B, 3 - Lab C

though there is a suggestion of column length differences, and gaps in the columns hinting that the equal variability assumption may be in doubt. This minor difference in variability was hinted at within the exploratory analysis presented in Example 3.1.

Minitab has a macro accessible through the menu commands **Stat** $\triangleright$ **ANOVA** $\triangleright$ **Residual Plots** for production of residual model diagnostics. Output 3.6 shows the plots produced by this macro for the recovery experiment of Example 3.1. The main plots of interest are the 'Residuals *vs.* Fits' plot (bottom right) and the 'Normal Plot of Residuals' (top left). The fits plot, however, exhibits a similar pattern to the laboratory residual plot in Output 3.5 and so results in the same interpretation. The normal plot looks reasonably linear, indicative that normality for the lasalocid sodium data may be acceptable.

The outcome of the diagnostic checking appears to be that the reported sodium data may not conform to equal variability but do to normality. A variance stabilising transformation of such data may be worthy of consideration in future experiments when such a response is measured.

Output 3.6 *Diagnostic checking: normal and fits plots for Example 3.6*

Select **Stat** ▷ **ANOVA** ▷ **Residual Plots** ▷ for *Residuals*, select **RESI1** and click **Select** ▷ for *Fits*, select **FITS1** and click **Select** ▷ click **OK**.

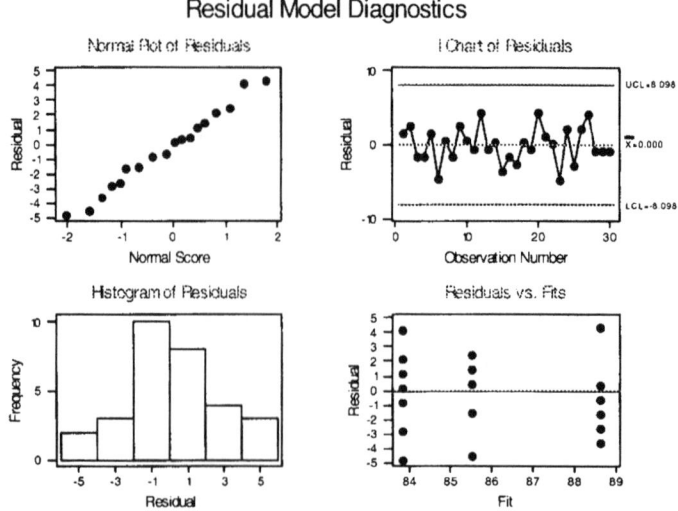

From Examples 3.1 to 3.6, we can see that the steps in the analysis associated with a CRD experimental structure are model specification, exploratory data analysis, inferential data analysis through treatment effect test and follow-up to pinpoint treatment differences, and diagnostic checking. Through use of these steps, a full and comprehensive analysis of the data collected can be forthcoming. On the basis of such an approach, we can say that the lasalocid sodium recovery experiment of Example 3.1 shows difference between the laboratories ($p < 0.05$) with laboratory B providing significantly different measurements from the other two laboratories ($p < 0.05$). Only laboratory A appears to be reporting measurements near to target though all laboratories report acceptably consistent measurements. There is also a suggestion that transformation of the response may be worth considering in future comparable studies.

Exercise 3.1

A chemical engineer is studying a newly developed polymer to be used in removing toxic wastes from water. Experiments were conducted at five different temperatures I to V representing incrementally increasing temperature. The response measured was the percentage of impurities removed by the treatment. Does temperature affect percent removal?

Temperature I is the presently used temperature for this removal process. Is the average percentage removed at this temperature different from the average of the other four temperature settings?

Temperature	I	II	III	IV	V
	40	36	49	47	55
	35	42	51	49	60
	42	38	53	51	62
	48	39	53	52	63
	50	37	52	50	59
	51	40	50	51	61

Source: J.S. Milton and J.C. Arnold, 'Introduction to Probability and Statistics', 2nd Edn., McGraw-Hill, New York, 1990: reproduced with the permission of The McGraw-Hill Companies.

4 UNBALANCED CRD

Discussion in Sections 2 and 3 has centred on explaining the analysis components of a balanced CRD (equal numbers of measurements for each treatment). In some cases, experimental constraints may dictate that unequal numbers of measurements be made for each treatment, as illustrated in Table 3.6 for an experiment comparing three analytical methods. All the outlined analysis components, and software information, in Sections 2 and 3 apply equally to an unbalanced CRD, though modification of related formulae is necessary to cater for the imbalance in design structure.

Application of the *ANOVA principle* to an unbalanced CRD still rests on the split of response variability into treatment and residual elements. The associated ANOVA table format, illustrated in Table 3.7, is unchanged though some minor changes occur in degrees of freedom and sum of squares calculations.

General treatment differences are still assessed by the *F* test statistic (3.1). Specific treatment comparisons, by means of multiple comparisons

Table 3.6 *Unbalanced CRD structure*

Analytical method	A	B	C
	x	x	x
	x	x	x
	x	x	x
	x	x	x
		x	x
		x	

x, denotes a reported measurement

Table 3.7 *General ANOVA table for unbalanced CRD*

Source	df	SS	MS
Treatments $k-1$		$\mathrm{SSTr} = \sum\limits_{j=1}^{k}\dfrac{T_j^2}{n_j} - \dfrac{\left(\sum_{i=1}^{n}\sum_{j=1}^{k}X_{ij}\right)^2}{\sum_{j=1}^{k}n_j}$	$\mathrm{MSTr} = \mathrm{SSTr}/(k-1)$
Residual	$\sum n_j - k$	SSRes (by subtraction)	$\mathrm{MSRes} = \mathrm{SSRes}/[\sum n_j - k]$
Total	$\sum n_j - 1$	$\mathrm{SST} = \sum\limits_{i=1}^{n}\sum\limits_{j=1}^{k}X_{ij}^2 - \dfrac{\left(\sum_{i=1}^{n}\sum_{j=1}^{k}X_{ij}\right)^2}{\sum_{j=1}^{k}n_j}$	

k, number of treatments; n, number of observations per treatment; T_j, sum of the responses for treatment j; X_{ij}, the experimental response for unit i within treatment j; df, degrees of freedom; SS, sum of squares; MS, mean square

or linear contrasts, must also be considered with appropriate modifications to account for the imbalance. For the SNK follow-up, calculation of the least significant range (3.2) is modified to

$$W_r = q_{\alpha,r,resdf}\sqrt{\frac{MS\ Res}{2}(1/n_i + 1/n_j)} \qquad (3.5)$$

for the comparison of treatments in positions i and j in the ordered list of means. In a similar way, the $100(1-\alpha)\%$ confidence interval for a linear contrast L, equation (3.3), must also be modified to

$$\text{estimate of } L \pm t_{\alpha/2,resdf}\sqrt{\left(\sum_j \frac{c_j^2}{n_j}\right) MS\ Res} \qquad (3.6)$$

to account for the unequal numbers of observations associated with each treatment.

5 USE OF CRD IN COLLABORATIVE TRIALS

The CRD structure is often used as the basis of *collaborative trials* (CT) of analytical laboratories when validating an analytical procedure using reference material. Interest, in such a quality assurance study, lies in assessing method precision rather than accuracy through estimation of the within laboratory (residual) and between laboratory (treatment) variation. The former measures the sampling variation due to differences between experimental units (sample material) assigned to the laboratories for testing, while the latter is essentially the measurement variance incorporating possible instrumental errors.

The within laboratory variation is denoted σ_0^2 and is estimated by the residual mean square (MSRes term in the ANOVA table). The between laboratory variation is denoted by σ_1^2 and is estimated as

$$\sigma_1^2 = (MSTr - MSRes)/n \qquad (3.7)$$

If σ_0^2 is considerably less than σ_1^2, then variation in measurements is primarily due to differences between the laboratories. If σ_0^2 is considerably in excess of σ_1^2, then the F test statistic will be non-significant, hinting that response variation is primarily due to measurement error and not laboratory differences. If $\sigma_0^2 \approx \sigma_1^2$, then just as much variation in the response is caused by measurement error as difference between the laboratories tested.[8]

From these variation estimates, we can assess the precision of the method of analysis in terms of repeatability and reproducibility values.[9] The standard deviation of repeatability S_r is expressed as $\sqrt{\sigma_0^2}$ while the standard deviation of the reproducibility S_R is given as $\sqrt{(\sigma_0^2 + \sigma_1^2)}$ with both representing properties of the method. Generally, it is expected that $S_R{}^2$ will be greater than $S_r{}^2$ so formal statistical testing of the results is generally unnecessary. Upper 95% confidence limits can be expressed as $r = 2.8s_r$ for repeatability and $R = 2.8s_R$ for reproducibility. The former provides a rational basis for assessing whether replicate measurements from a standard method are similar enough, while the latter provides a similar base but in respect of replication between laboratories. In addition, in food and drug analysis, Horwitz's generalisation of $\sigma_r/\sigma_R \approx 0.6$ is often used to assess measurement precision.

Example 3.7

Consider the reproducibility study described in Example 3.1. From the design structure and ANOVA table presented in Output 3.3, we know that $n = 10$, MSTr $= 59.233$, and MSRes $= 6.241$.

Based on this, we can estimate σ_0^2 as 6.241 and σ_1^2 as $(59.233 - 6.241)/10 = 5.2992$ from equation (3.7). Since σ_0^2 marginally exceeds σ_1^2, we can conclude that there is a suggestion that response variation is caused more by measurement error than by differences between the laboratories. This suggests error plays a major part in the response measurements reported by the laboratories as highlighted when model fit was discussed in Section 3.5.

[8] Laboratory of the Government Chemist, VAMSTAT Software, Anova, A collaborative trial, 1–5.
[9] P. Brereton and H. Wood, *Anal. Proc.*, 1992, **29**, 186.

The standard deviation estimates for repeatability and reproducibility are $S_r = \sqrt{\sigma_0^2} = 2.498$ and $S_R = \sqrt{(\sigma_0^2 + \sigma_1^2)} = 3.397$, respectively. In this case, Horwitz's generalisation of σ_r/σ_R can be estimated as 2.498/3.397 = 0.735 which suggests measurement precision is in doubt in this food analysis study.

6 RANDOMISED BLOCK DESIGN (RBD)

The experimental error in a CRD based experiment arises from the difference between responses of experimental units (samples of material) within treatments, the within group variation. The influence of this error can be reduced further by considering blocking the experiment and testing each treatment in every block. Introduction of a blocking factor can help explain more of response variation and thereby improve the likelihood of detecting true treatment differences if they exist.

Each block represents a replication and the corresponding structure is referred to as a *randomised block design* (*RBD*). The blocks generally correspond to homogeneous groups, the units of which are similar to one another, *e.g.* batches of material or analysts. An RBD is essentially an extension of the paired sample experimentation concept described in Chapter 2, Section 8.

To illustrate its underlying philosophy, consider a study into the effect on reaction time of a chemical process of four catalysts using five batches of chemical. Such an experiment could be blocked by batch, as variation between batches could substantially affect reaction time. The first batch of chemical could be split into four equal parts to form one block with the four catalysts randomly assigned to a part of the split batch. The next batch would be split similarly to form the next block and this would continue until all five batches had been split and the catalysts assigned accordingly. Table 3.8 illustrates this design layout. We should note that the batches are randomised within the blocks, *i.e.* blocking represents a restriction on randomisation and so an RBD is not a CRD.

Table 3.8 *Possible design layout for reaction time experiment*

	Catalyst	A	B	C	D
	1	Part 3	Part 2	Part 1	Part 4
	2	Part 4	Part 3	Part 2	Part 1
Batch	*3*	Part 1	Part 2	Part 4	Part 3
	4	Part 2	Part 4	Part 3	Part 1
	5	Part 3	Part 1	Part 2	Part 4

Table 3.9 *Randomised block design (RBD) structure*

	Treatments	1	2	.	.	k
	1	X_{11}	X_{12}	.	.	X_{1k}
Blocks	2	X_{21}	X_{22}	.	.	X_{2k}
	.	.	.	.	.	.
	.	.	.	.	.	.
	n	X_{n1}	X_{n2}	.	.	X_{nk}

X_{ij}, *measurement made on the experimental unit within block i receiving treatment j*

6.1 Response Data

The experimental layout for an RBD structure comprising k treatments (or factor levels) and n blocks is shown in Table 3.9 where each treatment is tested on each block of material. The structure is such that there are k experimental units in each block providing a total of kn experimental units as in the CRD. The difference between the two designs lies in the inclusion of the blocking factor thought to influence the response but not requiring full assessment. Again, the structure is such that only one treatment factor is being investigated.

6.2 Model for the Measured Response

For a recorded observation X_{ij} within an RBD experiment, the suggested response model structure is

$$X_{ij} = \mu + \beta_i + \tau_j + \varepsilon_{ij}$$

where μ, τ_j and ε_{ij} are as for the CRD (see Section 2.2), and β_i defines the effect of block i, assumed constant for all units within block i, but which could differ for units within different blocks (often a fixed component). This additive response model again specifies that response variation is explained by the effect of the controlled factors (blocks and treatments) plus the effect of uncontrolled factors (error). The inherent *assumptions* are again equal response variability for all treatments and normality of response.

6.3 Exploratory Data Analysis (EDA)

As with previous illustrations, initial examination of data plots and summaries should be considered to gain initial insight into what the data specify about the experimental objectives.

6.4 ANOVA Principle, Test Statistics, and Follow-up Analysis

As in the CRD, we need to determine if treatment differences are of statistical significance. Again, the ANOVA principle provides the background to this aspect of the data analysis providing the mechanisms for test statistic construction and derivation based on response model specification. The general procedure involves specification of hypotheses, derivation of an ANOVA table and treatment test statistic, and decision on which hypothesis best reflects the evidence the data are providing.

Hypotheses for an RBD-based experiment can be specified in the same way as those for the CRD, *i.e.* H_0: no treatment difference versus H_1: treatment difference.

As with the CRD, application of the *ANOVA principle* involves decomposing response variation into the sources of variation specified in the RBD model namely, blocks, treatments, and residual. This gives rise to the general ANOVA table shown in Table 3.10 where,

$$SSTotal\,(SST) = SSBlocks\,(SSBl) + SSTreatment\,(SSTr) + SSResidual\,(SSRes)$$

The MS terms represent variance components and are, as with the CRD, part of the statistical theory underpinning experimental designs. Determination of the importance of the treatment effect is again based on the ratio of treatment variance to residual variance while block effect assessment is based on the ratio of block variance to residual variance.

Table 3.10 *General ANOVA table for an RBD experiment*

Source	df	SS	MS
Blocks	$n-1$	$SSBl = \dfrac{\sum_{i=1}^{n} B_i^2}{k} - \dfrac{\left(\sum_{i=1}^{n}\sum_{j=1}^{k} X_{ij}\right)^2}{kn}$	$MSBl = SSBl/(n-1)$
Treatments	$k-1$	$SSTr = \dfrac{\sum_{j=1}^{k} T_j^2}{n} - \dfrac{\left(\sum_{i=1}^{n}\sum_{j=1}^{k} X_{ij}\right)^2}{kn}$	$MSTr = SSTr/(k-1)$
Residual	$(k-1)(n-1)$	SSRes (by subtraction)	$MSRes = SSRes/[(k-1)(n-1)]$
Total	$kn-1$	$SST = \sum_{i=1}^{n}\sum_{j=1}^{k} X_{ij}^2 - \dfrac{\left(\sum_{i=1}^{n}\sum_{j=1}^{k} X_{ij}\right)^2}{kn}$	

n, number of blocks; k, number of treatments; B_i, sum of responses for block i; X_{ij}, experimental response for block i and treatment j; T_j, sum of responses for treatment j; df, degrees of freedom; SS, sum of squares; MS, mean square

From the statistical theory underpinning experimental designs, the *treatment effect test statistic* is specified as

$$F = MSTr/MSRes \qquad (3.8)$$

with degrees of freedom $df_1 = (k-1)$ and $df_2 = (n-1)(k-1)$, *i.e.* treatment degrees of freedom and residual degrees of freedom. Essentially, this is the same test statistic as occurred in the CRD. The block effect test statistic is specified as

$$F = MSBl/MSRes \qquad (3.9)$$

with $df_1 = (n-1)$ and $df_2 = (n-1)(k-1)$ degrees of freedom though formal use of this test is not strictly necessary.

To decide whether the evidence within the data points to acceptance or rejection of the null hypothesis, either of the two standard *decision rules* can be used provided that software producing p values has been used for ANOVA table determination. The *test statistic approach* is based on the same form of decision rule as previously, *i.e.* test statistic < critical value ⇒ accept H_0 (no significant treatment differences detected). The *p value approach* depends on software usage and operates as previously, *i.e.* p value > significance level ⇒ accept H_0.

It is also useful in an RBD analysis to carry out a *block effect check* using the p value for the blocking factor. If it is less than 0.1, then use of blocking has been beneficial to explanation of response variation whereas if greater than 0.1, inclusion of the blocking factor has not been as beneficial as hoped.

As with the CRD, this treatment F test is simply a general test of treatment differences with further analysis, through *standard error plot* (Section 3.1), *multiple comparison* (Section 3.2), or *linear contrasts* (Section 3.3) necessary to pinpoint specific treatment differences if they are found to exist. The final check in RBD analysis is the *diagnostic checking* of the residuals.

In summary, we can see that the steps in the analysis associated with an RBD experimental structure are model specification, exploratory data analysis, inferential data analysis through treatment effect test and follow-up to pinpoint treatment differences, and diagnostic checking. Use of these analysis components, which are the same as those considered for a CRD based experiment, enables a full and comprehensive analysis of the collected data to be forthcoming.

Example 3.8

Malarial patient blood sulfadoxine measurements, in $\mu g \ ml^{-1}$, were determined from finger-prick blood specimens by three analytical methods: HPLC, a filter photometer, and a spectrophotometer, the latter two corresponding to colorimetric assay methods. Blood was sampled from eight patients where the patients acted as the blocking element in an RBD structure ($k = 3$, $n = 8$) with samples split equally between the analysis methods. The purpose of the experiment was to compare the analytical methods for any difference between them. Also of interest was a comparison between HPLC and the colorimetric assay methods. The recorded data are presented in Table 3.11.

Table 3.11 *Blood sulfadoxine measurements for Example 3.8*

	Method	HPLC	Filter photometer	Spectrophotometer
	1	11.2	10.5	9.8
	2	32.4	41.3	39.5
	3	54.8	64.1	50.3
Patient	4	19.8	26.2	24.3
	5	12.8	17.2	19.1
	6	32.8	40.4	39.6
	7	6.5	4.7	5.2
	8	19.7	26.2	24.7

Source: M.D. Green, D.L. Mount and G.D. Todd, *Analyst (Cambridge)*, 1995, **120**, 2623.

Data entry

Data entry in Minitab for RBD data follows the same principles outlined in Example 3.1 for CRD data though now three columns of data need to be entered: response (C1), block codes (C2), and treatment codes (C3). Blood sulfadoxine data entry into column C1 can be achieved through block by block input as the result for (patient 1, HPLC) then the result for (patient 1, filter photometer) then the result for (patient 1, spectrophotometer) then the result for (patient 2, HPLC) and so on.

The patient codes (1 to 8) are entered into column C2 and as each patient has three results associated with them, each code must appear three times. Based on the form of response data entry, we need to enter the codes as three 1s then three 2s then three 3s and so on. Method codes are entered into column C3 and as each has been tested on eight samples, each code must appear eight times. From the mode of response data entry, we need to enter the codes as 1 then 2 then 3 and repeat this pattern seven times more to produce codes relevant to response data entry. The menu commands **Calc** ▷ **Set Patterned Data** can be used to simplify codes entry.

Response model
As the design structure is that of an RBD, then the model for the response will be,

$$\text{blood sulfadoxine} = \mu + \beta_i + \tau_j + \varepsilon_{ij}$$

$i = 1, 2, \ldots, 8$ (eight patients) and $j = 1, 2, 3$ (three procedures). The term β_i defines patient effect, τ_j the analytical method effect, and ε_{ij} the error effect. In addition, we assume that blood sulfadoxine measurements are normally distributed and that variability in these measurements is similar for all three procedures.

Hypotheses
The null and alternative hypotheses for the method effect within this RBD experiment are H_0: no difference in blood sulfadoxine with analytical procedure against H_1: difference in blood sulfadoxine with analytical procedure.

Exploratory data analysis
Exploratory analysis will be based on the Minitab plots and summaries presented in Output 3.7 where Method 1 refers to HPLC, Method 2 to filter photometer, and Method 3 to spectrophotometer. The plot highlights similar spaced out results for each method with no obvious difference between them. Consistency of measurement differs slightly indicating similarity between methods though the range covered by all is high. The means show HPLC least and spectrophotometer highest. *RSD*s are 66.1%, 66.8%, and 58.7% highlighting large but similar variability in analytical method results.

ANOVA table
The Minitab output presented in Output 3.8 contains the ANOVA table for the blood sulfadoxine data and conforms to the general pattern shown in Table 3.10 with response variation assumed to be affected by patient source of sample, analytical method, and residual error. The table shows that SSTotal = SSPatient + SSMethod + SSResidual with the MS terms derived as the ratio SS/DF for each of the patient, method, and residual effects as per the underlying theory.

Output 3.7 *Initial summaries of blood sulfadoxine data for Example 3.8*

Menu commands as Output 3.1.

```
Character Dotplot

Method
1
                  .     . .      . .          :                    .
        +---------+---------+---------+---------+---------+-------Sulfad
Method
2
           .        .       .       :           :                .
        +---------+---------+---------+---------+---------+-------Sulfad
Method
3
          .    .         . .           :            .
        +---------+---------+---------+---------+---------+-------Sulfad
        0        12        24        36        48        60
```

Menu commands as Output 3.2

Descriptive Statistics

Variable	Method	N	Mean	Median	TrMean	StDev	SEMean
Sulfad	1	8	23.75	19.75	23.75	15.71	5.55
	2	8	28.83	26.20	28.83	19.27	6.81
	3	8	26.56	24.50	26.56	15.60	5.51

Variable	Method	Min	Max	Q1	Q3
Sulfad	1	6.50	54.80	11.60	32.70
	2	4.70	64.10	12.18	41.08
	3	5.20	50.30	12.13	39.57

Test statistic
Equation (3.8) specifies the method F test statistic as the ratio of MSTr to MSRes. The figure of 5.06 in the F column for the method effect in the ANOVA table of Output 3.8 has been evaluated as this ratio (51.71/ 10.23). Again, both test statistic and p value approaches are available for testing treatment differences statistically.

Using the *p value approach* alone, we have the p value for the method effect F test as 0.022. This is below the significance level of 5% ($\alpha = 0.05$) implying rejection of H_0 and the conclusion that it appears blood sulfadoxine measurements differ according to analytical procedure used ($p < 0.05$).

Output 3.8 *ANOVA table for Example 3.8*

Select **Stat** ▷ **ANOVA** ▷ **Balanced ANOVA** ▷ for *Responses*, select **Sulfad**
and click **Select** ▷ select the **Model** box, enter **Patient Method** ▷ for *Storage*,
select **Residuals** and select **Fits** ▷ click **OK**.

Analysis of Variance (Balanced Designs)

Factor	Type	Levels	Values							
Patient	fixed	8	1	2	3	4	5	6	7	8
Method	fixed	3	1	2	3					

Analysis of Variance for Sulfad

Source	DF	SS	MS	F	P
Patient	7	5887.75	841.11	82.25	0.000
Method	2	103.43	51.71	5.06	0.022
Error	14	143.17	10.23		
Total	23	6134.34			

Block effect

Though the objectives of this experiment do not require analysis of the
patient (blocking) effect, it is advisable to assess its effectiveness in order
to confirm whether blocking has been beneficial. From Output 3.8, the
patient *F* test statistic is 82.25 (ratio 841.11/10.23) with *p* value quoted as
0.000 approximately, *i.e.* $p < 0.0005$. This implies that there is evidence
of a substantial difference between the patients chosen, and that
blocking has been beneficial to the explanation of response variation.

Multiple comparison

To pinpoint the detected method differences, we must validate these
differences on a statistical basis. The SNK multiple comparison procedure, at the $\alpha = 0.05$ (5%) significance level, will again fulfill this role.
The design structure provides us with $n = 8$ and $k = 3$, and from the
ANOVA table in Output 3.8, we have resdf = 14 and MSRes = 10.23.
The null hypothesis to be tested in this follow-up is H_0: no difference in
blood sulfadoxine between analytical methods compared. The method
mean sulfadoxine levels and the table of differences in means, the test
statistics for the method comparisons, are presented in Table 3.12.

  The corresponding 5% critical values W_r, stemming from evaluation
of equation (3.2), are shown in Table 3.13 with the *q* values read from
Table A.5 for the combination of terms (df = 14, $r = 2$, 3). Using the
principles outlined in Example 3.4, we can assess the method differences
for any which show statistical difference.

Table 3.12 *SNK test statistics for Example 3.8*

	Mean	HPLC 23.75	Spectrophotometer 26.56	Filter photometer 28.83
HPLC	23.75	–	2.81	5.08 *
Spectrophotometer	26.56		–	2.27
Filter photometer	28.83			–

*, difference statistically significant at the 5% significance level

Table 3.13 *Critical values for W_r for SNK multiple comparison of Example 3.8*

r	2	3
$q_{0.05,r,14}$	3.03	3.70
W_r	3.43	4.18

For the comparison of HPLC and spectrophotometer, the corresponding test statistic from Table 3.12 is 2.81. These methods are in positions $i = 1$ (1st) and $j = 2$ (2nd) in the ordered list. The critical value to use for comparison of this test statistic corresponds to $r = 2 + 1 - 1 = 2$ (adjacent), *i.e.* use $W_2 = 3.43$ from Table 3.13. As test statistic is less than critical value, we accept H_0 and indicate that there appears insufficient evidence of a statistical difference in sulfadoxine measurement between these procedures ($p > 0.05$).

For the HPLC and filter photometer comparison ($i = 1, j = 3, r = 3$), the test statistic is 5.08. The associated critical value is 4.18. As test statistic exceeds critical value, we conclude significant difference between these methods ($p < 0.05$). For the comparison of spectrophotometer and filter photometer ($i = 2, j = 3, r = 2$), the test statistic is 2.27 and associated critical value is 3.43 leading to the conclusion that no statistical difference in sulfadoxine measurement exists between these methods ($p > 0.05$).

From these follow-up results, we can deduce that only the HPLC measurements differ from the filter photometer results on a statistical basis and that no other statistical differences between the methods can be detected.

Contrast of HPLC against colorimetric methods
The investigator was also interested in comparing whether HPLC and colorimetric methods differ from one another. This comparison can be expressed in contrast form as

$$L = \text{HPLC} - (\text{filt} + \text{spectro})/2.$$

The null hypothesis for this contrast is H_0: $L = 0$, *i.e.* mean sulfadoxine for HPLC is not different from mean level of the other two methods. This contrast is estimated as $23.75 - (28.83 + 26.56)/2 = -3.945$ using the method means specified in Output 3.7. The negative value of the contrast estimate implies that HPLC indicates less sulfadoxine, on average, compared with the other two methods.

We know that resdf = 14, $n = 8$, and MSRes = 10.23 (from Output 3.8). The 95% confidence interval (3.3), using a critical of $t_{0.025,14} = 2.145$ from Table A.1, is given by

$$-3.945 \pm (2.145)\sqrt{(1 + (-1/2)^2 + (-1/2)^2)\frac{10.23}{8}} = -3.945 \pm 2.971$$

which results in the interval $(-6.916, -0.974)$ µg ml^{-1}. The interval does not contain zero so we can conclude that average sulfadoxine measurement for HPLC differs statistically from that associated with the colorimetric methods ($p < 0.05$). The interval values suggest HPLC results can be lower by anything from 0.974 µg ml^{-1} to 6.916 µg ml^{-1}, on average, compared with the colorimetric method results.

Diagnostic checks
Lastly, we require to analyse response model residuals. The plotting of residuals (error estimates) against the methods, shown in Output 3.9, shows that all columns appear reasonably similar in length indicating that the equal variability assumption appears acceptable. There is, however, a hint of outliers in the second and third columns which may be distorting the picture.

Output 3.9 *Diagnostic checking: analytical method plot for Example 3.8*

Menu commands as Output 3.5.

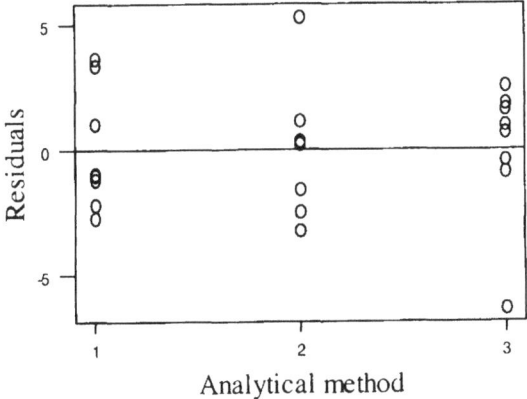

1 - HPLC, 2 - Filt Photom, 3 - Spectroph

Output 3.10 *Diagnostic checking: fits and normal plot for Example 3.8*

Menu commands as Output 3.6.

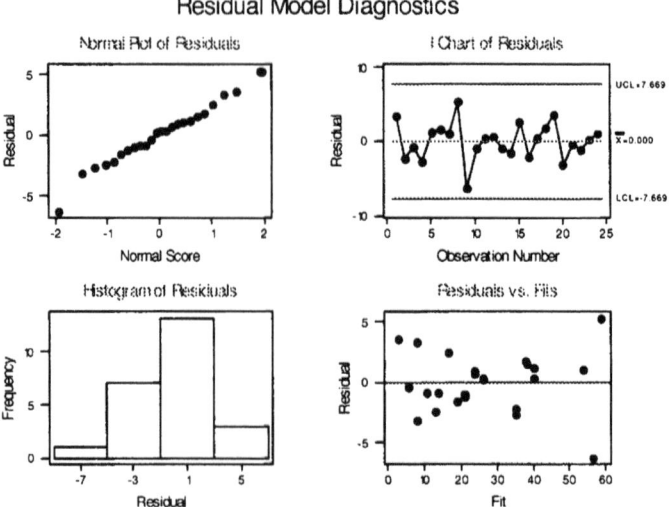

Residual Model Diagnostics

Output 3.10 shows the 'Residuals *vs.* Fits' plot (bottom right) and the 'Normal Plot of Residuals' (top left). The former shows a butterfly pattern suggestive of inappropriate model fit. The normal probability plot, on the other hand, exhibits a relatively straight line pattern but two outliers are again suggested. Normality of blood sulfadoxine measurements appears acceptable.

Summarising the analyses of the blood sulfadoxine data, we can conclude that the analytical procedures differ ($p < 0.05$) and are producing widely differing results. It appears also that HPLC looks to be generating different data from the colorimetric methods ($p < 0.05$). The outcome of the diagnostic checking appears to be that both assumptions, equal variability and normality, are satisfied by the sulfadoxine measurements reported.

Exercise 3.2

An industrial laboratory experiment was carried out to compare different procedures for the measurement of selenium in urine. The general procedure consisted of wet oxidising the organic matter, separating the selenium by distillation and precipitation, dissolving the metal and titrating the selenium with 0.001N sodium thiosulphate. Because the oxidation reaction was slow, four variations B to E of these steps were devised for comparison with the general procedure, A. Since it is not easy to prepare and maintain a reservoir of this physiological material, a randomised block design was used blocking on each subject's 24-hour urine specimen. Analysis of each specimen (block) in the experiment was carried out whenever time was available in the laboratory. The results of the experiment are given below and represent mg Se per 100 ml sample. Do the procedures differ? Compare average performance of procedure A (general procedure) with the average performance of its variants.

Wet oxidation procedure		*A*	*B*	*C*	*D*	*E*
	1	27.8	26.0	26.7	24.0	28.6
	2	21.6	16.8	19.5	18.5	19.6
	3	17.5	11.9	13.7	10.5	16.6
	4	28.9	25.8	24.6	24.1	29.3
Subject	5	26.6	25.7	24.4	24.9	28.6
	6	36.8	34.4	34.6	31.4	36.3
	7	20.4	16.1	19.0	14.1	19.4
	8	23.9	19.7	18.9	18.3	21.5
	9	30.8	25.9	27.1	24.2	28.6

Source: G.T. Wernimont, 'Use of Statistics to Develop and Evaluate Analytical Methods', Association of Analytical Chemists, Arlington, Virginia, 1985: reproduced with the permission of AOAC International, Inc.

7 ADDITIONAL ASPECTS OF ONE FACTOR DESIGNS

7.1 Missing Observations in an RBD Experiment

In experiments, observations may be lost due to contamination of experimental material or instrument malfunction. When this occurs within an RBD based experiment, we can still analyse the remaining data using the outlined analysis techniques by estimating the missing value in such a way as to minimise the residual mean square. For one missing observation, we can estimate it, rounded to conform to the accuracy and nature of the response data, as

$$X' = \frac{nB' + kT' - G'}{(n-1)(k-1)} \tag{3.10}$$

where B' is the total of the block containing the missing observation, T' the total of the treatment containing the missing observation, and G' is the overall total of all recorded observations. The residual degrees of freedom and total degrees of freedom in the ANOVA table are each reduced by one to reflect the loss of information the estimation process gives rise to. The *General Linear Models* (*GLM*) technique, a multiple regression based procedure for unbalanced designs, is an alternative to this estimation process.

7.2 Efficiency of an RBD Experiment

The efficiency of an RBD structure relative to a CRD structure for the same experimental units can be estimated as

$$E = \left[\frac{(n-1)MSBl + n(k-1)MS\,Res}{(kn-1)MS\,Res} \right] * 100\% \qquad (3.11)$$

If E is greater than 100%, the RBD is more efficient in comparing treatment means and blocking has provided a more sensitive test of treatment effects. E less than 100% implies that there has been a loss in sensitivity by using the RBD structure and the precision of the treatment comparisons has been reduced. For Example 3.8, $E = 2572\%$ which readily confirms the appropriateness of including subjects as a blocking factor. Checking the blocking effect p value, as highlighted in Example 3.8, provides a comparable conclusion (see Section 6.4).

8 POWER ANALYSIS IN DESIGN PLANNING

In Chapter 2, Section 9, the power analysis[10] aspect of experimentation was introduced and its importance in design planning explained. When using formal design structures, the same concepts and principles apply in order to ensure that the best design, commensurate with the experimental objective(s), is being implemented. In experimental designs, when testing a null hypothesis of no difference between treatments, the ability to detect the alternative hypothesis that at least one treatment differs when it is correct is of fundamental importance. Without this ability, the treatment effect statistical test will lack power (probability of detecting a false null hypothesis) and be unable to answer the question being investigated adequately. We must therefore be sure that the statistical test planned can detect chemical differences in the data if such exist.

In CRD and RBD based experiments, power analysis can be approached in a number of ways. Most are based on using the non-central F distribution characterised by the numerator and denominator degrees of freedom of the relevant F test statistic (treatment f_1, residual f_2), and a third parameter referred to as the *non-centrality parameter*.

[10] M.W. Lipsey, 'Design Sensitivity: Statistical Power for Experimental Research', Sage Publications, Newbury Park, London, 1990

Table 3.14 *Degrees of freedom for treatment tests in one factor designs*

	f_1	f_2
CRD	$k-1$	$k(n-1)$
RBD	$k-1$	$(k-1)(n-1)$

Table 3.14 summarises the relevant degrees of freedom for one factor designs while Table A.6 provides power estimates for the non-central $F(f_1, f_2)$ distribution for planned tests at the 5% significance level.

To use these tables in power analysis, we require to specify a *minimum detectable difference* δ between the treatments which, if detected, would provide the investigator with chemical, or scientific, proof of treatment differences. This difference essentially reflects the smallest numerical difference we expect to detect between two population means, *i.e.* the smallest difference between any two treatments. Using this pre-set estimate, we compute the term ϕ (phi)

$$\phi = \sqrt{\frac{n\delta^2}{2k\sigma^2}} \tag{3.12}$$

which is related to the non-centrality parameter. In equation (3.12), n is the number of observations per treatment, δ the minimum detectable difference, k the number of treatments, and σ^2 is the residual mean square estimate (MSRes estimate).

Examining Table A.6, we can see that the larger the value of ϕ, then the larger the power, and from equation (3.12), we detect that ϕ can increase by increasing n, increasing δ, reducing k, or decreasing the residual mean square estimate. Specification of these parameters, together with the significance level ($100\alpha\%$) for the proposed treatment F test, enables this power analysis procedure to be implemented to assess the viability of a proposed one factor experiment.

8.1 Power Estimation

For power estimation, the parameters n, δ, k, σ^2, and the significance level ($100\alpha\%$) of the planned statistical test of treatments need to be specified. Using these pre-set values, we evaluate ϕ from equation (3.12) and use Table A.6 for the characteristics combination f_1, f_2, and ϕ to estimate the power of the treatment effect test at the 5% significance level. Ideally, as in Chapter 2 Section 9, a power estimate in excess of 80% is desirable.

Example 3.9

Consider a CRD based experiment involving the comparison of four treatments using a total of five samples for each treatment. We wish to estimate the power of the treatment effect F test, at the 5% significance level, for a minimum detectable difference of 6.75. The estimate for the residual mean square, from previous information, is assumed to be 7.05. We want to estimate the power of the planned F test for treatment differences to assess if it exceeds the target of 80%.

Information
This is a CRD experiment with $k = 4$ treatments and $n = 5$ samples per treatment. From Table 3.11, we have $f_1 = 4 - 1 = 3$ and $f_2 = 4(5 - 1) = 16$. We also know that the minimum detectable difference is $\delta = 6.75$, the estimate of residual mean square $\sigma^2 = 7.05$, and that the treatment effect test is to be carried out at the 5% ($\alpha = 0.05$) significance level.

Estimate of ϕ
From equation (3.12), we estimate ϕ as $\sqrt{\dfrac{(5)(6.75)^2}{(2)(4)(7.05)}} = \sqrt{4.039} = 2.01$

Power estimation
An extract from Table A.6 based on $f_1 = 3$ and $f_2 = 16$ is presented in Table 3.15. Given that ϕ is estimated to be 2.01, it can be seen that the power is estimated to be between 85.6% and 91.8%. As $\phi = 2.01$ is near the tabulated entry for ϕ of 2.0, we could assume power is approximately 86%. As this is in excess of the target of 80%, it would appear that the proposed design structure is acceptable if the specified level of treatment difference is likely to exist and it is desired to prove this statistically.

Table 3.15 *Extract from Table A.6 for Example 3.9*

ϕ	0.5	1.0	1.2	1.4	1.6	1.8	2.0	2.2	2.6	3.0
Power %	10.1	28.8	40.3	53.0	65.6	76.7	85.6	91.8	98.0	99.7

Exercise 3.3

The concentration, in ppm, of chromium in a body of water near an industrial waste disposal site is to be investigated by a regional water company using an RBD structure. The body of water is to be sampled at five different locations (increasing distance from site) and at four depths at each location with depth acting as the blocking factor. The water company is interested in detecting a minimum location effect of 7.85 ppm at the 5% significance level. Estimate the power of the proposed significance test, assuming a residual mean square estimate of 8.65.

8.2 Sample Size Estimation

Power analysis calculations are most often reversed to estimate the size of the sample (n) required in an experiment to ensure a treatment effect F test power of at least 80%. In other words, we want to determine the number of replicate measurements necessary to ensure the experiment has a good chance of detecting significant treatment differences, if such are occurring in the chemical outcome data. Estimation is based on trial and error using an initial guess and refining it until a power close the desired figure is reached. Example 3.10 will provide a suitable illustration.

Example 3.10

Before being used in production, chemical raw material is often checked for purity. For chemical powders, for instance, powder samples are heated and the residual ash content, which represents the impurities, weighed. The purity, expressed as a percentage, is then determined by

$$purity = [1 - (ash\ weight/initial\ weight)]*100\%.$$

A pharmaceutical company wishes to compare the purities of a chemical powder that it receives from its four regular suppliers within a CRD structure. A batch of the powder is to be randomly selected from the deliveries of each supplier and its purity ascertained. The quality assurance manager is unsure as to how many measurements to make on each selected batch but is of the opinion that a treatment difference of at least 4.24% would be sufficient to suggest that purity differed with supplier sufficiently to give grounds for further investigation. The manager wishes the planned supplier test to be carried out at the 5% significance level and to have a power of at least 90%. Past data suggest a residual mean square estimate of 5.54 would be appropriate. How many measurements should be taken for each supplier?

Information
We have $k = 4$ suppliers ($f_1 = 3$), minimum detectable difference $\delta = 4.24\%$, significance level of the test of 5% ($\alpha = 0.05$), residual mean square estimate $\sigma^2 = 5.54$, and power of at least 90%.

Estimation of n

Initially, consider $n = 8$. For this case, $f_2 = 4(8-1) = 28$ and equation (3.12) becomes,

$$\phi = \sqrt{\frac{(8)(4.24)^2}{(2)(4)(5.54)}} = 1.80$$

Consulting Table A.6 for $f_1 = 3$ and $f_2 = 28$, power is estimated to be 81.4% which, though acceptable, is lower than the required power. We must raise n and check power again.

Trying $n = 10$, we have $f_2 = 4(10-1) = 36$ and

$$\phi = \sqrt{\frac{(10)(4.24)^2}{(2)(4)(5.54)}} = 2.01$$

which provides a power estimate of at least 90.4% (Table A.6, $f_1 = 3, f_2 = 36$). This satisfies the power stipulation. It is recommended, therefore, that each batch of material supplied be tested at least 10 times.

Exercise 3.4

A food analyst wished to be set up a CRD based experiment to assess the butyric acid content of three commercially produced butter-based biscuits. The analyst is unsure of how many samples of each biscuit type to test but is prepared to assume that if a minimum difference of 0.6 mg ml^{-1} in the mean butyric acid content could be detected, it would provide sufficient evidence to conclude that butyric acid content differed with biscuit type. The general significance test is planned to be carried out at the 5% significance level with power of at least 80%. A residual mean square estimate of 0.21 is available from past comparable studies. The analyst initially believes four biscuits of each type is a reasonable number of samples for assessment. Is four sufficient or are more necessary?

Most experimenters neglect power analysis within the planning phase of an experiment primarily because they do not give enough thought to formulating a minimum treatment difference which, if detected, would prove something scientifically justifiable in respect of the investigation. Its omission is often also because experimenters do not appreciate the importance of power and sample size estimation within design planning to ensure the statistical tests to be applied are capable of detecting significant effects if they exist. Other mechanisms of carrying out power analysis include the use of operating characteristic curves[11] and the specification of treatment effects for all treatments being compared.[12] Retrospective power analysis after data collection and analysis is often also appropriate.[13]

9 DATA TRANSFORMATIONS

The underlying equal variance (homoscedastic) and normality assumption, required by parametric inference procedures, may not always be satisfied by experimental data. Often, response data may be inappropriate for normality to be assumed (skewed, count data), *e.g.* copper levels in blood and latoxin levels in peanuts, or they may be such that the variances associated with the treatments are not constant and seem to vary with the magnitude of the treatment mean (heteroscedastic). Data re-coding is useful in such cases to enable the non-normality, non-constant variance, and non-additivity to be corrected before implementing inferential data analysis, the techniques of which may depend on these specifications being valid in the response data. Re-coding, or data transformation, essentially changes the scale of the response data and may also help to improve the fit of a specified model to experimental data. Several transformations exist; the most commonly used are described briefly in Box 3.4.

If small numbers of observations have been collected, the range can be used in preference to the standard deviation when assessing whether data transformation is necessary. Data analysis, in such cases, proceeds as it would if the actual response data were being used. The inferences reached are then extrapolated to the original experimental investigation for problem context inference. Example 3.11 which follows will be used to illustrate the principle of data transformations.

Example 3.11

Consider the data presented in Table 3.16 which represent the determinations using AAS of combined sulphur in rubber samples, expressed in mg. The summaries for the collected data are also presented in Table 3.16 and show that the ranges, replacing the standard deviation due to the small number of observations, are not constant for all samples implying that the response data are heteroscedastic. We need, therefore, to assess which transformation to apply to the collected data to enable the property of similar variability (homoscedasticity) across all samples to be satisfied.

[11] Reference 2, pp. 120–129.
[12] W.P. Gardiner, 'Statistics for the Biosciences', Prentice Hall, Hemel Hempstead, 1997.
[13] Reference 3, pp. 194–195.

Box 3.4 *Commonly applied data transformations*

Logarithm. If the standard deviation is proportional to the mean for each treatment group, *i.e.* standard deviation over mean (RSD) is approximately constant (data exhibit heteroscedasticity), then we can use the logarithmic transformation $\log_{10}(X)$, or $\log_{10}(X+1)$ if the data contain zeros. For skewed data, this can produce transformed data which are symmetric. This transformation is used often in laboratory experimentation to re-scale factor settings for the purpose of results presentation, *e.g.* concentration of a solution and density of organisms in a growth experiment.

Square root. If the experimental data are such that the variance is proportional to the mean for each treatment grouping, *i.e.* variance over mean approximately constant, then the square root transformation $\sqrt{X}$, or $\sqrt{(X+0.5)}$ if the data are low and contain many zeros, should be applied. Such a data transformation is useful to make the group variances independent of the means. Transforming response measurements in this way is appropriate for count data, *e.g.* radiation counts and number of abnormal cells in viral cultures.

Arcsine (angular). When response data are in the form of probabilities, proportions or percentages, we can apply the arcsine transformation $\sin^{-1}(\sqrt{X})$ where X lies between 0 and 1. Typical data of this type may be proportion of cells inhibited by toxic substances and percentage of respiring bacteria in sewage samples.

Reciprocal. For each data grouping, if the standard deviation is proportional to the square of the mean, then we can use the reciprocal transformation $1/X$ to transform the data.

Table 3.16 *Combined sulphur measurements for Example 3.11*

Sample	A	B	C
	0.24	0.16	0.07
	0.50	0.32	0.15
Mean	0.37	0.24	0.11
Range	0.26	0.16	0.08

Source: Reprinted from W. Puacz, W. Szahun and M. Kopras, *Talanta*, 1995, **42**, 1999–2006 with kind permission of Elsevier Science – NL, Sara Burgerhartstraat 25, 1055 KV Amsterdam, The Netherlands.

We first check the ratio of (range/mean) providing values 70.3, 66.7, and 72.7 for samples A, B, and C respectively. These values are roughly constant so the logarithmic transformation would appear appropriate. We will use $\log_{10}(X) + 1.5$ to ensure the transformed data have positive values. Applying this transformation results in the re-coded data presented in Table 3.17. The summary figures for these coded data show that variability, as measured through the range, is now more similar across all samples.

Table 3.17 *Transformed data for Example 3.11*

Sample	A	B	C
	0.830	0.704	0.345
	1.199	1.005	0.676
Mean	1.0395	0.8545	0.5105
Range	0.319	0.301	0.331

We can see from the coded results of Example 3.11 how transforming data can improve the properties of the data enabling the assumptions of a planned inference procedure to be more readily satisfied.

10 LATIN SQUARE DESIGN

RBDs do not eliminate all the possible variation stemming from block-to-block operations of treatments. For example, in an experiment involving comparison of three methods of measuring the toxicity of soil samples from a land-fill site, analysts could be used as a blocking factor to account for a potentially extraneous source of measurement variability. It may also be that each experiment is time consuming with only one capable of being completed by each analyst in a day. The day effect could be included as a second blocking factor in order to account for a further source of extraneous variation and to provide a more efficient design structure for the planned experiment.

A *Latin Square* (LS) design structure[14] would be suitable in such a case as it enables one treatment factor and two blocking factors to be accounted for. Table 3.18 provides an illustration of this design structure for an experiment to assess four treatments, A to D, accounting for two blocking factors. Within the illustrated structure, one blocking factor is assigned to the columns, the second to the rows, and the treatments in a way that ensures each occurs once only in each

[14] Reference 1, pp. 269–285.

row and each column. The design is square with number of rows = number of columns = number of treatments. An LS design can be further extended to a *Graeco-Latin Square design* by inclusion of a third blocking factor. Again, assignment of treatments must ensure that they appear once only with respect to all blocking factor levels. Analysis of data from such design structures follows comparable procedures to those outlined in the RBD illustration of Example 3.8.

Table 3.18 *Example of Latin square design structure for four treatments A to D*

| | | Block 1 | | | |
		1	2	3	4
	1	B	C	A	D
Block 2	2	D	A	B	C
	3	C	B	D	A
	4	A	D	C	B

11 INCOMPLETE BLOCK DESIGNS

Sometimes, we may be forced to design an experiment in which we must sacrifice some balance to perform the experiment.[15] For example, suppose a comparison is to be made of the purity of three different batches of a chemical product. There are three analysts available on any given day with the product being of such complexity that an analyst can do just two analyses per day. Completion of all batch comparisons in a single day is achievable provided each analyst analyses just two of the three batches, as illustrated in the design structure displayed in Table 3.19. Carrying out the analyses on a single day would help to eliminate time, storage, and other factors which could affect the outcome to the experiment.

Table 3.19 *Example of incomplete block design structure*

| | | Analyst | | |
		A	B	C
	1	x	x	–
Batch	2	x	–	x
	3	–	x	x

x, denotes an analysis measurement

---

[15] Reference 1, pp. 306–324.

The design layout of Table 3.19 is called a *Balanced Incomplete Block Design* (*BIBD*) where the number of treatments tested in each block is less than the number of treatments being compared. The illustrated structure is a special case of an incomplete block design in which the inherent balance of the design has been retained. Other incomplete block designs include *Youden Squares* and *Lattice Designs* with the *General Linear Models* (*GLM*) technique providing the base for determination of the relevant analysis elements.

CHAPTER 4

Factorial Experimental Designs for Chemical Experimentation

1 INTRODUCTION

The ANOVA concepts and principles for one factor designs, described in Chapter 3, can be readily extended to investigations where the effect on a response of two or more controlled factors needs to be assessed. Multi-factor experimentation occurs often within chemical processes when a chemical outcome may depend on many factors such as pH, temperature, instrumentation conditions, and chemical concentrations. Many, or a combination of, such factors may influence the response with experimentation used to determine how this may be occurring.

For example, the fluorescence intensity of an aluminium complex may depend on quantity of the dyestuff Solochrome Violet RS (SVRS) and pH. By studying these two factors simultaneously within a two factor experiment, we could assess how they influence the fluorescence of the complex, either independently or in combination. This measurement of independent and dependent factor influences (main effects and independent effects) is best obtained using *factorial design structures*. Simplicity of practical implementation, more effective use of experimental responses, and provision of more relevant response information are the three features which make factorial structures better experimental bases than *one factor designs* (CRD, RBD) or *one-factor-at-a-time (OFAT) experimentation*, both of which are inefficient mechanisms for multi-factor experimentation, especially if estimation of interaction effects is important.

The drawback of one factor designs is that they are only capable of assessing the effect of a single factor on a chemical outcome while, in most scientific experimentation, it is likely that more than one factor will influence a response either independently or, more likely, in combination. In addition, they only represent a slice from a factorial with the conclusions reached depending heavily on the slice chosen to be investigated. In other words, if the assessment of interaction effects is overlooked, then the inferences obtained may not be as relevant as hoped for.

OFAT experimentation is the classical approach to chemical experimentation. It involves taking each factor in turn and testing it at different levels with all other factors held constant. An optimal level is then chosen for the tested factor and the process continued until all factors have been assessed and their optimal level ascertained. A major drawback of the OFAT approach is that specification of an optimal factor combination is due more to chance than design. This is because determination of optimal depends on the combination of settings tested and in OFAT experiments, not all potential treatment combinations are tested. In addition, such experimentation does not enable a wide range of experimental conditions to be assessed and is also unable to estimate factor interaction as it assumes each factor works independently of every other. Use of OFAT approaches is unlikely to provide a true picture of the phenomena being studied in multi-factor experiments.

Independent influence of factors on a chemical response is highly unlikely in most chemical experiments. Factors are likely to interact in their effect on a response in such a way that the difference in mean response for each level of a factor varies across the levels of another factor, *i.e.* the factors work together, not separately, in their influence on the measured response. Factorial experimentation attempts to address this point though OFAT experimentation is often adopted in multi-factor experiments when, through adequate planning, simple factorial structures would be more advantageous as they can provide more relevant scientific information more efficiently.

In factorial-based experiments, interest lies in assessing how much influence, if any, each controllable factor has on the response and whether the tested factors work independently or in combination. The former refers to the concept of a *main effect* which measures the change in the response resulting from changing the levels of a factor. Factors working in combination reflect the concept of an *interaction effect*, a measure of the differing effect on the response of different factor-level (treatment) combinations. In other words, interaction is a measure of the combined effect of two or more factors where this effect cannot be explained by the sum of the individual factor effects. It is this concept which factorial designs are primarily geared to assessing and estimating.

By way of an illustration, consider an experiment to assess the effect of three mobile phase factors on ion-interaction HPLC retention times of the analyte simazine. The factors to be tested are the methanol concentration, pH, and flow rate. Each factor is to be investigated at three levels, meaning that the design structure would be that of a $3 \times 3 \times 3$ factorial. By including replication of the treatment combinations, we would be able to provide estimation of the interdependency of the factors (factor interaction) and how this affects, or does not affect, retention time.

The concept of interaction is best explained graphically as the plots in Figure 4.1 illustrate for a response of absorbance in respect of two factors, temperature and chemical concentration. The response lines, representing lines joining associated treatment combination means, in both plots are not parallel indicating non-uniformity of response across the treatment combinations. In Figure 4.1A, we can see that absorbance varies markedly across concentration as temperature changes. For low temperature, the trend is downward while for high temperature, it is upward. Mid temperature provides a more quadratic effect. This form of interaction plot illustrates the case where interaction is occurring as a difference in the direction of the responses, *i.e.* non-uniformity of response order. Figure 4.1B, on the other hand, highlights a magnitude difference form of interaction, *i.e.* uniformity of response order but magnitude differences. The plot shows that the temperature trends are linear with same temperature order at each concentration though the difference in mean absorbance varies as concentration changes. Both plots in Figure 4.1 highlight that the factors are not acting independently but are inter-dependent. Absence of interaction would be shown by simple parallel lines indicative of a simple additive, rather than interactive, effect when changing factor levels.

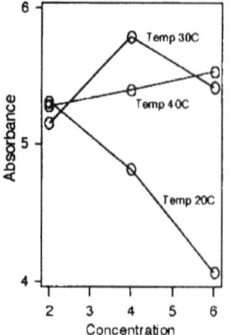 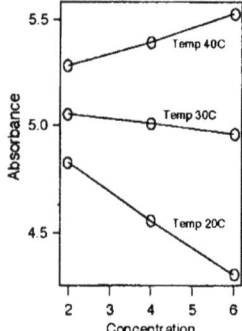

Figure 4.1 *Illustration of interaction plots*

Experimental factors in factorial designs are generally specified at a number of different levels. For example, temperature: 50 °C, 70 °C, and 90 °C; pH: 4, 7, and 10; acid levels: 2 ml, 2.5 ml, and 3 ml; and analysts: A, B, and C. These qualitative or quantitative 'levels' reflect the different factor settings to be controlled in the experiment. It is the effect of these levels on a measured response, separately and in combination, which factorial design structures are designed to assess and estimate. Analysis of the data collected from such experiments is on similar lines to that outlined for one factor designs though follow-up analysis is generally more graphical than statistical in its basis.

To understand the principles of factorial experimentation better, consider the following example. It will be used to illustrate the *OFAT* approach, the *factorial-based* approach, and the contrast between them.

Example 4.1

An experiment is to be conducted into the effect of temperature and pressure on the yield of a chemical reaction. It has been decided to base the experiment on three levels of each factor as shown in Table 4.1. The purpose of the experiment is to determine an optimum treatment combination which maximises yield. How do we do that? How do we set up the experiment to answer the specified objectives?

Table 4.1 *Factors and levels for reaction experiment of Example 4.1*

Level	Temperature	Pressure
low	50 °C	600 mm Hg
medium	75 °C	750 mm Hg
high	100 °C	900 mm Hg

Using one-factor-at-a-time (OFAT) experimentation
In this form of experimentation, each factor is taken in turn and tested at its three levels with all other factors held at constant level. Once the factor chosen has been optimised (best level for maximum yield), it is then held constant at that level and the next factor tested. After investigating each factor in turn, it is hoped that the best (optimal) combination of factors producing the optimal response is reached.

For OFAT, there are six starting points, one of the three temperatures or one of the three pressures. Choice of starting point can greatly affect the outcome which may, in fact, not be the true optimal. This is primarily because OFAT experiments cover only a fraction of the experimental region (limited number of treatment combinations tested) and so only provide a limited amount of information on factor effects. In addition, no account is taken of interaction effects as factors are assumed independent, leading to such experimentation having low precision.

For this illustration, suppose we set initial temperature at 50 °C. We would then measure the reaction yield for each pressure setting at say, two replicates. Suppose 900 mmHg was best. Next, we would set pressure at 900 mm Hg and measure reaction yield at each temperature for two replicates. Suppose 50 °C appeared best. On this basis, we would suggest that (50 °C, 900 mmHg) was the optimal combination. Within this experiment, however, only five treatment combinations have been tested, these being (50 °C, 600 mmHg), (50 °C, 750 mmHg), (50 °C, 900 mmHg), (75 °C, 900 mmHg), and (100 °C, 900 mmHg). A total of 12 experiments would be required as the combination (50 °C, 900 mmHg) is repeated twice.

From the design structure presented in Table 4.1, we can see that four factor combinations, (75 °C, 600 mmHg), (75 °C, 750 mmHg), (100 °C, 600 mmHg), and (100 °C, 750 mmHg), have not been tested. Though an 'optimum' has been reached, we cannot be sure it is correct because of the four untested combinations.

Using Factorial-based experimentation

For the planned experiment, a factorial-based structure would test all nine possible treatment combinations covering all the planned experimental region. The inclusion of replication would enable the interaction effect to be estimated. If we consider replicating twice, this would mean a total of $3 \times 3 \times 2 = 18$ experiments, six more than the OFAT experiment would require. However, the benefits of ensuring full coverage of the experimental region (all treatment combinations tested) and the ability to estimate interaction effects far outweigh the drawbacks that the marginal increase in amount of experimentation provides.

2 TWO FACTOR FACTORIAL DESIGN WITH N REPLICATIONS PER CELL

Consider two factors A and B which can be tested at a and b levels respectively, providing a total of ab treatment combinations to which the experimental units require to be randomly assigned. Generally, each treatment combination is experimented on n times (n replicate measurements) in order to provide sufficient data for efficient effect estimation. It is this replication which enables the interaction effect to be estimated. This two factor design will be used as the basis for the description of the underlying concepts and analysis procedures appropriate to factorial based experiments. Blocking in such designs can also occur but will not be discussed.

2.1 Experimental Set Up

The experimental set up for a two factor factorial experiment requires each factor-level combination to be run n times (n replications). Suppose we have three levels for factor A, four levels for factor B, and three replications are to be carried out at each factor–level combination. The set up for such an experiment is illustrated in Table 4.2 and would require $3 \times 4 \times 3 = 36$ separate experiments to be conducted. The specified treatment combinations would be run in randomised order.

Table 4.2 *Illustration of the experimental structure for a 3x4 factorial with 3 replicates*

		Replications		
Factor A	*Factor B*	*1*	*2*	*3*
Level 1	Level 1	x	x	x
Level 1	Level 2	x	x	x
Level 1	Level 3	x	x	x
Level 1	Level 4	x	x	x
Level 2	Level 1	x	x	x
Level 2	Level 2	x	x	x
Level 2	Level 3	x	x	x
Level 2	Level 4	x	x	x
Level 3	Level 1	x	x	x
Level 3	Level 2	x	x	x
Level 3	Level 3	x	x	x
Level 3	Level 4	x	x	x

x, denotes a response measurement

For two factor experimentation, it may appear that the associated design structure is similar to that of an RBD. This is not the case. An RBD is used to investigate the effect of one treatment factor using blocking as a means of excluding extraneous variation from the response to provide a more sensitive test of treatment effect. A factorial design, on the other hand, does not sacrifice information on any factor enabling the effect of all factors and their interactions to be fully assessed.

2.2 Model for the Measured Response

As with the one factor designs of Chapter 3, the first step in the analysis is the specification of a response model for the chemical response. Denoting X_{ijk} to be the measured response on the kth replicate of experimental units receiving factor A level i and factor B level j, we specify the response model, assuming additivity of effects, to be

$$X_{ijk} = \mu + \alpha_i + \beta_j + \alpha\beta_{ij} + \epsilon_{ijk}$$

where $i = 1, 2, \ldots, a, j = 1, 2, \ldots, b$, and $k = 1, 2, \ldots, n$.

The term μ refers to the grand mean (assumed constant for all factor–level combinations) with α_i (alpha) representing the effect on the response of factor A at level i (assumed constant for all units receiving factor A at level i but may differ for units receiving a different treatment combination) and β_j (beta) corresponding to the effect on the response of factor B at level j (assumed constant for all units receiving factor B at level j but may differ for units receiving a different treatment combination). The interaction term $\alpha\beta_{ij}$ measures the combined effect on the response of factor A level i with factor B level j while ϵ_{ijk} again represents the uncontrolled variation (the within treatment combination effect). The *assumptions* underpinning factorial designs are the same as those for one factor designs, *i.e.* equal response variability across the factors, normality of response, and additivity of effects.

2.3 Exploratory Data Analysis (EDA)

As in all previous data handling illustrations, simple exploratory data analysis through plots and summaries should be considered to help gain initial insight into what they specify concerning the experimental objectives.

2.4 ANOVA Principle and Test Statistics

The ANOVA principle again provides the basis of the statistical analysis aspect of factorial designs by providing the mechanisms for test statistic construction and derivation based on response model specification and the underpinning statistical theory. The procedures involved mirror those of the CRD and the RBD structures.

Hypotheses associated with factorial designed experiments define a test of absence of effect versus presence of effect. Since there are three possible effects to investigate, we can test the main effects of factor A and of factor B, and the interaction effect A × B. The associated null hypotheses for each of these cases is as follows:

main effect of factor A: H_0: no factor A effect occurring ($\alpha_1 = \alpha_2 = \ldots = \alpha_a = 0$),
main effect of factor B: H_0: no factor B effect occurring ($\beta_1 = \beta_2 = \ldots = \beta_b = 0$),
interaction effect A × B: H_0: no interaction effect occurring ($\alpha\beta_{11} = \alpha\beta_{12} = \ldots = \alpha\beta_{ab} = 0$),

The corresponding alternative hypotheses simply define presence of the effect being assessed.

The *ANOVA principle* within a two factor factorial experiment splits response variation into the four sources of variation associated with the specified response model. Again, sum of squares (SS) and mean square (MS) terms form the basis of this with the main SS term, Total SS, expressed as,

$$SSTotal\ (SST)\ =\ SSA + SSB + SSInteraction\ (SSAB) + SSResidual\ (SSRes)$$

The calculation aspects and general form of ANOVA table for a two factor factorial experiment are presented in Table 4.3. The obvious difference from the corresponding ANOVA table for an RBD experiment (Table 3.7) lies in the replacement of the block and treatment effects by the separate factor and interaction effects.

Table 4.3 *General ANOVA table for a two factor factorial experiment with n replicates*

Source	df	SS	MS
A	$a-1$	$SSA = \dfrac{\sum_{i=1}^{a} A_i^2}{bn} - \dfrac{\left(\sum_{i=1}^{a}\sum_{j=1}^{b}\sum_{k=1}^{n} X_{ijk}\right)^2}{abn}$	MSA = SSA/df
B	$b-1$	$SSB = \dfrac{\sum_{j=1}^{b} B_j^2}{an} - \dfrac{\left(\sum_{i=1}^{a}\sum_{j=1}^{b}\sum_{k=1}^{n} X_{ijk}\right)^2}{abn}$	MSB = SSB/df
A × B	$(a-1)(b-1)$	SSAB (see below)	MSAB = SSAB/df
Residual	$ab(n-1)$	SSRes (by subtraction)	MSRes = SSRes/df
Total	$abn-1$	$SST = \sum_{i=1}^{a}\sum_{j=1}^{b}\sum_{k=1}^{c} X_{ijk}^2 - \dfrac{\left(\sum_{i=1}^{a}\sum_{j=1}^{b}\sum_{k=1}^{n} X_{ijk}\right)^2}{abn}$	

$$SSAB = \sum_{i=1}^{a}\sum_{j=1}^{b} AB_{ij}^2 - SSA - SSB + \dfrac{\left(\sum_{i=1}^{a}\sum_{j=1}^{b}\sum_{k=1}^{n} X_{ijk}\right)^2}{abn},$$

A_i, sum of the responses for level i of factor A; a, number of levels of factor A; b, number of levels of factor B; n, number of replications of each factor–level combination; X_{ijk}, response measurement for experimental unit k receiving factor A level i and factor B level j; B_j, sum of the responses for level j of factor B; AB_{ij}, sum of the responses for the treatment combination of factor A level i and factor B level j; df, degrees of freedom; SS, sum of squares; MS, mean square

The *test statistics* for the three effects associated with a two factor factorial design are again specified as a ratio of two mean squares, the MS for factor being tested and MS for the residual. For the *factor A effect*, the ratio of the factor A variance to the residual variance provides the test statistic, *i.e.*

$$F = MSA/MSRes \qquad (4.1)$$

with degrees of freedom $df_1 = a - 1$ and $df_2 = ab(n - 1)$. For the *factor B effect*, the test statistic is constructed as the ratio of the factor B variance to the residual variance, *i.e.*

$$F = MSB/MSRes \qquad (4.2)$$

with degrees of freedom $df_1 = b - 1$ and $df_2 = ab(n - 1)$. For the *interaction effect*, the test statistic is based on the ratio of the interaction variance to the residual variance, *i.e.*

$$F = MSAB/MSRes \qquad (4.3)$$

with degrees of freedom $df_1 = (a - 1)(b - 1)$ and $df_2 = ab(n - 1)$.

Essentially, the *F* ratios (4.1) to (4.3) measure the importance of each factorial effect. Small values will indicate that the tested effect explains little of the variation in the response, while large values will be indicative of significant effect and that the factor, or interaction, effect being tested has a significant influence on the response. It is this influence we are interested in detecting if it is occurring within the data. Use of either the *test statistic approach* or the *p value approach* is feasible here. Generation of requisite ANOVA table and test statistics can be achieved in Minitab through the same Balanced ANOVA dialog window used for one factor designs (see Figure 3.3).

As three *statistical tests* can be carried out in a two factor factorial design, it is necessary to decide which should be carried out first. In general, the interaction significance test should be carried out *first* as it is important to detect factor interaction if it exists since this means that the two controllable factors influence the response in combination and not independently. Presence of a significant interaction effect can make testing of main effects (test of factors A and B) unnecessary unless the interaction is orderly (a similar pattern of results across all factor–level combinations) when a test on main effects can be meaningful.

Example 4.2

To determine formaldehyde levels in aqueous solution, measurements of the optical absorbance, at 570 nm, of the reaction product formed with chromotropic acid (CTA) in the presence of sulphuric acid (H_2SO_4) can be obtained. To find the optimum conditions for carrying out the complex reaction, a factorial experiment was run using three fixed levels of each acid. The design matrix and recorded absorbance measurements are provided in Table 4.4. The objectives of the study were to determine the amounts of sulphuric acid and chromotropic acid that produce the greatest absorbance for a level of formaldehyde of 2 ppm and to understand the effects of changing the amounts of these two acids on the absorbance in the region of the optimum.

Table 4.4 *Optical absorbance data for Example 4.2*

Sulphuric Acid (ml)	Chromotropic Acid (ml)	Recorded absorbance values	
		Replication 1	Replication 2
2.5	0.1	0.524	0.538
2.5	0.3	0.455	0.509
2.5	0.5	0.386	0.428
2.8	0.1	0.515	0.516
2.8	0.3	0.583	0.575
2.8	0.5	0.545	0.537
3.1	0.1	0.526	0.530
3.1	0.3	0.534	0.545
3.1	0.5	0.554	0.551

Source: G.T. Wernimont, 'Use of Statistics to Develop and Evaluate Analytical Methods', Association of Analytical Chemists, Arlington, Virginia, 1985: reproduced with the permission of AOAC International, Inc.

Two factors are being controlled: level of sulphuric acid and level of chromotropic acid. The design is such that factor interaction can be assessed as each factor-level combination has been replicated twice. The basic design is therefore a 3×3 factorial with two replicates so $a = 3$, $b = 3$, and $n = 2$. The response is the optical absorbance of the reaction mixture and the basic question being assessed is 'do the two acids affect absorbance and if so, how?'.

Minitab data entry
Data entry in Minitab for factorial designs follows the principles for one factor designs with response data in one column and codes for the factor levels in second and subsequent columns. Repetition of factor-level codes is necessary to account for the replication of factor-level combinations. We enter the absorbance data into column C1, based on

the order of data presentation in Table 4.4, as the two replicates for (sulphuric 2.5, chromotropic 0.1) then the two replicates for (sulphuric 2.5, chromotropic 0.3) then those for (sulphuric 2.5, chromotropic 0.5) and so on.

Three levels of H_2SO_4 have each been experimented on six times so each of the codes 1, 2, and 3 must appear six times in column C2. Based on the form of response data entry, these codes will correspond to six 1s then six 2s then six 3s. We have also experimented on each level of CTA on six occasions. From the form of response data entry, we must enter the codes into column C3 as two 1s then two 2s then two 3s repeated on three occasions. The menu commands **Calc** ▷ **Set Patterned Data** can be used to aid entry of the codes.

Response model
Based on this factorial design structure, the model for the absorbance response is

$$absorbance = \mu + \alpha_i + \beta_j + \alpha\beta_{ij} + \epsilon_{ijk}$$

with i = 1, 2, 3 (three sulphuric acid levels), j = 1, 2, 3 (three chromotropic acid levels), and k = 1, 2 (two replicates). The term α_i measures the effect of sulphuric acid on absorbance, β_j the effect of chromotropic acid on absorbance, $\alpha\beta_{ij}$ the interaction effect of the two acids, and ϵ_{ijk} the uncontrolled variation. Again, model specification indicates the belief that the response is affected by both controlled and uncontrolled variation. We assume that the absorbance measurements have equal variability for both acids and that they are normally distributed.

Hypotheses
The null hypothesis for each possible factorial effect is as follows:

interaction H_0: no sulphuric acid × chromotropic acid interaction effect on absorbance,
sulphuric acid H_0: no sulphuric acid effect on absorbance,
chromotropic acid H_0: no chromotropic acid effect on absorbance.

Exploratory data analysis
Initial data plots and summaries for both factors are presented in Outputs 4.1 and 4.2, the former for the H_2SO_4 factor and the latter for the CTA factor.

For H_2SO_4, there appears an obvious difference between the levels tested. At 2.5 ml (Sulph 1), absorbance is low and inconsistent. At 2.8

ml (Sulph 2) and 3.1 ml (Sulph 3), absorbance is higher though similar at both levels with 3.1 ml providing the more consistent measurements. The means highlight the same effect with 2.8 ml least and the other levels similar. The *RSD*s are 12.7%, 5.3%, and 2.2%, respectively, providing further evidence of the precision differences in the absorbance measurements hinted at in the dotplots. Initially, 3.1 ml may be best as mean absorbance is high, though not the highest, and precision is high (minimum *RSD*).

Output 4.1 *Initial summaries of absorbance data of Example 4.2 for sulphuric acid levels tested*

Select **Graph** ▷ **Character Graphs** ▷ **Dotplot** ▷ *for Variables*, click **Absbnce** and click **Select** ▷ select **By variable**, click the empty box, select **Sulph**, and click **Select** ▷ click **OK**.

```
Character Dotplot

Sulph
1            .         .         .          .  .   .
    -------+---------+---------+---------+---------+---------Absbnce
Sulph
2                                    :      . .        . .
    -------+---------+---------+---------+---------+---------Absbnce
Sulph
3                                         ... . ..
    -------+---------+---------+---------+---------+---------Absbnce
       0.400     0.440     0.480     0.520     0.560     0.600
```

Select **Stat** ▷ **Basic Statistics** ▷ **Descriptive Statistics** ▷ for *Variables*, select **Absbnce** and click **Select** ▷ select **By variable**, click the empty box, select **Sulph**, and click **Select** ▷ for *Display options*, select **Tabular form** ▷ click **OK**.

Descriptive Statistics

Variable	Sulph	N	Mean	Median	TrMean	StDev	SEMean
Absbnce	1	6	0.4733	0.4820	0.4733	0.0601	0.0245
	2	6	0.5452	0.5410	0.5452	0.0288	0.0118
	3	6	0.54000	0.53950	0.54000	0.01161	0.00474

Variable	Sulph	Min	Max	Q1	Q3
Absbnce	1	0.3860	0.5380	0.4175	0.5275
	2	0.5150	0.5830	0.5157	0.5770
	3	0.52600	0.55400	0.52900	0.55175

For CTA, a similar pattern emerges except in reverse order of level. At 0.1 ml (Chrom 1) absorbance measurements are high and consistent while at 0.5 ml (Chrom 3), they appear more spread out and lower. The highest absorbances appear to be occurring at 0.3 ml (Chrom 2). The means specify an order of 0.5 ml, 0.1 ml, and 0.3 ml with distinct differences between each level. The *RSD*s are 1.7%, 8.8%, and 14.7%, respectively, for 0.1 ml, 0.3 ml, and 0.5 ml highlighting the large difference in measurement precision and that 0.1 ml of CTA appears to provide the most consistent results. Because of the differences in both mean and consistency, choice of best level, on an initial basis, may be difficult as no level stands out as best in respect of both summary characteristics.

Output 4.2 *Initial summaries of absorbance data of Example 4.2 for chromo-tropic acid levels tested*

Menu commands as Output 4.1.

```
Character Dotplot

Chrom
1
                                                     :  ... .
          -------+---------+---------+---------+---------+---------Absbnce
Chrom
2
                              .              .      . .        . .
          -------+---------+---------+---------+---------+---------Absbnce
Chrom
3
           .          .                          . . ..
          -------+---------+---------+---------+---------+---------Absbnce
            0.400     0.440     0.480     0.520     0.560     0.600
```

Menu commands as Output 4.1.

Descriptive Statistics

Variable	Chrom	N	Mean	Median	TrMean	StDev	SEMean
Absbnce	1	6	0.52483	0.52500	0.52483	0.00868	0.00354
	2	6	0.5335	0.5395	0.5335	0.0470	0.0192
	3	6	0.5002	0.5410	0.5002	0.0736	0.0301

Variable	Chrom	Min	Max	Q1	Q3
Absbnce	1	0.51500	0.53800	0.51575	0.53200
	2	0.4550	0.5830	0.4955	0.5770
	3	0.3860	0.5540	0.4175	0.5518

ANOVA table
Output 4.3 contains the ANOVA table for the absorbance data produced using the ANOVA procedure in Minitab. The symbol '|' in the Menu commands presented is a shorthand specification of the factorial model while the Options sub-menu enables the interaction means to be evaluated and printed off. The Storage sub-menu is to enable the residuals, or error estimates, and predicted values, fits, to be stored in new columns for later use in diagnostic checking.

Output 4.3 *ANOVA table for Example 4.2*

Select **Stat** ▷ **ANOVA** ▷ **Balanced ANOVA** ▷ for *Responses*, select **Absbnce** and click **Select** ▷ select the **Model** box and enter **Sulph|-Chrom**.

Select **Options** ▷ select the **Display means for (list of terms)** box and enter **Sulph*Chrom** ▷ click **OK**.

For *Storage*, select **Residuals** and select **Fits** ▷ click **OK**.

Analysis of Variance (Balanced Designs)

Factor	Type	Levels	Values		
Sulph	fixed	3	1	2	3
Chrom	fixed	3	1	2	3

Analysis of Variance for Absbnce

Source	DF	SS	MS	F	P
Sulph	2	0.0192623	0.0096312	33.66	0.000
Chrom	2	0.0035893	0.0017947	6.27	0.020
Sulph*Chrom	4	0.0166973	0.0041743	14.59	0.001
Error	9	0.0025755	0.0002862		
Total	17	0.0421245			

MEANS

Sulph	Chrom	N	Absbnce
1	1	2	0.53100
1	2	2	0.48200
1	3	2	0.40700
2	1	2	0.51550
2	2	2	0.57900
2	3	2	0.54100
3	1	2	0.52800
3	2	2	0.53950
3	3	2	0.55250

The ANOVA table presented mirrors the general form in Table 4.3 with the controlled components (Sulph, Chrom, Sulph*Chrom) and uncontrolled components (Error) adding to explain the Total variability in the optical absorbance data. The MS terms are again derived as the ratio SS/DF. The SS term for sulphuric acid (Sulph) is much larger than that for CTA (Chrom) implying possibly greater effect of H_2SO_4 on reaction product absorbance.

Interaction test

Equation (4.3) specifies that the interaction *F* test statistic is the ratio of the interaction MS term and the Residual MS term. From the ANOVA table in Output 4.3, this is expressed as 0.0041743/0.0002862 resulting in $F = 14.59$ as specified in row 'Sulph*Chrom' and column 'F'. With both test statistic and *p* value available, we can use either decision rule mechanism for the interaction test.

For the *test statistic approach*, we know $F = 14.59$ with degrees of freedom $df_1 = 4$ (interaction df) and $df_2 = 9$ (residual df) and associated 5% critical value of $F_{0.05,4,9} = 3.63$ (Table A.3). For the *p value approach*, the interaction *p* value is estimated as 0.001 (row 'Sulph*Chrom', column 'P'), a low value indicating rejection of the null hypothesis and the conclusion that the evidence implies that H_2SO_4 and CTA interact in their effect on absorbance of the reaction product ($p < 0.05$).

Model fit

As described in Chapter 3, Section 3.5, we can estimate the error ratio (SSRes/SST)% to help assess adequacy of model fit. For this experiment, we have this ratio as (0.0025755/0.0421245)% = 6.1%, a low value, signifying that the specified absorbance response model is a reasonable explanation of what affects the absorbance of the reaction product.

Unfortunately, the interaction test only enables us to conclude on the general effect of factor interaction on a response. It provides no information on how this effect is occurring within the response data so we must consider further analysis to pinpoint and interpret this effect.

3 FOLLOW-UP PROCEDURES FOR FACTORIAL DESIGNS

3.1 Significant Interaction

A significant interaction, such as found in Example 4.2, indicates that the levels of one factor result in a difference in the response across the range of levels of the other factor, *i.e.* the different treatment combinations do not give rise to similar response measurements. We know this effect is occurring but not how it is occurring in respect of the treatment combinations tested. We need, therefore, to analyse the interaction effect further to pinpoint how it is manifesting itself within the chemical responses.

Normally, for fixed levels of the factors (model I experiment), it is sufficient to plot the interaction means at each level of one factor for each level of the other factor in an *interaction*, or *profile*, *plot* of the treatment combination means, such as those illustrated in Figure 4.1. By inspection of the plot, we can identify the trends and patterns within the data which explain why significant interaction was detected and so provide a means of explaining how interaction is manifested within the response data. In addition, assessment of an optimal treatment combination should also be carried out in light of the conclusions reached from the plot analysis. Minitab has an interaction plot macro (%Interact) which can produce interaction plots. The menu command access is via **Stat ▷ ANOVA ▷ Interactions Plot** which produces the dialog window in Figure 4.2. The usual selecting, checking, and filling in of boxes applies.

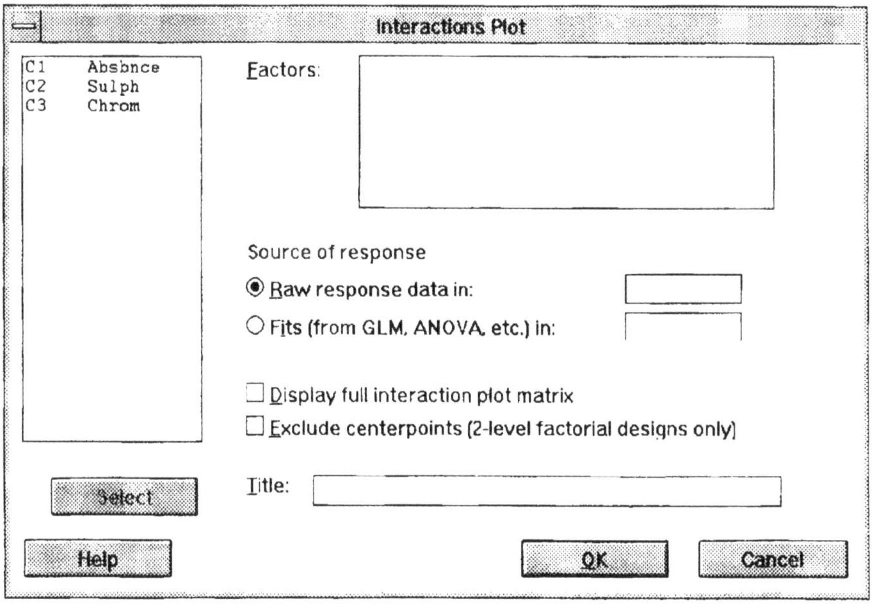

Figure 4.2 *Interactions plot dialog window in Minitab*

We could consider a formal multiple comparison of the means of each treatment combination. Based on a simple $a \times b$ factorial design, this would produce a total of $ab(ab - 1)/2$ possible pairings to assess. Even for small a and b, this can be too numerous to consider. Hence, the adoption of simple graphical approaches in the assessment of interaction effects.

Example 4.3

In Example 4.2, interaction between the acids was found to have a statistically significant effect on the absorbance of the reaction product. To assess this further, we need to produce an interaction plot to investigate the trends and patterns in the interaction effect in more detail. Output 4.4 illustrates the Minitab produced interaction plot using the dialog window of Figure 4.2 with the annotation at the side of the plot providing a plot legend to help aid interpretation.

Output 4.4 *Interaction plot of absorbance means*

Select **Stat** ▷ **ANOVA** ▷ **Interactions Plot** ▷ for *Factors*, enter **Sulph Chrom** ▷ for *Source of response*, select the **Raw response data in** box, select **Absbnce**, and click **Select** ▷ select the **Title** box and enter **Interaction plot of absorbance means** ▷ click **OK**.

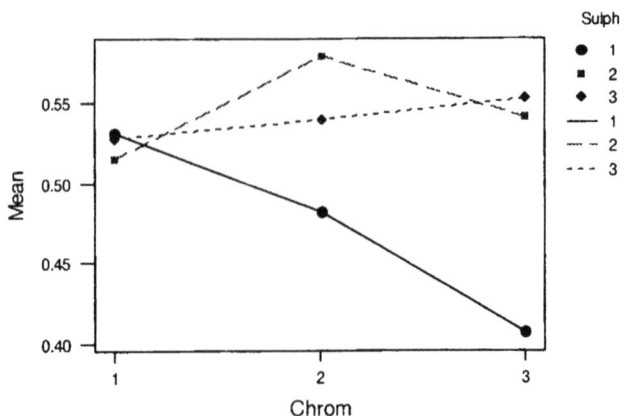

Interaction plot of of absorbance means

Interaction plot
The presence of non-parallel lines in the interaction plot is indicative of the occurrence of an interaction effect and that the interaction is reflecting a non-uniform order of response as illustrated in Figure 4.1A. The patterns differ for each level of H_2SO_4 tested. For low level, (Sulph 1, 2.5 ml), absorbance diminishes markedly as amount of CTA is increased while for medium level (Sulph 2, 2.8 ml), the effect is more

peaked. High levels of H_2SO_4 (Sulph 3, 3.1 ml) produce similar results for all levels of CTA though a slight increase can be seen. The pattern, in respect of levels of CTA, differs with level. It would appear therefore that increase in H_2SO_4 suggests increased absorbance while increased CTA produces more inconsistency in absorbance measurements, CTA 0.1 ml (Chrom 1) being best in this regard.

Optimal combination

To aid the conclusion, we should construct a simple table of the best results found and assess them. Such a table is presented in Table 4.5 based on the four highest results from the interaction plot, means from Output 4.3 and ranges from the original absorbance measurements

Table 4.5 *Possible optimal results for mean absorbance for Example 4.3*

H_2SO_4	CTA	Mean absorbance	Range (max − min)
2.8 ml (2)	0.3 ml (2)	0.579	0.008
2.8 ml	0.5 ml (3)	0.541	0.008
3.1 ml (3)	0.3 ml	0.5395	0.011
3.1 ml	0.5 ml	0.5525	0.003

Combination (H_2SO_4 2.8 ml, CTA 0.3 ml) gives highest mean absorbance, while combination (H_2SO_4 3.1 ml, CTA 0.5 ml) has a lower mean and lower range. The latter combination, though giving a lower mean absorbance, may be providing more consistent results than the former. Which combination to choose as best in this case depends on whether attainment of an optimal response and/or good precision in response is the more important. Ideally, we want both aspects of the optimal satisfied though in many cases, we may have to play one off against the other, the concept of *trade-off*, when concluding on a best combination of factors.

3.2 Non-significant Interaction but Significant Factor Effect

If interaction is not significant statistically, the evidence points to the factors acting independently on the response. On this basis, we must test the separate main effects for evidence of effect. If such effects are significant, follow-up checks based on multiple comparisons can be carried out to complete the statistical analysis of the response data. Main effect, or standard error, plots could also be considered. Planned comparisons of interaction effects and main effects through linear contrasts or orthogonal polynomials must still be assessed irrespective of whether any, some, or all of the factorial effects are shown to be significant.

3.3 Diagnostic Checking

As factorial design structures are based on comparable assumptions to one factor designs, inclusion of diagnostic checking as an integral part of the data analysis is important. Plotting residuals (measurements minus predicted values) against the factors tested and against model fits check for equality of response variability and adequacy of model. Normality of the measured response is again checked by a normal plot of residuals with a straight line ideal. Inappropriateness of these assumptions can be overcome through either a non-parametric routine (see Chapter 6, Section 7) or data re-scaling (see Chapter 3, Section 8).

Example 4.4

Residual plots for the absorbance experiment of Example 4.2 are given in Outputs 4.5 and 4.6, the former providing the factor plots and the latter, the fits and normal plots.

Factor plots
The H_2SO_4 plot in Output 4.5 has obvious column length differences as does the CTA plot. This, together with the evidence of gaps in some columns of the plots, hints strongly of unequal response variability across the acid levels tested. The exploratory analysis of Example 4.2 also highlighted this effect.

Output 4.5 *Diagnostic checking: sulphuric acid and chromotropic acid plots*

Produced using the **Layout** feature which enables the residual plots to be produced side by side and the **Graph ▷ Plot** menu (see Outputs 3.5 and 3.10) including the use of the **Title** sub-menu.

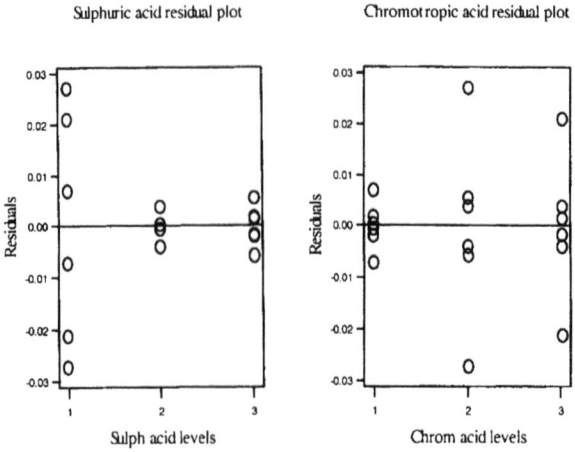

Fits and normal plots

The fits plot, in Output 4.6, shows a distinct bunching pattern after absorbance 0.5 indicative of inadequate model fit especially for absorbance values below 0.5. The symmetry pattern present is due solely to the fact that each treatment combination was replicated twice. The normal plot shows an S-shaped pattern hinting at non-normality for the absorbance response.

Output 4.6 *Diagnostic checking: normal and fits plots*

Menu commands as Output 3.6.

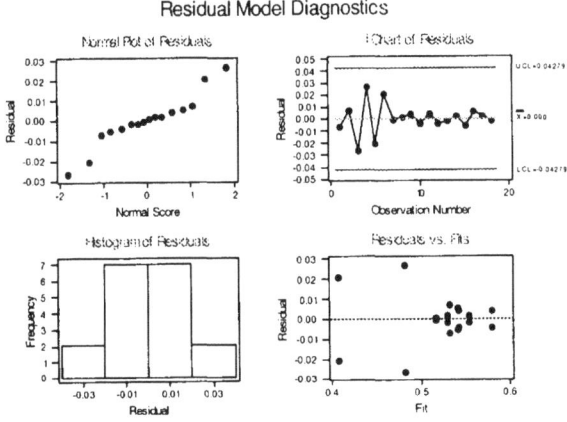

Residual Model Diagnostics

The analysis points to H_2SO_4 and CTA having different forms of influence on reaction product absorbance. Statistically, interaction is significant ($p < 0.01$) and the interaction plot shows the non-uniform nature of this effect. Optimal absorbance appears to occur using H_2SO_4 2.8 ml or 3.1 ml and CTA 0.3 ml or 0.5 ml. Perhaps, future experiments should investigate around these levels. Diagnostic checking suggests that equal variability and normality of the absorbance response data may not be acceptable and that, in future experiments, some form of data transformation to stabilise variability and normality effects may need to be considered prior to data analysis.

In practical investigations, we would generally carry out all the illustrated analyses in a single coherent manner so providing a procedure for the full analysis of the factorial design data. Conclusions, as ever, may feed other experiments. In the absorbance experiment illustrated, this could mean conducting further experiments around the 'best' combination to home in further on the optimal. Inclusion of further factors, such as temperature and pH, could be considered to understand further how factors affect absorbance. Such sequential experimentation is the basis of most practical chemical experimentation.

3.4 Overview of Data Analysis for Two Factor Factorial Designs

Exploratory data analysis, inferential data analysis, and diagnostic checking should all be part of the data analysis of factorial experiments. The remaining analysis aspects are dependent on whether or not interaction is significant. The absorbance example provides an outline of what to do when interaction is statistically significant. If interaction is not statistically significant, then it appears that the two factors affect the response independently, and we should therefore assess each factor separately through the formal F test and appropriate follow-up, if necessary.

Exercise 4.1

An experiment was conducted to test the effect of both temperature (°C) and duration (minutes) on the output of a particular chemical process. Both factors were tested at each of three fixed levels. The data collected refer to the percent yield of chemical and are presented below. How do temperature and duration affect yield? Current practice is based on a temperature of 40 °C. Is there a significant difference between the results for this temperature and the other two temperatures tested?

Duration	20 min			30 min			40 min		
30 °C	41	40	42	50	53	53	44	44	44
Temperature 40 °C	50	50	48	52	51	50	52	53	52
50 °C	50	49	41	43	42	44	53	55	54

Source: E.R. Dougherty, 'Probability and Statistics for the Engineering, Computing and Physical Sciences', Prentice Hall, Englewood Cliffs, New Jersey, 1990: reproduced with the permission of Prentice Hall, Inc.

4 POWER ANALYSIS IN TWO FACTOR FACTORIAL DESIGNS

In factorial designs, it is often necessary to ask 'How many replications will be required?' Estimation of the amount of replication is important in the planning phase of a factorial experiment since replication can make a specified treatment difference easier to detect statistically. As with the outlined principles of Chapter 2, Section 9 and Chapter 3, Section 8, information on likely experimental outcomes is necessary to enable such estimation to take place. Again, this entails the experi-

menter specifying the minimum detectable difference δ that they expect
to detect if differences exist within the data. For interaction analysis,
this corresponds to the minimum difference expected to be detected
between any two treatment combinations. For main effects analysis, δ
defines the minimum difference expected to be detectable between two
factors.

We use the same procedure as shown in Chapter 3, Sections 8.1 and
8.2 with ϕ re-defined as

$$\phi = \sqrt{\frac{n'\delta^2}{2k'\sigma^2}} \qquad (4.4)$$

where k' refers to the number of levels of the effect being examined and
n' is used to define the number of observations to be recorded at each of
these levels. This expression is a general one appropriate for all three
testable effects in an $a \times b$ factorial experiment. Further clarification of
how k' and n' are defined for each effect and the related degrees of
freedom for its associated test is provided in Table 4.6. Replication
estimation, as in Chapter 3, Section 8.2, is found by estimating n'
iteratively until a satisfactory power estimate is obtained. Generally, for
the interaction effect, the number of replications should be small so
estimates of $n = 2$, $n = 3$, and $n = 4$ are appropriate starting points for
replication estimation.

Table 4.6 *Summary information for power analysis in a two factor factorial*

Effect	k'	n'	f_1	f_2
factor A	a	bn	$a-1$	$ab(n-1)$
factor B	b	an	$b-1$	$ab(n-1)$
interaction A $\times$ B	ab	n	$(a-1)(b-1)$	$ab(n-1)$

*a, number of levels of factor A; b, number of levels of factor B; n, number of replications
of each factor–level combination*

Example 4.5

As part of a food analysis study, four independent laboratories are to
be provided with three similar samples of a food product, each of which
is to be analysed a number of times. Past quality assurance studies
suggest that a minimum difference of 7.31 ng g^{-1} would be indicative
of a sample/laboratory interaction. It is planned to test the interaction
effect at the 5% significance level with power of at least 85%. A residual
mean square estimate of 1.22 is available from past QA studies.

Estimate the number of replicate measurements of each sample each laboratory should perform to satisfy the specified experimental constraints.

Information
We have $a = 4$ laboratories, $b = 3$ samples, minimum detectable difference $\delta = 7.31$ ng g^{-1} for the interaction effect, significance level of 5%, residual mean square estimate $\sigma^2 = 1.22$, and power of at least 85%. From Table 4.6, we can specify that $n' = n$, $k' = (4)(3) = 12$, $f_1 = (4-1)(3-1) = 6$, and $f_2 = (4)(3)(n-1) = 12(n-1)$.

Estimation of n
Initially, consider $n' = n = 2$. For this case, $f_2 = 12(2-1) = 12$ and equation (4.4) becomes,

$$\phi = \sqrt{\frac{(2)(6.5)^2}{(2)(12)(1.22)}} = 1.70$$

Consulting Table A.6 for $f_1 = 6$ and $f_2 = 12$, power is estimated as 70%, marginally too low for adequate experimentation. Increasing replications to three ($n' = n = 3$, $f_2 = 24$, $\phi = 2.08$), power is estimated as approximately 97%. Based on this, the advice would be that at least three replicate measurements would be necessary.

Operating characteristic curves,[1] specification of all interaction and factor effects,[2] and simple confidence interval principles[3] could also be used for power analysis.

Exercise 4.2

A factorial experiment is being planned to investigate the effect of a quantity of SVRS and level of pH on the fluorescence intensity of an aluminium complex. Three levels of SVRS and three pH settings are to be used, and replicate measurements of each treatment combination are planned to enable factor interaction to be tested. It has been suggested that a minimum detectable difference of 105.2 in the fluorescence intensity would be indicative of evidence for the presence of an

[1] D.C. Montgomery, 'Design and Analysis of Experiments', 4th Edn., Wiley, New York, 1997, pp 249–251.
[2] W.P. Gardiner, 'Statistics for the Biosciences', Prentice Hall, Hemel Hempstead, 1997.
[3] L. Davies, 'Efficiency in Research, Development and Production: The Statistical Design and Analysis of Chemical Experiments', The Royal Society of Chemistry, Cambridge, 1993, pp. 51–52.

interaction effect. The interaction test is planned to be carried out at the 5% significance level with power of at least 85%. Based on a residual mean square estimate of 464, estimate the number of replications of the treatment combinations necessary to meet the experimental constraints specified.

5 OTHER FEATURES ASSOCIATED WITH TWO FACTOR FACTORIAL DESIGNS

5.1 No Replication

When $n = 1$ in a two factor factorial experiment, there is only one observation (single replicate) for each factor-level combination. In this case, the response model does not contain an interaction term because interaction and error are confounded, *i.e.* measured by the same sum of squares term. This means there are insufficient data with which to estimate these two effects independently. The response model is therefore the same as that of an RBD though interpretation of the terms relate to those for a two factor factorial. The purpose of the factorial experiment is to test for differences in each main effect, unlike the RBD which is constructed to test for treatment effects using blocking to provide a more sensitive test.

5.2 Unequal Replications per Cell

In some factorial designs, it may not be possible to have equal amounts of replication at each treatment combination due to experimental constraints. Loss of replicate measurements at certain combinations could also occur. In these instances, the design is referred to as *unbalanced*, as unequal numbers of observations occur at certain treatment combinations. The General Linear Models (GLM) estimation procedure, used for incomplete block designs, provides the means of obtaining the ANOVA table and related test statistics in such a case.

For unbalanced factorial designs, there also exist special sum of squares terms called *Type I*, *Type II*, *Type III*, and *Type IV*. Depending on the experimental objectives, one of these SS types should be used as the basis of the statistical test component of the analysis. Choice of SS type to use is important as each, with the exception of Type III, examines a marginally different model hypotheses structure from that specified for a balanced design.[4,5]

---

[4] R.G. Shaw and T. Mitchell-Olds, *Ecology*, 1993, **74**, 1638.
[5] R.O. Kuehl, 'Statistical Principles of Research Design and Analysis', Duxbury Press, Belmont California, 1994, pp. 197–207.

6 METHOD VALIDATION APPLICATION OF TWO FACTOR FACTORIAL

Interference studies[6] are often used in the initial phases of analytical method validation with a two factor factorial design structure used to assess the effect of interference on a chemical outcome. The study experiment is based on adding a possible interferent to the 'normal' matrix both with and without the analyte being present. A possible design structure for such a study is shown in Table 4.7 and corresponds to a 2×2 factorial design. If $B \neq A$, we say there is evidence of background interference reflecting translational, or additive, interference. If $(D - C) \neq (B - A)$, we say there is evidence of a matrix effect which reflects the presence of rotational, or multiplicative, interference.

Table 4.7 *2×2 factorial structure for an interference study*

		Interferent level absent	high
Analyte	*absent*	A	B
	present	C	D

7 THREE FACTOR FACTORIAL DESIGN WITH N REPLICATIONS PER CELL

Two factor factorial designs can be easily extended to investigate three or more factors. For example, the chromatographic response function (CRF), a summation term of the individual resolutions between pairs of peaks in HPLC, may be affected by three factors: the concentration of citric acid (A), the proportion of methanol (B) (methanol:water), and the concentration of acetic acid (C) in the mobile phase. It could be that different factor combinations may result in different CRF measurements (interaction effects present) with experimentation enabling this to be investigated.

For such an experiment involving three factors A, B, and C, we would set each factor at levels *a*, *b* and *c* respectively providing *abc* treatment combinations and carry out replicate determinations of the related chemical response at each combination. The effects that can be tested from such an experiment are the main effects A, B and C, the second order interactions $A \times B$, $A \times C$ and $B \times C$, and the third order interaction $A \times B \times C$, *i.e.* seven possible effects.

---

[6] Laboratory of the Government Chemist, VAMSTAT Software, The quality of analytical data, Method validation: Preliminary tests, 2.

7.1 Model for the Measured Response

The model for an experimental observation from a three factor factorial is given by

$$X = \mu + \alpha + \beta + \alpha\beta + \gamma + \alpha\gamma + \beta\gamma + \alpha\beta\gamma + \varepsilon$$

where α is the factor A effect, β the factor B effect, $\alpha\beta$ the $A \times B$ interaction effect, γ the factor C effect, $\alpha\gamma$ the $A \times C$ interaction effect, $\beta\gamma$ the $B \times C$ interaction effect, $\alpha\beta\gamma$ the $A \times B \times C$ interaction effect, and ε the error effect. As previously, response model specification consists of both controlled elements (factors and their interactions) and uncontrolled elements (error, unexplained variation). The assumptions associated with this response model are as defined for all previous experimental designs (equality of response variability across all factors, normality of response).

7.2 ANOVA Principle and Test Statistics

Again, the ANOVA principle underpins the statistical analysis aspects of a three factor factorial design through its provision for test statistic construction and derivation based on response model specification and underpinning statistical theory. The associated procedures mirror and extend those shown previously for other formal design structures.

Hypotheses essentially specify a test of absence of effect (H_0) versus presence of effect (H_1) for all seven possible effects.

The underlying *ANOVA principle* again splits response variation into controlled and uncontrolled elements from which the general sum of squares expression of

$$SST = SSA + SSB + SSAB + SSC + SSAC + SSBC + SSABC + SSRes$$

emerges. This expression underpins the general ANOVA table illustrated in Table 4.8 where all aspects of the table are as previously described for the two factor factorial design.

Table 4.8 *General ANOVA table for a three factor factorial experiment with n replicates*

Source	df	SS	MS
A	$a-1$	SSA	MSA = SSA/df
B	$b-1$	SSB	MSB = SSB/df
AxB	$(a-1)(b-1)$	SSAB	MSAB = SSAB/df
C	$c-1$	SSC	MSC = SSC/df
AxC	$(a-1)(c-1)$	SSAC	MSAC = SSAC/df
AxC	$(b-1)(c-1)$	SSBC	MSBC = SSBC/df
AxBxC	$(a-1)(b-1)(c-1)$	SSABC	MSABC = SSABC/df
Residual	$abc(n-1)$	SSRes	MSRes = SSRes/df
Total	$abcn-1$	SST	

The *test statistics* for the seven possible effects are constructed, as before, as the ratio of two MS terms. When all factors are classified as fixed effects, this corresponds to all test statistics being derived as the ratio MSEffect/MSRes as shown in Table 4.9. When random effects occur, test statistic construction is based on different ratios of MS terms depending on which factors are fixed and which are random (mixed model).[7,8]

Table 4.9 *Test statistics for a fixed effects three factor factorial design*

Source	MS	F
A	MSA	MSA/MSRes
B	MSB	MSB/MSRes
A × B	MSAB	MSAB/MSRes
C	MSC	MSC/MSRes
A × C	MSAC	MSAC/MSRes
B × C	MSBC	MSBC/MSRes
A × B × C	MSABC	MSABC/MSRes

Again, we must have an order to the inferential tests to be carried out. Essentially, we test the three factor interaction A × B × C first with the other effects, two factor interactions and main effects, tested as appropriate depending on the outcome to the tests of higher order effects. For example, if the A × B × C interaction is not significant, we would then test all the two factor interactions. Suppose interactions A × B and A × C were found to be significant. We would not then carry out the main effect tests for A, B and C as the significance of all these factors has been adequately shown through the significance of the highlighted two factor interactions. At the end of any such analysis,

[7] Reference 5, pp. 220–250
[8] Reference 1, pp. 470–502.

however, we must be sure that all effects have been tested statistically and those found significant, analysed fully.

7.3 Overview of Data Analysis for Three Factor Factorial Designs

Appropriate analysis of the collected data can be carried out by extending the procedures and principles outlined previously (model, exploratory data analysis, inferential data analysis, interaction plots, contrasts, diagnostic checking). As with the two factor factorial design, the level of data analysis necessary for a three factor factorial depends on which, if any, interactions are significant. If the three factor interaction is significant, it is advisable to plot this as a two factor interaction at each level of the third factor and assess each interaction plot in itself and in comparison with the plots for the other factor levels. If the three factor interaction is not significant, we would then assess the two factor interactions using the same principles developed for the two factor design. The basic rule in these designs is to ensure that conclusions on the influence of all factors and their interactions are forthcoming through appropriate techniques (formal tests, interaction plots, multiple comparisons).

Example 4.6

When separating phenols by reverse phase HPLC, several factors are known to influence the separation. A simple study was set up to examine three factors in order to optimise the separation of 11 priority pollutant phenols on an isocratic HPLC (solvent composition cannot be altered during the chromatographic run). The chemical factors chosen were concentration of acetic acid (AA), proportion of methanol (MW), and concentration of citric acid (CA). Factors AA and CA are added to the mobile phase as they can reduce the degree of peak tailing, a severe phenomenon in the HPLC of phenols. The response chosen to be monitored is the chromatographic response function (CRF), a summation term of the individual resolutions between pairs of peaks, which will be high if the peaks are separated at the base-line and the degree of peak tailing is small. The design matrix and the CRF values obtained are presented in Table B.1 where each experimental run was conducted in a randomised order.

Data entry

Entry of CRF response and factor codes data into Minitab follows similar lines to the data entry principles explained in Example 4.2 with one column for the CRF responses and three columns for the factor codes. Repetition of codes will again be necessary to account for the replication of treatment levels. Entry of the CRF data in the way it is presented in Table B.1 can greatly simplify codes entry through the inherent factor patterning that it will exhibit. Codes entry can utilise the **Calc ▷ Set Patterned Data** menu commands.

Response model
Based on the factorial design structure utilised, we would specify the response model as

$$CRF = AA + MW + AA \times MW + CA + AA \times CA + MW \times CA + AA \times MW \times CA + error$$

AA is the contribution of the acetic acid effect to the response, MW the contribution of the methanol proportion, AA × MW the interaction effect of acetic acid and methanol proportion, and CA the contribution of the citric acid effect with the interaction terms involving CA similarly defined. Again, we assume that the CRF response has equal variability across all factor levels and is normally distributed.

Hypotheses
The hypotheses for all seven effects associated with this experiment are all based on the test of no effect on response (H_0) against effect on response (H_1).

Exploratory data analysis
Initial plots and summaries of the CRF data with respect to each factor, though not shown, suggested some interesting results. No obvious difference in effect of the two acids was apparent though, for citric acid, the consistency of data appeared to differ with concentration. The most obvious effect was shown with the methanol:water proportion which provided distinctly different results for the two levels tested with 80% giving markedly higher results compared with 70%. Initially, only the methanol proportion looked to be exerting a strong influence on CRF.

Statistical tests
The ANOVA table output from Minitab is presented in Output 4.7 based on fitting a three factor factorial model to the CRF response data. The symbol '|' in the Menu commands again provides a shorthand notation for a factorial model. The Options and Storage sub-menus are as previously explained.

Output 4.7 ANOVA table for Example 4.6

Select **Stat** ▷ **ANOVA** ▷ **Balanced ANOVA** ▷ for *Responses*, select **CRF** and click **Select** ▷ select the **Model** box and enter **AA|MW|CA**.

Select **Options** ▷ select the **Display means for (list of terms)** box and enter **MW*CA** ▷ click **OK**.

For Storage, select **Residuals** and select **Fits** ▷ click **OK**.

Analysis of Variance (Balanced Designs)

Factor	Type	Levels	Values		
AA	fixed	3	1	2	3
MW	fixed	2	70	80	
CA	fixed	3	2	4	6

Analysis of Variance for CRF

Source	DF	SS	MS	F	P
AA	2	0.4356	0.2178	2.72	0.093
MW	1	26.6944	26.6944	333.68	0.000
CA	2	0.6572	0.3286	4.11	0.034
AA*MW	2	0.0089	0.0044	0.06	0.946
AA*CA	4	0.3778	0.0944	1.18	0.353
MW*CA	2	1.6572	0.8286	10.36	0.001
AA*MW*CA	4	0.3444	0.0861	1.08	0.397
Error	18	1.4400	0.0800		
Total	35	31.6156			

MEANS

MW	CA	N	CRF
70	2	6	9.683
70	4	6	9.417
70	6	6	9.483
80	2	6	10.833
80	4	6	11.250
80	6	6	11.667

Looking at the p values, we can see that the three factor interaction is statistically insignificant ($p = 0.397$). The only significant interaction appears to be the one between methanol proportion and citric acid ('MW*CA') with a p value of approximately 0.001. The methanol effect (p approximately 0.000, $p < 0.0005$) and the citric acid effect ($p = 0.034$) are also statistically significant. The SS for the methanol proportion is by far the largest suggesting that it is the most important of the three factors in its influence on variation in the CRF response.

We can conclude, therefore, that CRF appears to be primarily influenced by the methanol and citric acid interaction though the acetic acid factor ('AA'), with a *p* value of 0.093, may also play a role though only a minor one by comparison.

Follow-up
According to the statistical evidence, only the methanol and citric acid interaction needs further assessment. The associated interaction plot is provided in Output 4.8 based on the interaction means shown in Output 4.7. A differing pattern is obvious according to methanol proportion used. For low level, CRF remains low for all concentrations of citric acid. A high methanol proportion produced not only higher CRF values but also an upward trend as level of citric acid increases.

Output 4.8 *Interaction plot for Example 4.6*

Menu commands as Output 4.4.

Methanol:water x Citric acid interaction plot

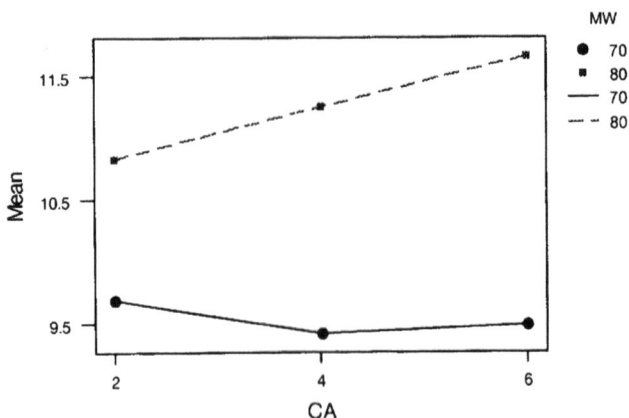

Highest CRF appears to occur at (MW 80%, CA 6 g dm^{-3}). Checking the original data (Table B.1) shows that this combination together with AC 0.004 mol dm^{-3} provided the highest set of CRF values which were also consistent (low range). Given these results, it could be that higher CRF measurements could be obtained at higher proportion of methanol and higher concentration of citric acid. This may be an area where future experimentation should focus in order to endeavour to optimise the HPLC system.

Diagnostic check

The residual plots for diagnostic checking, though not shown, were also checked and provided distinct column length differences in all the factor plots. The fits plot was shaped as a sideways V indicative of inadequate model fit. Normality of CRF response appeared acceptable. In light of these results, perhaps the CRF response requires to be transformed if being used as the chemical response in a subsequent experiment.

Summarising the analysis of Example 4.6, we can say that the methanol proportion appears to be the major factor influencing the CRF response. The only interaction to be statistically significant was that between methanol proportion and citric acid ($p = 0.001$). Assessment of this interaction showed that a combination of MW 80% and CA 6 g dm^{-3} appeared to be the best for maximum CRF, hinting that high methanol proportion and high concentration of citric acid could be the area for future experimentation. Diagnostic checking hints at CRF response rescaling as being worth considering for such future experiments.

Exercise 4.3

The measurement data below were collected from a recovery experiment investigating spectrophotometric determination of colourants in synthetic food mixtures. The mixtures comprised different proportions, measured as mg l^{-1}, of the colourants E-123 (Amaranth), E-124 (Poinceau 4R), and E-120 (Carminic acid). The data refer to percent recovery of Amaranth as measured at 549.7 nm in the first derivative ratio spectrum using Carminic acid (32 mg l^{-1}) as divisor. How does presence of different colourants affect recovery? Are any colourant interactions important?

		E-120			
E-123	E-124	12		24	
12	12	100.2	102.4	99.7	99.3
	24	91.8	92.5	101.9	102.3
	36	102.1	103.7	100.4	100.8
24	12	95.2	95.8	97.9	97.3
	24	102.6	102.1	100.6	99.8
	36	97.7	97.3	95.2	95.8
36	12	101.6	102.3	99.7	99.2
	24	100.9	100.6	98.7	98.3
	36	100.4	100.2	96.7	97.4

Source: Reprinted from J.J. Berzas Nevado, C. Guiberteau Cabanillas and A.M. Contento Salcedo, *Talanta*, 1995, **42**, 2043–2051 with kind permission of Elsevier Science-NL, Sara Burgerhartstraat 25, 1055 KV Amsterdam, The Netherlands.

8 OTHER FEATURES ASSOCIATED WITH THREE FACTOR FACTORIAL DESIGNS

8.1 Pseudo-*F* Test

For a *model II* (all factors random) or *mixed model* (at least one factor random and one fixed) factorial experiment, an exact *F* test for certain main effects may not be available by direct means through the ratio of two known mean square (MS) terms. This lack of available test statistic can be overcome by producing a *pseudo-F*, or *approximate-F*, test, the construction of which still uses the MS ratio principle, except that the denominator is expressed as a linear combination of other MSs within the design. The numerator MS remains that of the effect to be tested. For example, if A is fixed and B and C are random (mixed model experiment), then

$$F = MSA/(MSAB + MSAC - MSABC)$$

provides a pseudo-*F* test for testing the statistical significance of the fixed main effect A. It is therefore important in three factor factorial designs, if factor levels are chosen randomly, to use the underlying theory to ascertain the format of the test statistics required to help test the interaction and main effects.

8.2 Pooling of Factor Effects

In some factorial designs, particularly when more than two factors are being investigated, certain factor interactions may be known to be unimportant or can be assumed negligible in their effect. These 'negligible' effects can be combined with the residual to form a new estimator for the residual sum of squares (SSRes′) with greater degrees of freedom (Resdf′). For example, if in a three factor factorial experiment the interaction $A \times B \times C$ can be assumed negligible, we could combine it with the residual sum of squares to form a new residual sum of squares SSRes′ (SSABC + SSRes) and with new degrees of freedom Resdf′ (ABCdf + Resdf). Appropriate *F* test statistics for the remaining model components can then be derived based on the new residual mean square estimate of MSRes′ = SSRes′/Resdf′.

9 NESTED DESIGNS

Nested design structures arise in situations when replicate measurements are made on the same experimental unit. They superficially resemble the

RBD structure except that the levels of the second factor are not common to all treatments. Generally, the levels of this second factor, often representing a subject effect, are randomly selected whereby the levels are similar but not identical for the different levels of the main factor. This structure is used as the basis of *co-operative trials* in which laboratories are provided with a series of specimens for analysis and interest lies in estimating the level of random and systematic error in the system.[9,10]

To illustrate the concept of a nested design, consider a study set up to investigate the analysis results reported by three laboratories on the level of lead in a supplied sample. In each laboratory, the sample is to be analysed, in duplicate, by three technicians. A possible design structure is illustrated in Table 4.10. The technicians carrying out the analysis are obviously not the same as they are attached to the laboratory undertaking the analysis. Based on this, we say the technicians are *nested* within the laboratories with the design structure referred to as a *two factor nested design*. The response model would be,

lead measurement = lab + technician(lab) + error

Essentially, the nested term represents the technician and technician/laboratory interaction effects, so a two factor nested design has similar underlying features to a two factor factorial.

Table 4.10 *Possible design structure for the lead study*

Laboratory		A			B			C	
Technician	*1*	*2*	*3*	*4*	*5*	*6*	*7*	*8*	*9*
	x	x	x	x	x	x	x	x	x
	x	x	x	x	x	x	x	x	x

x denotes a lead measurement

Appropriate test statistics for nested designs can be constructed from the related MS terms for each effect once it is known which factor is fixed and which is random. Analysis of data from nested designs follows a similar pattern to that outlined for the design structures considered in this chapter. In co-operative trials, however, the objective is more to estimate and assess the three sources of variation, these being the variation between replicates (residual), the variation between technicians (the nested effect), and inter-laboratory variation. Often, these

[9] Analytical Methods Committee, *Analyst (London)*, 1987, **112**, 679.
[10] Analytical Methods Committee, *Analyst (London)*, 1989, **114**, 1489.

may also be a function of the analyte concentration which can affect the analysis carried out.

10 REPEATED MEASURES DESIGN

A *repeated measures design* can arise in a longitudinal study, such as a clinical trial and an inter-laboratory study when preparing precision statements where several factors are experimented on and measurements are collected at regular time points.[11] It is such that both factorial and nested effects appear in the same experimental structure and is a powerful design with application in observational studies, in particular. The associated data analysis involves techniques appropriate to both factorial and nested designs.

For example, an experiment is to be set up to study the effect of three treatment regimes on the level of copper in the blood of animals grazing near a toxic waste disposal site. Six randomly selected animals are to be assigned to each regime and their level of copper measured, in duplicate, on a monthly basis over the trial period. The design structure would follow that presented in Table 4.11 where the animals assigned to each regime differ and must therefore be considered as nested within regime. The main factors within this illustration are regime (factor A), animal (factor B) nested within regime, and time (factor C). The other terms, $A \times C$ and $B(A) \times C$, correspond to interaction effects.

Table 4.11 *Experimental layout for the blood copper study*

Treatment regime	Animal	Time 1		2		.	6	
A	1	x	x	x	x	.	x	x
A	2	x	x	x	x	.	x	x
.	.	.	.	.	.			
A	6	x	x	x	x	.	x	x
B	7	x	x	x	x	.	x	x
B	8	x	x	x	x	.	x	x
.	.	.	.	.	.			
B	12	x	x	x	x	.	x	x
C	13	x	x	x	x	.	x	x
C	14	x	x	x	x	.	x	x
.	.	.	.	.	.			
C	18	x	x	x	x	.	x	x

x denotes a measurement of level of copper

The model for the response of copper measurement would be

[11] Reference 5, pp. 499–521.

$$\text{Cu measurement} = A + B(A) + C + A \times C + B(A) \times C + \text{error}$$

where the model parameters are as defined for factorial and nested design models. Based on this model, there are a number of tests of interest: regime effect, nested animal effect, time effect, the interaction between regime and time, and the interaction of the nested animal effect and time. When only one replicate measurement is made at each level of factor C (each time point), the latter interaction cannot be estimated and so no test of this effect is available. Factor B, in general a subject factor, is mostly classified as a random effect while C is assumed fixed meaning such designs are generally of mixed model type.

11 ANALYSIS OF COVARIANCE

Analysis of covariance (ANCOVA) combines both ANOVA and regression techniques and is used when wishing to control one, or more, extraneous, or confounding, variables (covariates) that may influence the measurable response, e.g. completely uniform material may not be available at the start of an experiment. Covariates can be included in an experimental design structure at little extra cost in terms of increased experimentation and enable adjustments to be made to the response for chance differences between the groups due to the random assignment of non-uniform experimental units to the treatment groupings. ANCOVA can help to increase the power of an experiment provided a high correlation exists between the selected covariate(s) and the response enabling the final analysis to more precisely reflect the influence of the effects being tested.[12]

To apply this design, the experimenter must anticipate those attributes that may affect the proposed response variable prior to experimentation. Appropriate measurements of the response and covariate(s) can then be obtained with the response adjusted to eliminate the effect of the covariate(s) so providing more relevant, and meaningful, response data.

---

[12] Reference 5, pp. 562–584.

CHAPTER 5

Regression Modelling in the Chemical Sciences

5.1 INTRODUCTION

Applied chemical experimentation is not always aimed at comparative experiments, such as comparing analytical procedures or comparing laboratories. Often, we may be more interested in ascertaining whether the measured response and the experimental factors can be modelled by a mathematical relationship. The concept of *least squares regression* plays a fundamental role in this modelling aspect of experimentation by providing the estimation technique for determining the form of relationship best suited to the patterning exhibited by the experimental data. Regression, therefore, represents a modelling tool for the derivation of the relationship between a dependent variable Y and any number of independent (controlled) variables, the Xs, based on collected experimental data.

For example, the solubility of a chemical may be related to the temperature at which solubility is measured. Through regression, it may be possible to build a model of solubility as a function of temperature, with such a model used to predict solubility at specified temperatures. In a chemical process, the yield of a reaction product may be related to factors such as process operating temperature, quantity of catalyst, humidity, pH, and time. By application of regression, it may be possible to build a model of yield as a function of these characteristics to provide a mechanism for predicting yield which could aid process optimisation or process control. In 'supervised' analysis of multivariate data, we may be interested in relating concentration of a chemical to absorbance measurements at particular wavelengths through a multiple regression model to produce a mechanism for prediction of concentration.

Most regression models are based on simple mathematical functions specified in terms of parameters which require to be estimated from the collected experimental data. Simple models which occur often in chemical experimentation include the *linear model*

168

$$Y = \alpha + \beta X$$

used in linear calibration, the *non-linear model*

$$Y = \alpha e^{\beta x}$$

often used to model exponential decay or growth, and the *quadratic model,*

$$Y = \alpha + \beta_1 X + \beta_2 X^2$$

Data plots exhibiting trends commensurate with such models are shown in Figure 5.1. Both positive and negative trends are feasible for linear models as Figure 5.1A demonstrates. The exponential plots of Figure 5.1B show the accelerated trend of such relationships while the quadratic plots of Figure 5.1C show varying trend in response as the experimental variable changes value. Each of the plots in Figure 5.1 shows a distinctly different trend in the data, highlighting that linearity of response data is not always going to be the case in regression modelling.

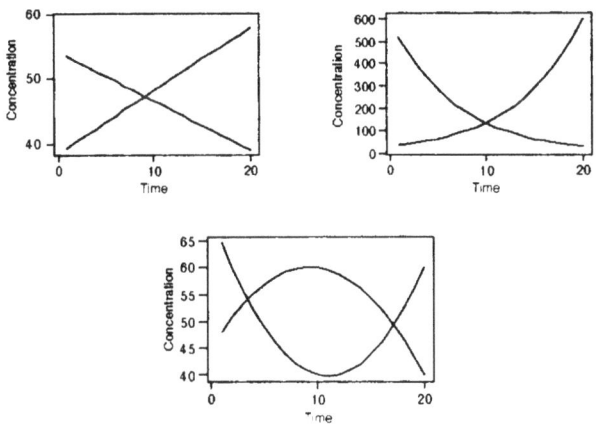

Figure 5.1 *Illustrations of data plots*

Linear modelling is widely used in many areas of chemistry. They include linear calibration, comparison of analytical accuracy[1] and detection of analytical bias,[2] specification of detection limits for analy-

[1] M. Thompson, *Analyst (London)*, 1982, **107**, 1169.
[2] B.D. Ripley and M.T. Thompson, *Analyst (London)*, 1987, **112**, 377.

tical procedures,[3] relating sources of variation to concentration of analyte in co-operative trials,[4] and relating reproducibility and repeatability to analyte concentration in collaborative trials. In many instances, however, an experimental response Y may depend on more than one X variable giving rise to such as the *multiple linear model*

$$Y = \alpha + \beta_1 X_1 + \beta_2 X_2 + \beta_3 X_3 + \beta_4 X_4$$

of Y as a function of four independent X variables. *Multiple linear regression* (MLR) can often be more beneficial than simple modelling through its ability to relate the measured response to many contributing variables as occurs in multivariate calibration.

The purpose of regression modelling is to use the experimental data to derive estimates of the constants α and β associated with a proposed regression model to provide a numerical form for the model. Once the form of model is estimated by the technique of *ordinary least squares* (OLS), we must carry out a full statistical and practical assessment of its applicability as a representative model of the process under investigation.

Irrespective of type of model fitted, we must always assess both its statistical and practical appropriateness. Statistical validity is obviously important but it is practical validity which is probably the most important aspect of regression modelling. If model predictions do not match the recorded responses well enough, then the fitted model may not be of much practical use for its intended purpose as a model of a chemical process. Hence, the need to check predictive ability of the fitted model as an integral part of the assessment of a regression model. The requisite steps in the analysis of regression models are summarised in Box 5.1 though not all are necessarily appropriate in the applications of regression methods within chemical experimentation.

For the case of two related variables X (controlled) and Y (recorded), we will primarily consider the linear model though the principles introduced will be equally applicable to non-linear and polynomial modelling. Applications of linear regression in the form of linear calibration and the comparison of two linear equations, of primary use in analytical experimentation, will also be introduced. The principles of linear regression will be extended to multi-variable situations where MLR can be applied. It must always be remembered, however, that derivation of any regression model depends, to a large extent, on the experimental data collected. A regression model is also

---

[3] M. Thompson, *Anal. Proc.*, 1987, **24**, 355.
[4] Analytical Methods Committee, *Analyst (London)*, 1987, **112**, 679.

Box 5.1 *Analysis steps in regression modelling*

Data plot. Plot the experimental data on a scatter diagram, if feasible, to ascertain what pattern is inherent in the data. Propose a model commensurate with the data pattern.

Parameter estimation. Obtain estimates for the parameters of the proposed model.

Statistical validity. Confirm the statistical validity of the fitted model through inferential data analysis (tests), estimation (confidence intervals), and assessment of the coefficient of determination.

Practical validity. Check the practical validity of the fitted model. Matching of recorded responses and predicted values implies good model fit though matching is never likely to occur at all data points.

Diagnostic checking. Use the residuals to check model construction and assumptions by plotting residuals in a similar way to that for experimental designs (see Box 3.3).

Conclusion. Is model fit acceptable? Is model interpretable chemically? Are there enough regressor variables?

---

only valid with respect to the range of values the experimental data cover, with good coverage improving the applicability of a well fitting regression model.

In this chapter, we will consider how best to choose an appropriate relationship to fit to experimental data, how to fit such a relationship using OLS, and how to assess its relevance in the context for which it is intended. Essentially, we ask whether the fitted relationship is a statistically appropriate one, is chemically appropriate, and is of practical use.

2 LINEAR REGRESSION

Linear regression corresponds to modelling the relationship between two variables by a simple straight line. For this form of simple modelling, we have two variables: X, the independent/controlled variable representing the set of experimental conditions, and Y, the response/dependent/recorded variable. Obviously the recorded values of Y depend on the input values of X. To obtain relevant data, an experiment is conducted using a set of X values to generate a set of recorded Y responses, as illustrated in Table 5.1. Essentially, we conduct an experiment at setting x_1 and record the response y_1. This would be repeated at the other X settings until we have collected a total of n pairs of observations. These observations become the data set on which we base the model building.

Table 5.1 *Typical data layout for a simple regression experiment*

X	x_1	x_2	x_3	.	.	x_n
Y	y_1	y_2	y_3	.	.	y_n

n, number of data points

2.1 Data Plot

For experiments involving only two variables X and Y, it is always advisable first to plot the data on a *scatter diagram* to assess the nature of the relationship between Y and X. In linear modelling, the plot will exhibit a monotonic linear pattern such as those illustrated in Figure 5.2. The trend will be in either a positive direction (increase in X providing increase in Y, line slopes upwards from left to right) or a negative direction (increase in X providing decrease in Y, line slopes downwards left to right). Simple data plotting is also useful when assessing the viability of a model suggested prior to experimentation, *i.e.* do the data conform to the suggested trend?

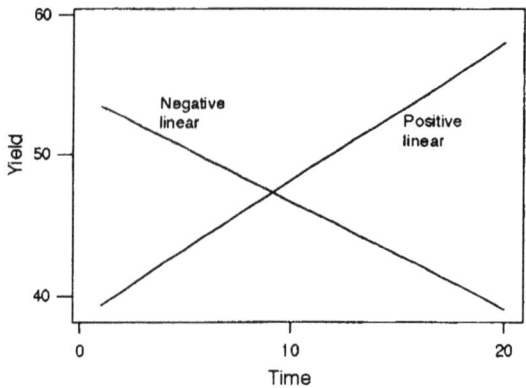

Figure 5.2 *Illustrations of linear data plots*

2.2 Simple Linear Model

The simple linear regression model is expressed as

$$Y = \alpha + \beta X + \epsilon$$

where α, the intercept coefficient and point where line cuts the y-axis, and β, the slope coefficient, are the population regression coefficients, and ε is the error term. As in the specification of a response model in experimental designs, this form of response model defines the response to be a function of the controlled variation, the X variable, and the uncontrolled variation, the error or noise. Hopefully, the latter will be minimal enabling the validity of the specified model to be readily shown.

The general *assumptions* underpinning regression modelling are that the X values are fixed and measured without error, and that the error ε is assumed normal with common error variance σ^2 for each X value. Assuming normality for the error means that the response Y can be assumed normal while the common variance assumption implies equal response variability at all X values.

2.3 Parameter Estimation

Estimation procedures for the intercept α and the slope β of the linear model, labelled a and b, are derived by applying the OLS estimation technique, a procedure which fits the 'best' fitting straight line to the data based on minimising the sum of squared deviations of the recorded responses from the model predictions. These deviations are referred to as the *errors*, or *residuals*, and denoting them by e_i, we can construct the sum of squared deviations as

$$S = \sum_{i=1}^{n} e_i^2 = \sum_{i=1}^{n} \left[y - (a + bx_i) \right]^2 \tag{5.1}$$

This expression is called the *residual*, or *error, mean square*.

A graphical illustration of this procedure is shown in Figure 5.3 with the residuals e_i represented by the vertical distances indicated. The data illustrated in Figure 5.3 specify a positive trend but do not fully conform to a perfect linear model as a straight line cannot pass through all points. OLS estimation attempts to fit the 'best' straight line through these data subject to minimising equation (5.1) across all data points resulting in the fitted line passing close to some points but missing others.

Minimising equation (5.1), using partial differentiation, results in the normal equations

$$na + b_1 \sum x = \sum y$$
$$a \sum x + b_1 \sum x^2 = \sum xy$$

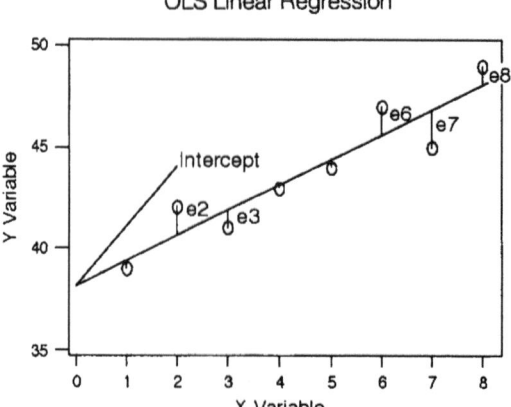

Figure 5.3 *Linear least squares regression*

with the condition that $\sum e_i = 0$, *i.e.* positive and negative errors add to zero. Solving these equations provides the solutions for a and b:

$$b = \frac{\sum_{i=1}^{n} x_i y_i - \frac{\sum_{i=1}^{n} x_i \sum_{i=1}^{n} y_i}{n}}{\sum_{i=1}^{n} x_i^2 - \frac{\left(\sum_{i=1}^{n} x_i\right)^2}{n}} = \frac{S_{XY}}{S_{XX}} \tag{5.2}$$

$$a = \frac{\sum_{i=1}^{n} y_i - b \sum_{i=1}^{n} x_i}{n} \tag{5.3}$$

where $\sum x_i = x_1 + x_2 + \ldots + x_n$, $\sum y_i = y_1 + y_2 + \ldots + y_n$, $\sum x_i y_i = x_1 y_1 + x_2 y_2 + \ldots + x_n y_n$, $\sum x_i^2 = x_1^2 + x_2^2 + \ldots + x_n^2$, and the S terms refer to the 'corrected sum of squares'.

Excel has regression modelling and assessment facilities which can be used to provide the necessary information for full statistical and practical analysis of a fitted regression model. They are located in the Data Analysis procedures within the Tools menu. The associated dialog window is illustrated in Figure 5.4 with the usual filling in and checking boxes required before implementing the regression estimation process. 'Confidence interval 95%' means that all estimation information on regression parameters will be based on presentation of 95% confidence intervals.

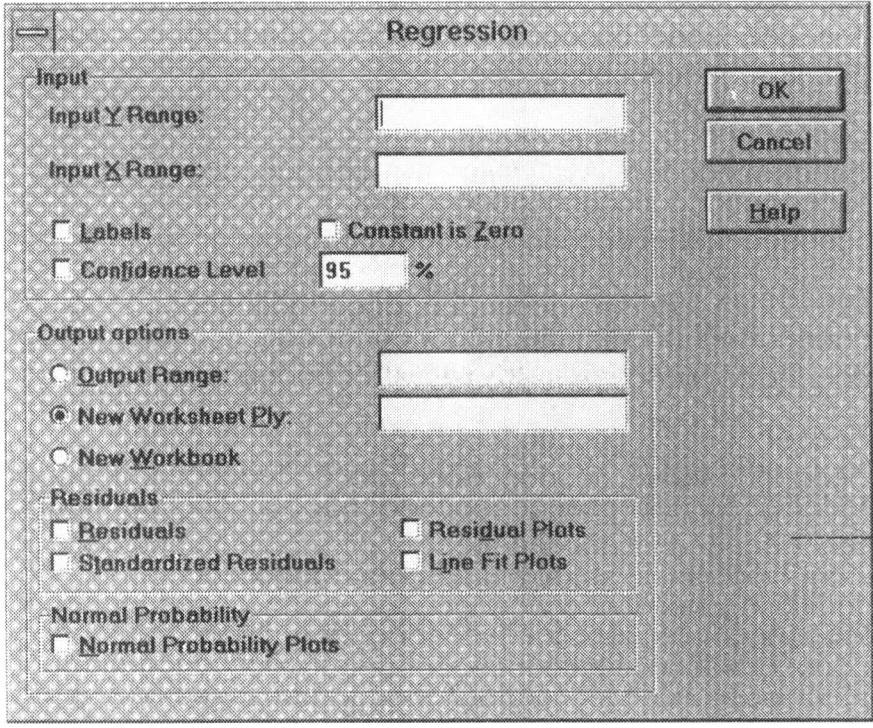

Figure 5.4 *Regression dialog window in Excel*

Example 5.1

Specially coated quartz piezoelectric crystals can be used as detectors to monitor SO_2 levels in gaseous emissions since adsorption of SO_2 can cause a decrease in the crystal's frequency of vibration. Following a standard measurement procedure, the frequency shift (Hz) was measured for crystal samples exposed to differing SO_2 concentrations (ppm) providing the data in Table 5.2. It has been suggested that frequency shift may be a linear function of SO_2 concentration.

Table 5.2 *Frequency shift data for Example 5.1*

SO_2 (ppm)	5	10	15	20	25	30	35	40	45	50	55	60	65	70	75
Freq(Hz)	45	55	67	75	90	101	113	122	143	150	159	177	192	195	215

Source: D. McCormick and A. Roach, 'Measurement, Statistics and Computation', ACOL Series, Wiley, Chichester, 1987: reproduced with the permission of HMSO.

Data entry
Data entry in Excel consists of entering the X and Y data into separate columns of the spreadsheet.

Model
Linearity has been proposed for the frequency shift data so

$$\text{frequency shift} = \alpha + \beta\ SO_2\ \text{concentration} + \varepsilon$$

becomes the proposed model. The SO_2 concentration settings are assumed fixed and measured without error while the frequency shift data are assumed approximately normally distributed.

Data plot
Output 5.1 contains an Excel scatter plot of frequency shift data in Table 5.2. The plot is supportive of there being a positive linear trend in

Output 5.1 *Scatter plot of frequency shift data for Example 5.1*

Data in cells A1:B16, labels in cell 1 of each column.
  Select **ChartWizard**. Click cell **D1**.
  Chart Wizard Step 1 of 5: for *Range*, click and drag across cells **A1:B16** ▷ click **Next**.
  Chart Wizard Step 2 of 5: select **XY (Scatter)** ▷ click **Next**.
  Chart Wizard Step 3 of 5: select **XY(Scatter) chart 2** ▷ click **Next**.
  Chart Wizard Step 4 of 5: click **Next**.
  Chart Wizard Step 5 of 5: for *Add a Legend?*, select **No** ▷ select the **Chart Title** box and enter **Plot of Frequency Shift Against SO2 Concentration** ▷ select the **Axis Titles Category (X)** box and enter **SO2 Concentration** ▷ select the **Axis Titles Value (Y)** box and enter **Frequency Shift** ▷ click **Finish**.
  Re-specification of plot as Output 2.1.

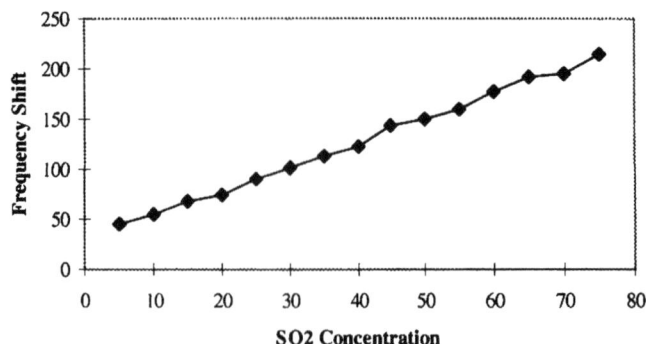

Plot of Frequency Shift Against SO2 Concentration

the data. As the points lie close to a straight line, it would appear that a linear model could produce an acceptable fit.

Parameter estimation

Excel's regression tool (see Figure 5.4) produces an output as shown in Output 5.2 which contains all the necessary regression statistics for statistical assessment of the fitment of a linear model. The additional plot selections in the menu procedure are used to produce practical and diagnostic check plots which will be addressed later (see Example 5.3).

Output 5.2 *Linear regression output for Example 5.1*

Y data in cells B1:B16 (label in B1) and X data in cells A1:A16 (label in A1).

Select **Tools** ▷ **Data Analysis** ▷ **Regression** ▷ click **OK** ▷ for *Input Y Range*, enter **B1:B16** ▷ select **Input X Range** box and enter **A1:A16** ▷ select **Labels** ▷ select **Output Range**, click the empty box, and enter **A18** ▷ select **Residual Plots** ▷ select **Line Fit Plots** ▷ select **Normal Probability Plots** ▷ click **OK** ▷ click the mouse button.

SUMMARY OUTPUT

Regression Statistics

Multiple R	0.9983
R Square	0.9966
Adjusted R Square	0.9963
Standard Error	3.2868
Observations	15

ANOVA

	df	SS	MS	F	Significance F
Regression	1	41237.15714	41237.16	3817.09	0.0000
Residual	13	140.4428571	10.8033		
Total	14	41377.6			

	Coefficients	Standard Error	t Stat	P-value	Lower 95%	Upper 95%
Intercept	29.514	1.7859	16.53	0.0000	25.6560	33.3726
SO2	2.427	0.0393	61.78	0.0000	2.3423	2.5120

The column 'Coefficients' specifies that the OLS estimation equations, (5.2) for slope and equation (5.3) for intercept, provide as $b = 2.427$ and

a = 29.514, respectively. The fitted linear model is therefore frequency shift = 29.514 + 2.427 SO_2. The intercept of 29.514 suggests that, when SO_2 concentration is zero, frequency shift will be of the order of 29.5 Hz. The slope estimate of 2.427 implies that, for every 1 ppm change in SO_2 concentration, the frequency shift of the crystal can change by 2.427 Hz.

Model extrapolation must be treated with caution as it can involve considering the fit of a model beyond the range of X values on which the model has been built. Excessive extrapolation must be avoided as it can result in nonsensical model interpretations, *e.g.* negative absorbance or negative concentration.

3 ASSESSING THE VALIDITY OF A FITTED LINEAR MODEL

The most important aspects in the assessment of a fitted regression model are the checks of its statistical and practical validity. The former is necessary to provide the objective basis for model assessment as only statistically valid models can be of any practical benefit. Validity of modelling data by simple relationships, however, does not just rest on statistical validity. The fitted model must also provide predicted values commensurate with the recorded experimental responses. In other words, if the model predictions are not reliable, how useful a model is it? Any fitted regression model must, in essence, be appropriate both statistically and practically if it is to be used as a suitable description of experimental data.

Since regression is based on model specification, the *ANOVA principle* of splitting response variability into explainable and unexplainable components can again be applied. The former refers to variation due to the regression model fitted, SSRegn, and the latter the residual variation, SSRes. Sum of squares (SS) and mean square (MS) terms again underpin this approach with the main SS term, SSTotal, expressed as,

$$SSTotal\ (SST) = SSRegression\ (SSRegn) + SSResidual\ (SSRes)$$

Table 5.3 *General ANOVA table for linear regression*

Source	df	SS	MS
Regression	1	SSRegn = $bS_{XY} - (\sum y)^2/n$	MSRegn = SSRegn/df
Residual	$n-2$	SSRes (by subtraction)	MSRes = SSRes/df
Total	$n-1$	SST = $\sum y^2 - (\sum y)^2/n$	

n, number of data points; df, degrees of freedom; SS, sum of squares; MS, mean squares; S_{XY}, corrected sum of squares [see equation (5.2)]

The calculation aspects and general ANOVA table for the simple linear model are presented in Table 5.3.

3.1 Statistical Validity of the Fitted Regression Equation

Checking the statistical validity of a fitted regression model can be achieved by carrying out a simple *hypothesis test*. For a linear model, several procedures can be utilised though all provide the same conclusion in respect of statistical validity. Box 5.2 illustrates the statistical checks available for linear model assessment, all based on standard decision rule mechanisms. All statistical elements described are usually produced in software by default.

Box 5.2 *Statistical validity checks for linear modelling*

t Test of β This form of model check assesses the two-sided hypotheses H_0: $\beta = 0$ and H_1: $\beta \neq 0$. The test statistic is specified as

$$t = b/s_b \tag{5.4}$$

based on $(n - 2)$ degrees of freedom.

Confidence interval for β. A $100(1 - \alpha)\%$ confidence interval estimate for the slope parameter β is given by

$$b \pm t_{\alpha/2, n-2} s_b \tag{5.5}$$

enabling assessment of the coefficient estimate to be considered.

F test of linear model. This test, based on the use of ANOVA principles, assesses the null hypothesis H_0: X and Y not linearly related. Based on the response variation split discussed previously, we specify the test statistic to be the ratio of regression variance to residual variance, *i.e.*

$$F = MSRegn/MSRes \tag{5.6}$$

with derivation of the MS terms as shown in Table 5.3. This test statistic is based on degrees of freedom df_1 = Regression df = 1 and df_2 = Residual df = $n - 2$.

Coefficient of determination. Once a model has been fitted, it is useful to estimate the variation in the recorded response accounted for by the X variable(s) and the fitted model to help assess the model's accountability. Such a summary is provided by the coefficient of determination

$$R^2 = SSRegn/SSRes \tag{5.7}$$

High values in excess of 80% to 90% are indicative of a good statistical fit, while for lower R^2 values, either the fitted model is inappropriate or other X variables may require to be included in a re-specified model.

The term s_b in equation (5.4) defines the standard error of the slope estimate b and is specified as

$$s_b = s_{y/x} / \sqrt{S_{XX}} \qquad (5.8)$$

where

$$s_{y/x} = \sqrt{[SSRes/(n-2)]} \qquad (5.9)$$

is the estimate of the error standard deviation σ (the regression standard deviation), and S_{XX} is the corrected sum of squares for X defined in equation (5.2). The divisor of $(n-2)$ in equation (5.9) can be similarly explained in terms of degrees of freedom as occurred for the standard deviation expression (2.3). In linear modelling, two parameters are estimated and so there are only $(n-2)$ independent pieces of information available for estimation. Hence, the use of $(n-2)$ as divisor for the regression standard deviation.

Acceptance of the null hypothesis in the t test for β, using test statistic (5.4), suggests that there appears no evidence that as X varies, Y exhibits a linear trend significantly different from a horizontal line. In other words, the variation in Y appears not to be explainable by a linear trend in X. For the linear model, we have $F = t^2$, *i.e.* the F test statistic (5.6) equals the square of the t test statistic (5.4). The F test is the best mechanism for assessing statistical validity of model fit as it is applicable in all regression modelling cases.

When using the coefficient of determination (5.7) within model assessment, it is often more appropriate to use R^2 adjusted for degrees of freedom, denoted R^2_{adj}, rather than the straight R^2. R^2_{adj}, expressed as

$$R^2_{adj} = 100 \left(\frac{(n-1)(R^2/100) - k}{n-k-1} \right) = 100 \left(\frac{(n-1)SSRegn - (k)SSTotal}{(n-k-1)SSTotal} \right)$$

$$(5.10)$$

takes account of the number of observations recorded (n) and the number of X terms in the fitted model (k), and will always be lower than R^2 though often not by much. This coefficient provides a better measure of explained variation in that it can account for the addition, or deletion, of new variables more adequately. Use of R^2_{adj} will be illustrated in more detail when MLR is considered (see Section 5.8) as its use in simple regression is generally not necessary.

Example 5.2

Having obtained the fitted linear model for the frequency shift data of Example 5.1, we now need to assess its statistical validity, carrying out all tests at the 5% significance level. To do this, we will use the Excel regression information presented in Output 5.2.

Hypotheses
Since we have fitted a linear model to the experimental data, the hypotheses we require to test are H_0: SO_2 concentration and frequency shift not linearly related ($\beta = 0$) against the alternative H_1: SO_2 concentration and frequency shift linearly related ($\beta \neq 0$).

t Test of β
The t test statistic (5.4) is determined as 61.78 (Output 5.2, row 'SO2', column 't Stat') with associated p value of 0.0000, *i.e.* $p < 0.00005$ (Output 5.2, row 'SO2', column 'P-value'). Based on testing at the 5% significance level, we can see that H_0 will be rejected, resulting in the conclusion that the linear model of frequency shift as a function of SO_2 concentration appears statistically valid ($p < 0.05$).

Confidence interval for β
We know that the slope estimate is $b = 2.427$ with standard error $s_b = 0.0393$ (Output 5.2, row 'SO2', column 'Standard Error'). A 95% confidence interval for β, using equation (5.5) with critical value $t_{0.025,13} = 2.160$, is (2.342, 2.512) Hz. As the interval does not contain 0, this provides further back-up of the statistical validity of the fitted model. In addition, as the interval limits do not differ markedly, the slope estimate provided can be considered as being of acceptable accuracy. From this interval, we can state that the rate of change of frequency shift with SO_2 concentration is likely to lie between 2.342 and 2.512 Hz per concentration change of 1 ppm.

F test
The F test statistic (5.6) can be found in the ANOVA table in Output 5.2 as 3817.09 (row 'Regression', column 'F') with associated p value of 0.0000, *i.e.* $p < 0.00005$ (row 'Regression', column 'Significance F'). Based on degrees of freedom $df_1 = 1$ ('Regression df') and $df_2 = 13$ ('Residual df') and 5% significance level, the critical value for test statistic comparison is given by $F_{0.05,1,13} \approx 4.60$ [Table A.3, (1, 14) degrees of freedom]. As test statistic exceeds critical value, we again conclude that the fitted linear model appears statistically valid ($p < 0.05$).

Coefficient of determination
The 'Regression Statistics' presented in Output 5.2 indicate that $R^2 =$ 99.7% ('R Square') and $R^2_{adj} = 99.6\%$ ('Adjusted R Square'). Since both are well above 80%, we can safely say that the linear model in terms of SO_2 concentration accounts for almost all of the variation in the frequency shift response data. This implies that the statistical fit of the linear model appears valid.

From this illustration, we can see that all tests of statistical validity provide the same statistical conclusion on model fit. This will always be the case in linear modelling and in most other forms of regression modelling. In practice, we should adopt a strategy of only considering two of these mechanisms as the foundation for the assessment of statistical validity. In most regression modelling, the F test and coefficient of determination are the most appropriate mechanisms to utilise.

3.2 Practical Validity

Once statistical validity of a fitted model has been assessed, the next step in model assessment rests with checking practical validity. This comparison of model predictions with the recorded experimental responses is a vital and important component in the analysis of any fitted regression model as it enables model validity to be confirmed at the level of its intended use, *i.e.* as a prediction tool. An appropriate model will obviously result in reliable predictions that are close to the corresponding recorded measurements. An inappropriate model will result in inaccurate predictions which differ markedly from the recorded response measurements. Examination of prediction accuracy should generally take the form of a print and plot of the recorded responses and model predictions. Procedures for determination of confidence intervals for predictions also exist.[5]

3.3 Diagnostic Checking

Diagnostic checking of the residuals (responses – model predictions) should be the final part of model analysis to assess model fit and validity of model assumptions. The checks, as before, are based on simple residual plots and can provide further valuable information about model effects not previously suspected.

A plot of the residuals against the controlled variable(s) (the Xs)

[5] L. Ott, 'An Introduction to Statistical Methods and Data Analysis', 4th Edn., Duxbury Press, Belmont, California, 1993, pp. 519–522.

should be examined for unaccounted trends or patterns indicative of suspect model fit (additional terms required). Figure 5.5A highlights just such a case where an obvious quadratic trend is apparent in the residual plot, suggesting that inclusion of a quadratic term in X, expressed as X^2, could improve model fit by removing this unexplained trend. A plot, such as that shown in Figure 5.5B, shows a different form of patterning suggesting unequal variation as the level of the X variable changes. In such a case, weighted least squares, or response data transformation, could be considered to overcome the highlighted trend in the data.

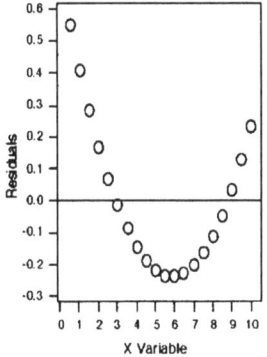 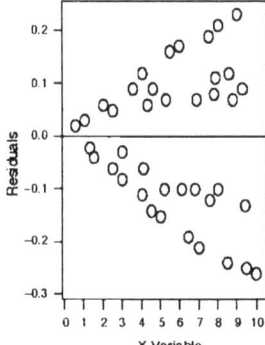

Figure 5.5 *Illustration of residual plots*

Normality of response can be assessed through a normal plot of the residuals, as described in Chapters 3 and 4 in respect of experimental designs. In Excel's Regression analysis tool, however, the normal probability plot default is a normal plot of the response recordings and not of the residuals. Interpretation of the produced plot is unaltered with linearity of points, the ideal trend.

Example 5.3

In Example 5.2, we found that the fitted linear model for frequency shift as a function of SO_2 concentration was statistically valid. We must now check its practical validity and carry out appropriate diagnostic checking to finalise the analysis of the fitted model. In Excel, this can be achieved by using the 'Line Fit', 'Residual', and 'Normal Probability' plot facilities within Excel's Regression analysis tool (see Figure 5.4) as Outputs 5.3 to 5.5 illustrate.

Practical validity
The print of the predicted values ('Preds') indicates that they appear close to the measured frequency shift measurements with the only differences occurring in the second decimal place. This would suggest acceptable fit for the model. The plot confirms practical validity with the experimental responses and model predictions coinciding almost exactly at most SO_2 settings.

Output 5.3   *Observed and predicted frequency shift for Example 5.3*

SO2	Freq	*Preds*
5	45	41.65
10	55	53.79
15	67	65.92
20	75	78.06
25	90	90.19
30	101	102.33
35	113	114.46
40	122	126.6
45	143	138.74
50	150	150.87
55	159	163.01
60	177	175.14
65	192	187.28
70	195	199.41
75	215	211.55

SO2 Line Fit Plot

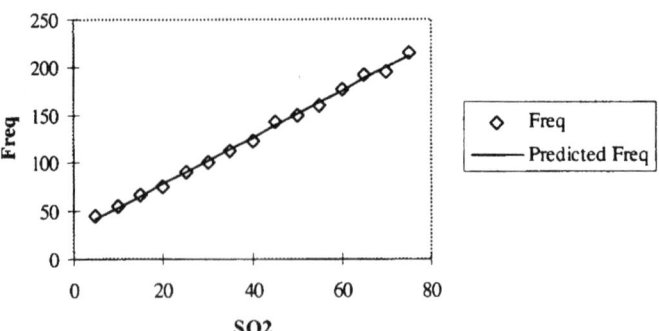

Diagnostic checking
The SO_2 residual plot in Output 5.4 does not fall into any obvious pattern though for low SO_2, there appears to be a downward trend in the residuals. There is a suggestion, though not very strong, of a quadratic trend and that inclusion of a quadratic term in SO_2 could improve model fit still further.

Output 5.4 *Diagnostic checking: SO₂ concentration residual plot*

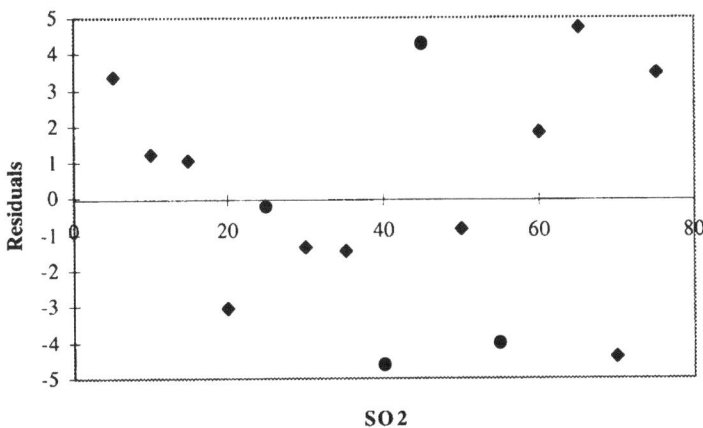

SO2 Residual Plot

The normal probability plot in Output 5.5 shows an ideal linear trend indicative of normality of frequency shift response being acceptable. The plot is based on the response as the *y* axis and an *x* axis based on standard normal percentiles of the ordered response data. A fuller explanation of the underlying principles of normal plots of response data is provided in Chapter 6 Section 9.1.

Output 5.5 *Diagnostic checking: normal plot of frequency data for Example 5.3*

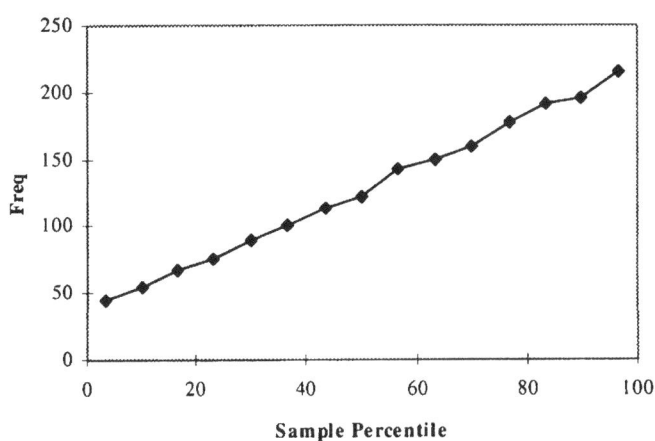

Normal Probability Plot

Pulling together all of the analyses carried out in Examples 5.1 to 5.3, we can conclude that a linear model of frequency shift as a function of SO_2 concentration is valid statistically and also, practically. The accuracy of the predicted values is good, though diagnostic checking hints at a possible unexplained trend suggestive that a quadratic model of frequency shift may be worth considering. This frequency shift example illustrates how all the analysis aspects provide a comprehensive picture, from a variety of angles, of the viability of a linear model for response data.

Exercise 5.1

An experiment was conducted into the solubility of the steroid *testosterone* in supercritical CO_2 at a temperature of 35 °C and varying pressures, in kg cm^{-2}. The mole fraction solubilities recorded are shown as mole fraction x 10^{-7}. It is thought that solubility and pressure may be linearly related. Carry out a full assessment of this assertion.

Pressure	84	101	116	132	149	164	181	200	215	231
Solubility	65.7	87.2	131	212	256	302	329	367	411	478

Source: J.R. Dean, M. Kane, S. Khundker, C. Dowle, R.L. Tranter and P. Jones, *Analyst (Cambridge)*, 1995, **120**, 2153.

4 FURTHER ASPECTS OF LINEAR REGRESSION ANALYSIS

The mechanisms illustrated for linear model assessment represent the main tools for the assessment of this form of regression modelling. Other analysis methods can be considered though they only assess specific aspects of model fit. This section will explain some of these approaches, the application of which depends on the requirements of the modelling being undertaken.

4.1 Specific Test of the Slope against a Target

Usage of this particular test could occur in a situation where we may wish to compare a fitted linear equation of absorbance in terms of concentration for a particular analyst with the known reference equation. Essentially, we would wish to compare the slope of the analyst's fitted equation against that of the reference equation. This can be achieved easily using a modified form of the slope test (5.4) as Box 5.3 summarises.

Box 5.3 *Slope test for comparison with a specified target*

Hypotheses. The hypotheses for this test are specified as H_0: $\beta = \beta_0$ and H_1: $\beta \neq \beta_0$ where β_0 refers to the target for comparison. A one-sided alternative may also be considered if appropriate to the assessment objectives.

Test statistic. The test statistic, representing a modification of the numerator of equation (5.4), is expressed as

$$t = (b - \beta_0)/s_b \qquad (5.11)$$

based on $(n-2)$ degrees of freedom where s_b is as defined in equation equation (5.8).

Decision rule. As all previous illustrations of the t test.

4.2 Test of Intercept

Within linear regression analysis, we can also carry out a statistical test of the intercept as Box 5.4 illustrates. Testing the intercept is valid if it is part of the goal of the experimentation to compare the estimate obtained from experimental data against a target value. For example, in linear calibration, it may be of interest to test the intercept against a target of 0 if it is expected that the linear calibration curve should pass through the origin. It should be noted that statistical software output generally provides a test of $\alpha_0 = 0$ when a linear model with intercept is fitted to data.

Box 5.4 *Test of intercept*

Hypotheses. The null hypothesis for the intercept test is expressed as H_0: $\alpha = \alpha_0$ with one- or two-sided alternative as appropriate.

Test statistic. The test statistic is expressed as

$$t = (a - \alpha_0)/s_a \qquad (5.12)$$

based on $(n-2)$ degrees of freedom.

Decision rule. As all previous illustrations of the t test.

The term s_a in equation (5.12) defines the standard error of the intercept estimate a and is specified as

$$s_a = s_{y/x}\sqrt{(\sum X^2/S_{XX})} \qquad (5.13)$$

using equation (5.9) for $s_{y/x}$ and S_{XX} as defined in equation (5.2). A confidence interval for the intercept coefficient, based on a simple re-statement of test statistic (5.12), can also be considered, if relevant.

Example 5.4

Referring to the frequency shift illustration of Example 5.1, suppose a previous experiment had yielded a linear model of frequency shift = 25.176 + 2.372 SO_2 concentration. Interest may lie in assessing whether this relationship differs statistically from that obtained in Example 5.1, any difference possibly reflecting experimentation differences.

Information
From Output 5.2, we know $n = 15$, $a = 29.514$, $s_a = 1.7859$ (row 'Intercept', column 'Standard Error'), $b = 2.427$, and $s_b = 0.0393$. The targets for comparison are $\alpha_0 = 25.176$ and $\beta_0 = 2.372$.

Test of slope
The hypotheses we will test are H_0: $\beta = 2.372$ against the two-sided alternative H_1: $\beta \neq 2.372$. The test statistic (5.11) is $t = (2.427 - 2.372)/0.0393 = 1.40$.

Based on the critical value of $t_{0.025,13} = 2.160$ (Table A.1), we must accept H_0 and conclude that the slope of the fitted linear equation does not differ statistically from the slope estimate of the previous experiment ($p > 0.05$).

Test of intercept
The hypotheses to be tested are H_0: $\alpha = 25.176$ and H_1: $\alpha \neq 25.176$. The test statistic (5.12) is $t = (29.514 - 25.176)/1.7859 = 2.43$.

Based on the critical value of $t_{0.025,13} = 2.160$ (Table A.1), H_0 is rejected leading to the conclusion that the intercept of the fitted linear equation appears to differ statistically from the intercept estimate of the previous experiment ($p < 0.05$).

In conclusion, we can state that there is no difference in slope ($p > 0.05$) but a difference in intercept ($p < 0.05$). This suggests there is an additive difference between the two experimental data sets which suggests experimentation differences exist, e.g. experiments carried out by different analysts.

4.3 Linear Regression with No Intercept

When modelling a response Y as a linear function of an experimental factor X, it may be logical to fit the model through the origin $(0, 0)$. For example, in modelling the uptake of a chemical in cells with respect to concentration, it is likely that rate of uptake will be zero at zero concentration. In such a case, fitting a linear model of the form $Y = \beta X$ would appear to be more appropriate than fitting the standard linear model $Y = \alpha + \beta X$ which includes an intercept.[6]

Fitting a linear model with no constant simplifies calculation of the slope estimate to

$$b = \sum xy / \sum x^2$$

compared to equation (5.2) for the standard linear model. In addition, the degrees of freedom for residual and total in the associated ANOVA table change to $(n-1)$ and n respectively. Sum of squares calculations also change since correction for the mean is no longer necessary. Assessment of model validity would, however, follow the steps illustrated in Examples 5.1 to 5.3.

5 PREDICTING X FOR A GIVEN VALUE OF Y

In analytical experimentation, interest may often lie in estimating the value of the experimental X variable corresponding to some measurement of the response Y made on an unknown sample. Measurements may refer to a single observation or replicate observations. This application of regression principles is generally referred to as *calibration*, or *inverse prediction*. Figure 5.6 provides a graphical illustration of the principle of calibration for linear modelling. In this illustration, a linear model of signal as a function of concentration has been derived and a signal measurement made on a sample of unknown concentration. It is desired to use the linear relationship between signal and concentration to predict the concentration of the unknown sample.

Inverse prediction involves using empirical data and prior experimental knowledge to determine predictions for an unknown variable from available measurements of a response via some mathematical transfer function. For example, in the analysis of a toxic chemical compound that absorbs light in the UV wavelength region it may be possible to express UV absorbance as a linear function of concentration. Using this relationship, we may be able to estimate the concentration of a chemical sample based on its UV absorbance reading.

---

[6] Reference 5, pp. 504–505.

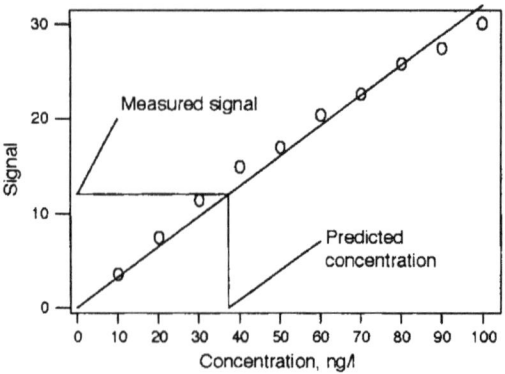

Figure 5.6 *Illustration of linear calibration*

Regression methods provide the necessary tools for calibration as well as enabling additional features of the calibration line to be assessed from a statistical perspective. These include assessing whether the fitted calibration curve passes through the origin (0, 0), comparing the slope of the fitted equation (the sensitivity) with a known target, and determining the accuracy (error) in a predicted X value. Box 5.5 summarises the analysis aspects of linear calibration though not all are appropriate in all calibration applications.

Example 5.5

Whole-blood plasma samples were taken from malaria patients receiving differing amounts of sulfadoxine and subjected to colorimetric assay. Absorbance readings were obtained from the processed samples with a spectrophotometer set at 560 nm. The collected data are presented in Table 5.4. The absorbance of a unknown blood sample was recorded in triplicate by the same technician. The readings obtained were 0.095, 0.089, and 0.093. We wish to estimate the sulfadoxine concentration of this specimen and determine a 95% confidence interval for the concentration estimate.

Table 5.4 *Absorbance measurements for Example 5.5*

Sulfadoxine concentration (μg ml^{-1})	5	10	25	50	100
Absorbance	0.018	0.031	0.072	0.128	0.224

Source: M.D. Green, D.L. Mount and G.D. Todd, *Analyst (Cambridge)*, 1995, **120**, 2623.

Box 5.5 *Analysis aspects for linear calibration*

Determination of calibration curve. Plot the experimental data, or appropriate transform of, on a scatter diagram to check for linearity. Determine the linear calibration curve and check its statistical significance with such required to be significant ($p < 0.05$) though highly significant ($p < 0.01$) is preferred.

Does the calibration line pass through the origin? Suppose the true calibration function should pass through the origin, *i.e.* intercept α should be 0. In calibration, we can assess whether this is true by carrying out the intercept test (5.12) based on $\alpha_0 = 0$ (see Box 5.4). This test is produced by default in most statistical software when using OLS to fit a linear model with intercept to experimental data.

Does the sensitivity of the calibration line differ from target? This check occurs when a target value for the slope of the calibration line has been specified prior to experimentation. The target may be a reference standard or a figure obtained previously. The general slope test (5.11) can be used for this comparison (see Box 5.3).

Estimation of the unknown X. Let be the measured response, or average of replicate responses. By re-expressing the fitted linear equation $Y = a + bX$, we can estimate the unknown X value as

$$\hat{X} = (\hat{Y} - a)/b \qquad (5.14)$$

provided b $B \neq 0$.

Confidence interval for estimated X value. An estimate of the error associated with the X prediction is given by

$$SE(\hat{X}) = \frac{s_{y/x}}{b}\sqrt{\frac{1}{m} + \frac{1}{n} + \frac{(\hat{Y} - \bar{Y})^2}{b^2 S_{XX}}} \qquad (5.15)$$

where $SE(\hat{X})$ defines the standard error of the predicted X value. The term $s_{y/x}$ is the regression standard deviation (5.9), b the least squares estimate of β (slope parameter), m the number of replicate response measurements made on the unknown test material, n the number of data pairs in the calibration experiment, $\bar{Y}$ the mean of the experimental responses, and S_{XX} the corrected sum of squares for X from equation (5.2). Using $SE(\hat{X})$, we can construct a $100(1-\alpha)\%$ confidence interval for the estimated X value as

$$\hat{X} \pm t_{\alpha/2,n-2}SE(\hat{X}) \qquad (5.16)$$

where $t_{\alpha/2,n-2}$ is the $100\alpha/2\%$ critical value from Table A.1.

Excel Outputs 5.6 to 5.8 contain the necessary graphical, regression, and summary information for this calibration example. Using these outputs, we want to assess the validity of a calibration line for the collected data and use this relationship, if statistically valid, to construct an approximate 95% confidence interval for the actual sulfadoxine concentration administered to the patient providing the unknown blood sample.

Data plot
The scatter plot in Output 5.6 shows a positive trend with increase in concentration resulting in increased absorbance. The pattern looks reasonably linear, enough to suggest that linear calibration would be appropriate for these data.

Output 5.6 *Scatter plot of absorbance measurements for Example 5.5*

Sulfadoxine concentration data in cells A1:A6 and absorbance data in cells B1:B6 (labels in row 1). Menu commands as Output 5.1.

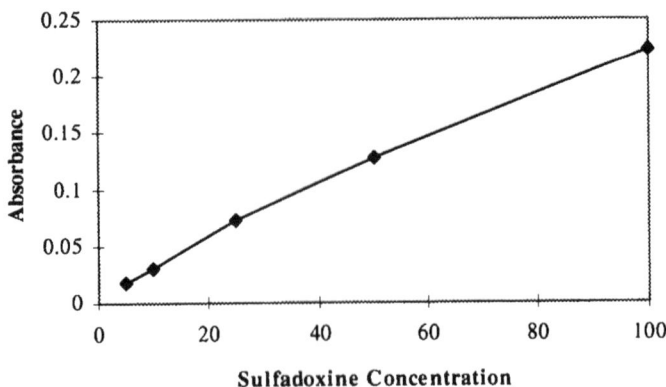

Absorbance against Sulfadoxine Concentration

Calibration line $Y = a + bX$
Given that the data plot suggests that a linear model may fit the absorbance data, we want to find this relationship and assess its statistical validity. The relevant Excel output is shown in Output 5.7.

Output 5.7 *Linear regression output for Example 5.5*

Sulfadoxine concentration data in cells A1:A6 and absorbance data in cells B1:B6 (labels in row 1). Menu commands as Output 5.2.

SUMMARY OUTPUT

Regression Statistics

Multiple R	0.9974
R Square	0.9949
Adjusted R Square	0.9932
Standard Error	0.0069
Observations	5

ANOVA

	df	SS	MS	F	Significance F
Regression	1	0.028139	0.028139	584.41	0.0002
Residual	3	0.000144	0.000048		
Total	4	0.028283			

	Coefficients	Standard Error	t Stat	P-value	Lower 95%	Upper 95%
Intercept	0.01251	0.0046	2.72	0.0725	−0.00213	0.02715
Concn	0.00216	0.000089	24.17	0.0002	0.00188	0.00244

The calibration line is estimated to be absorbance = 0.01251 + 0.00216 concentration. The p value for the linear model from the ANOVA table is estimated as 0.0002 ('Significance F') indicative of a highly significant linear fit ($p < 0.01$). R^2, at 99.5%, also suggests statistically valid fit. Use of this fitted linear calibration model to estimate the sulfadoxine concentration appears justified, on a statistical basis.

Estimation of unknown sulfadoxine concentration
From Output 5.7, we know that $a = 0.01251$ and $b = 0.00216$. The three absorbance readings ($m = 3$) are 0.095, 0.089, and 0.093. From these, we estimate $\hat{Y}$ as 0.0923, the average of the three readings. From the readings in Table 5.4, we would expect prediction of sulfadoxine concentration to be between 25 and 50 µg ml^{-1}. Equation (5.14) provides

$$\hat{X} = (0.0923 - 0.01251)/0.00216 = 36.940 \text{ µg ml}^{-1}$$

as the predicted concentration of sulfadoxine. This lies within the likely range so appears acceptable.

Confidence interval for estimated X concentration
To evaluate the standard error SE($\hat{X}$), equation (5.15), we need $s_{y/x}$, b, m, n, $\hat{Y}, \bar{Y}$, and S_{XX}. We know that $m = 3$, $n = 5$, and $\hat{Y} = 0.0923$. Output 5.7 provides $s_{y/x} = 0.0069$ ('Regression Statistics', row 'Standard Error') and $b = 0.00216$. We also need to estimate $\bar{Y}$ and $S_{XX} = \sum X^2 - (\sum X)^2/n$. $\bar{Y}$ and S_{XX} can be easily obtained within Excel as Output 5.8 demonstrates, providing values of $\bar{Y} = 0.095$ ('Mean Y') and $S_{XX} = 6030$ ('Sxx').

Output 5.8 *Summary data for Example 5.5*

Labels entered in cells A8:A10.
   In cell **B8**, enter = **COUNT(A2:A6)**. In cell **B9**, enter =**ROUND(AVERA-GE(B2:B6),3)** to generate mean of the Y responses. In cell **B10**, enter =**DEVSQ(A2:A6)** to generate the corrected sum of squares S_{XX}.

n	5
Mean Y	0.095
Sxx	6030

With all the terms for equation (5.15) now specified, we can evaluate it as,

$$SE(\hat{X}) = \frac{0.0069}{0.00216}\sqrt{\frac{1}{3} + \frac{1}{5} + \frac{(0.0923 - 0.095)^2}{(0.00216)^2(6030)}} = (3.1944)\sqrt{0.5336} = 2.3334$$

Now that we have all the necessary components, we can finally obtain the 95% confidence interval (5.16) for the predicted sulfadoxine concentration corresponding to the absorbance readings provided. Since we wish to determine a 95% confidence interval based on using $n = 5$ data points, the necessary critical value $t_{0.025,3}$ from Table A.1, will be 3.182. Given that $\hat{X} = 36.940$, $t_{0.025,5} = 3.182$, and SE($\hat{X}$) = 2.3334, equation (5.16) becomes

$$36.940 \pm (3.182)(2.3334) = 36.940 \pm 7.425 = (29.515, 44.365) \text{ µg ml}^{-1}$$

Based on this, we can say, with 95% confidence, that the sulfadoxine concentration administered to the patient who provided the unknown blood sample appears to lie between 29.515 and 44.365 µg ml^{-1}.

Calibration concepts can also be used to set detection limits for an analytical procedure. Such a limit represents the smallest measurement a procedure is capable of detecting. This involves determining the lowest predicted X value for which the 95% confidence interval does not include zero.[7] In other words, the limit of detection represents the X value, generally a concentration value, which can provide a response significantly different from the 'blank' response. There also exist procedures for non-linear and multivariate calibration. Fuller details on the application and usage of calibration in chemical experimentation can be found in Martens and Naes.[8]

Exercise 5.2

The data below were collected using FAES calibration at the 422.7 nm line. Assess the validity of a linear calibration curve for these data. An unknown specimen produced duplicate intensity measurements of 75.672 mV and 76.173 mV. Estimate the calcium concentration of this specimen and determine a 95% confidence interval for this estimate.

Ca (ppm)	0.0	0.1	0.2	0.4	0.6	0.8
Intensity (mV)	15.851	16.939	18.115	21.909	26.217	31.128

Ca (ppm)	2	4	6	8	10
Intensity (mV)	58.584	102.241	142.047	194.389	239.702

6 COMPARISON OF TWO LINEAR EQUATIONS

Comparing measurements from analytical procedures at only a few concentrations, using classical chemical experimental designs, may provide a misleading picture of their validity. As such procedures need to be valid across a wide range of concentrations, it is best to assess such data using regression methods. This is particularly the case if a new procedure is to be compared with a standard one over analyte concentrations that may vary substantially. Experimenting over a wide range of concentrations can also be useful when wishing to assess for bias within an analytical method tested on both standard and spiked material.

---

[7] J.C. Miller and J.N. Miller, 'Statistics for Analytical Chemistry', 3rd Edn, Ellis Horwood, Chichester, 1994, pp. 115–117.
[8] H. Martens and T. Naes, 'Multivariate Calibration', Wiley, Chichester, 1989.

The interest, in this type of experimentation, lies in ascertaining whether the relationships are similar for the compared procedures. Ideally, there should be no difference in the relationships reflecting that method does not affect measurement of experimental response. If a difference did occur, then we would need to investigate the methods further to ascertain why such an effect was detected. The comparison of the relationships on a statistical basis, involves checking whether a single line could explain all data sets, and checking for similarity of slope and intercept with differences in both or either signifying possible confounding effects.

Generally, in comparison experiments, linear relationships are fitted either to the original response data or to a transform of them providing data plots comparable to those presented in Figure 5.7. We want to be able to compare the fitted linear relationships, assuming the data conform to linear models, for each method tested in order to check whether there is any similarity in the pattern of the fitted relationships. Ideally, the fitted relationships should coincide but if not identical, there may still be some degree of similarity between them.

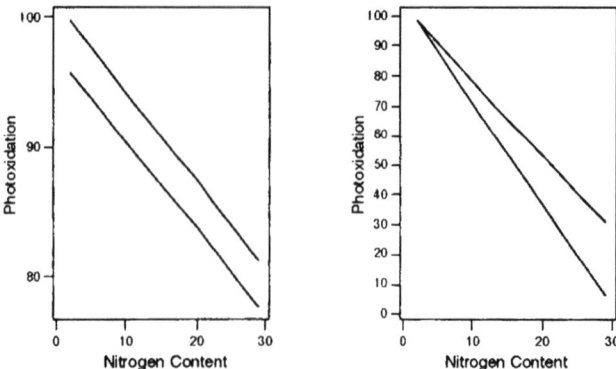

Figure 5.7 *Plots for the comparison of two linear equations*

Parallel lines with different intercepts, as shown in Figure 5.7A, reflect the presence of *translational bias* describing additive difference between the methods. Alternatively, as Figure 5.7B illustrates, the lines may converge at the y-axis (similar intercept, concurrent) though exhibiting differing slopes and, thereby, rates of response change. Such an illustration describes *rotational bias* where the method differences are multiplicative reflecting possible interference effects.

In such cases, data are generally available from at least two comparable data sets with the original data, or transform of, conforming to a linear relationship. The comparison procedure is based on the use of simple linear regression for each data set and multiple non-linear regression using indicator (dummy) variables. As with calibration, this analysis procedure is best described in stages to show how each aspect is put together.

The first step involves plotting the experimental data and assessing the scatter diagram for linearity and patterning. OLS estimation is then applied to determine the linear relationships for each data set with statistical validity of the fitted equations required to be significant ($p < 0.05$), though highly significant ($p < 0.01$) is preferable. Example 5.6 will be used as the basis of a demonstration of this procedure.

Example 5.6

As part of a chlorophyll calibration experiment, the data in Table 5.5 were collected. The data refer to the peak area from an HPLC chromatogram at various dilution levels, where a value of 0.01 means dilution to 1% of concentration of the original. Interest lay in comparing the HPLC peak areas at the two wavelengths in order to assess whether HPLC could be used to measure chlorophyll concentration in the absence of external standards. We want to assess whether these peak area data conform to linear patterns and whether any difference in the peak areas recorded exists through the patterns exhibited.

Table 5.5 *Peak area data for Example 5.6*

Dilution	0.06	0.12	0.18	0.24	0.30
Peak area at 430 nm	5.262	8.844	14.532	19.626	26.484
Peak area at 665 nm	4.278	7.182	11.742	15.768	20.790

Source: P.W. Araujo and R.G. Brereton, *Analyst (Cambridge)*, 1995, **120**, 2497.

Data plot
A plot of the collected data is presented in Output 5.9. We can see from the plot that both data sets look reasonably linear though they rise at marginally different rates, 430 nm changing at a higher rate. Initially, it appears that the two relationships appear to differ, appear to be non-parallel, and suggest that peak area measurement changes at a faster rate when measured at 430 nm.

Output 5.9 *Scatter plot of peak area data for Example 5.6*

Data in cells A1:C6, labels in cell 1 of each column.
  Select **ChartWizard**. Click cell **I1**.
  ChartWizard Step 1 of 5: for *Range*, click and drag from cell **A1:C6** ▷ click **Next**.
  ChartWizard Step 2 of 5: select **XY (Scatter)** ▷ click **Next**.
  ChartWizard Step 3 of 5: select **XY(Scatter) chart 2** ▷ click **Next**
  ChartWizard Step 4 of 5: click **Next**.
  ChartWizard Step 5 of 5: select the **Chart Title** box and enter **Peak Area Against Dilution** ▷ select the **Axis Titles Category (X)** box and enter **Level of Dilution** ▷ select the **Axis Titles Value (Y)** box and enter **Peak Area** ▷ click **Finish**.

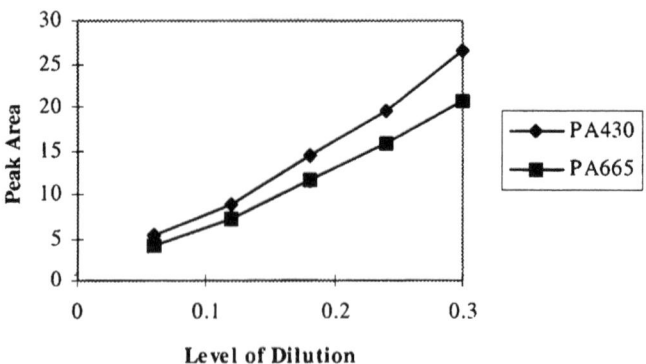

Peak Area Against Dilution

Regression equations
From the information presented for linear regression modelling (Example 5.1), we can use Excel to produce the necessary linear regression output for both data sets as shown in the summaries presented in Output 5.10. For *430 nm*, the fitted linear equation is peak area = $-1.0182 + 88.71$ dilution. For *665 nm*, the fitted linear equation is peak area = $-0.531 + 69.35$ dilution. Comparing the two equations, we see that the 430 nm line has a greater slope than the 665 nm line indicative of peak area changing at a faster rate and hinting at non-parallel lines. Intercepts also appear to differ providing further evidence that the two wavelengths appear to produce different peak area characteristics.

Statistical validity
Assessment of the statistical validity of both fitted equations will be based, for simplicity, on the R^2 ('R Square') and p values ('P-value')

Output 5.10 *Linear regression output for Example 5.6*

Menu commands as Output 5.1.

SUMMARY OUTPUT – 430 nm

Regression Statistics

Multiple R	0.9949
R Square	0.9898
Adjusted R Square	0.9864
Standard Error	0.9853
Observations	5

	Coefficients	Standard Error	t Stat	P-value
Intercept	−1.0182	1.0334	−0.99	0.3971
Dilution	88.7100	5.1932	17.08	0.0004

SUMMARY OUTPUT – 665 nm

Regression Statistics

Multiple R	0.9965
R Square	0.9930
Adjusted R Square	0.9907
Standard Error	0.6377
Observations	5

	Coefficients	Standard Error	t Stat	P-value
Intercept	−0.531	0.6689	−0.79	0.4853
Dilution	69.35	3.3612	20.63	0.0002

from the regression output in Output 5.10. For the *430 nm* line, R^2 = 99.0% and p = 0.0004 both indicative of a highly significant fit. For the *665 nm* line, the corresponding summaries are R^2 = 99.3% and p = 0.0002, also indicative of a highly significant linear fit. Such results show that both fitted linear equations are valid statistically ($p < 0.01$).

If the fitted linear equations for each data set are statistically valid, the second stage of the comparison procedure can be implemented. This involves merging the two data sets into a larger data set consisting of four variables Y, X_1, X_2, and X_1X_2. The first variable, Y, refers to all the recorded responses while the second, X_1, corresponds to all the X data. The third variable, X_2, is an indicator or dummy variable such that $X_2 = 1$ if an observation is from the first data set A and $X_2 = 0$ if from the second data set B. The final variable introduced is X_1X_2 which represents the interaction term between the variables X_1 and X_2. This merging process can be handled easily using Excel's Copy and Paste facilities as Output 5.11 demonstrates for the peak area data of Example 5.6.

Output 5.11 *Production of merged data for comparison of linear equations procedure*

Labels in cells D1 to G1.
   Copy and paste cells **B2:B6** to cell **D2**. Copy and paste cells **A2:A6** to cell **E2**.
Copy and paste cells **C2:C6** to cell **D7**. Copy and paste cells **A2:A6** to cell **E7**.
   In column F from cell F2, enter **1** five times then **0** five times.
   In cell **G2**, enter **=E2*F2**. Copy and paste this formula into cells **G3:G11**.

Y	X1	X2	X1X2
5.262	0.06	1	0.06
8.844	0.12	1	0.12
14.532	0.18	1	0.18
19.626	0.24	1	0.24
26.484	0.3	1	0.3
4.278	0.06	0	0
7.182	0.12	0	0
11.742	0.18	0	0
15.768	0.24	0	0
20.790	0.3	0	0

Once the merged data set is created, we model the merged data using the multiple non-linear regression model

$$Y = \alpha + \beta_1 X_1 + \beta_2 X_2 + \beta_3 X_1 X_2$$

The statistical information related to fitment of this model becomes the basis of the comparison test procedures.
   The first test carried out, *Test 1*, assesses whether a single line for the merged data or separate lines for each data set, best describe the relationships within the data. Box 5.6 summarises the steps associated with this test.

Box 5.6 *Test 1 for comparison of two linear equations*

Hypotheses. The null and alternative hypotheses for Test 1 are specified as H_0: no difference between single and separate lines ($\beta_2 = \beta_3 = 0$) and H_1: single line worse than two separate lines (not both $\beta_2 = 0$ and $\beta_3 = 0$).
Test statistic. The test statistic for Test 1 is an F test statistic expressed as

$$F = \frac{(SEQSS(X_2, X_1 X_2 | X_1)/2)}{MSRes} \qquad (5.17)$$

with degrees of freedom $df_1 = 2$ and $df_2 = n_A + n_B - 4$.
Decision rule. As all previous illustrations of the F test.

The null hypothesis in Test 1 is essentially saying that there appears no significant difference between the two data sets and that they exhibit comparable patterns. This similarity suggests that the two data sets can be safely described by one relationship through both sets instead of separate ones for each. The alternative hypothesis in Test 1 suggests that the data set differences are sufficient to suggest they cannot be assumed similar. In test statistic (5.17), MSRes refers to the residual mean square of the multiple non-linear regression equation, and n_A and n_B are the respective sample sizes for the two data sets.

SEQSS(X_2, $X_1 X_2 \mid X_1$) in equation (5.17) refers to the additional response variation (sum of squares) explained by adding X_2 and $X_1 X_2$ into the multiple non-linear model after X_1 has been fitted. It is such that

$$SEQSS(X_2, \ X_1 X_2 | X_1) = SEQSS(X_2 | X_1) + SEQSS(X_1 X_2 | X_1, \ X_2) \qquad (5.18)$$

where the former refers to the additional response variation explained by adding X_2 into the model after X_1 has been fitted, and the latter refers to the additional response variation explained by adding $X_1 X_2$ into the model after X_1 and X_2 have been fitted. Such terms are usually produced as part of the regression ANOVA table in most statistical software. In Excel, however, this is not the case.

To calculate SEQSS(X_2, $X_1 X_2 \mid X_1$) using Excel, we must determine, separately, the regression SS for the multiple non-linear model, SSRegn(X_1, X_2, $X_1 X_2$), and the regression SS for the linear model of Y as a function of X_1, SSRegn(X_1). Combining these as,

$$SSRegn(X_1, \ X_2, \ X_1 X_2) - SSRegn(X_1) \qquad (5.19)$$

will provide the required SEQSS term.

If the *null hypothesis is accepted* in Test 1, there is no need to go further as a single line adequately describes both data sets. This indicates that the two data sets exhibit very similar characteristics and suggests that the compared procedures produce comparable response measurements.

If the *null hypothesis is rejected* in Test 1, we must carry out a further test, *Test 2*, to pinpoint whether, given that separate lines are better than a single line, the separate lines are parallel, *i.e.* have similar slopes. Parallelism is reflected by the absence of the interaction term in the multiple non-linear model, *i.e.* $\beta_3 = 0$. Box 5.7 outlines the workings of this test procedure.

Box 5.7 *Test 2 for comparison of linear equations*

Hypotheses. The hypotheses for the parallelism test are specified as H_0: slopes same and lines parallel ($\beta_3 = 0$) and H_1: slopes different and lines not parallel ($\beta_3 \neq 0$).
Test statistic. The test statistic associated with Test 2 takes the form

$$F = SEQSS(X_1 X_2 | X_1, \ X_2)/MSRes \qquad (5.20)$$

based on $df_1 = 1$ and $df_2 = n_A + n_B - 4$.
Decision rule. As all previous illustrations of the F test.

The null hypothesis within Test 2 reflects that there may be only a magnitude difference between the lines and not a rate difference, while the alternative hypothesis suggests that both may be present. In test statistic (5.20), MSRes refers to the residual mean square of the multiple non-linear regression equation, and n_A and n_B are the respective sample sizes for the two data sets.

The SEQSS term in equation (5.20) refers to the additional response variation explained by adding $X_1 X_2$ into the model after X_1 and X_2 have been fitted. To calculate $SEQSS(X_1 X_2 \mid X_1, X_2)$ using Excel, we must determine the regression SS for the multiple non-linear model, $SSRegn(X_1, X_2, X_1 X_2)$, and the regression SS for the multiple linear model of Y as a function of X_1 and X_2, $SSRegn(X_1, X_2)$. Combining these as

$$SSRegn(X_1, \ X_2, \ X_1 X_2) - SSRegn(X_1, \ X_2) \qquad (5.21)$$

will provide the required SEQSS term.

If the *null hypothesis is accepted* in Test 2, the linear equations for each data set are classified as parallel (same slope) specifying that there appears not to be a difference in the rate of change of response between the data sets though there could be an additive one, manifested by a difference in intercept. An estimate of the population regression coefficient, β_c, underlying both b_A and b_B, is called the *common*, or *weighted*, regression coefficient and can be determined as

$$b_c = [(S_{XY})_A + (S_{XY})_B]/[(S_{XX})_A + (S_{XX})_B] \qquad (5.22)$$

where $S_{XY} = \sum xy - (\sum x)(\sum y)/n$, $S_{XX} = \sum x^2 - (\sum x)^2/n$, and subscript A or B refers to the relevant data set. The two regression equations, if parallel and different in intercept, can be written $Y_A = a_A + b_c X$ and $Y_B = a_B + b_c X$.

If the *null hypothesis is rejected* in Test 2, this suggests that the two data sets exhibit contrasting characteristics indicative of significant differences between them, *e.g.* procedure differences when comparing analytical procedures. In such cases, it may be that the two non-parallel lines intersect within the range of settings experimented on. The point of intersection of the two lines is (X_I, Y_I) where $X_I = (a_B - a_A)/(b_A - b_B)$, and $Y_I = a_A + b_A X_I$ or $Y_I = a_B + b_B X_I$.

Example 5.7

In Example 5.6, the statistical validity of the fitted linear equations for the peak area data sets was confirmed. We now want to carry out Test 1 and if necessary, Test 2, to assess the statistical similarity of the two data sets.

Regression SS information
We must first generate the necessary SEQSS terms for expressions (5.19) and (5.21). Excel's Regression tool can be used for this purpose by providing the ANOVA tables for the three underpinning models: full multiple non-linear model, Y as a function of X_1, and Y as a function of X_1 and X_2. The ANOVA tables appropriate to this example are summarised in Output 5.12. From this output, we find $SSRegn(X_1, X_2, X_1 X_2) = 478.9039$, $MSRes = 0.6888$, $SSRegn(X_1) = 449.6933$, and $SSRegn(X_1, X_2) = 472.1574$.

Output 5.12 *Regression output for SS determination in Example 5.7*

Select **Tools** $\triangleright$ **Data Analysis** $\triangleright$ **Regression** $\triangleright$ click **OK** $\triangleright$ for *Variable 1 Range*, enter **D1:D11** $\triangleright$ select the **Variable 2 Range** box and enter **E1:G11** $\triangleright$ select **Labels** $\triangleright$ select **Output Range**, click the box, and enter **A13** $\triangleright$ click **OK** $\triangleright$ click the mouse button.

Repeat for **E1:E11** and **E1:F11** changing the **Output Range** position accordingly. Output edited to produce only relevant summaries for comparison procedure.

Full multiple non-linear model
ANOVA

	df	SS	MS	F	Significance F
Regression	3	478.9039	159.6346	231.75	0.0000
Residual	6	4.1329	0.6888		
Total	9	483.0368			

Y as a function of X_1
ANOVA

	df	SS	MS	F	Significance F
Regression	1	449.6933	449.6933	107.89	0.0000
Residual	8	33.3435	4.1679		
Total	9	483.0368			

Y as a function of X_1 and X_2
ANOVA

	df	SS	MS	F	Significance F
Regression	2	472.1574	236.0787	151.9	0.0000
Residual	7	10.8794	1.5542		
Total	9	483.0368			

Test 1

This will assess whether the two peak area equations are similar enough to be explained by a single line. The test is based on the null hypothesis H_0: no difference between a single line for all the data and separate lines for the each wavelength ($\beta_2 = \beta_3 = 0$). We know that $n_A = 5 = n_B$.

From the Regression SS information and using equation (5.19), we obtain

$$SEQSS(X_2, \ X_1X_2|X_1) = 478.9039 - 449.6933 = 29.2106.$$

Using this together with the MSRes estimate of 0.6888, the F test statistic (5.17) becomes

$$F = (29.2106/2)/0.6888 = 21.20$$

based on degrees of freedom $df_1 = 2$ and $df_2 = 5 + 5 - 4 = 6$. The 5% critical from Table A.3 is $F_{0.05,2,6} = 5.14$. Since test statistic exceeds critical, it is obvious that H_0 is rejected and we conclude that the lines for the different wavelengths appear to differ sufficiently to imply that separate lines for each is better than a single line for both ($p < 0.05$). This result suggests that the peak area measurements at the two wavelengths are too dissimilar to be treated alike and must, therefore, be treated as different. As we have rejected H_0, we must now implement Test 2 to assess for parallelism.

Test 2
To finalise the comparison of the two data sets, we need, therefore, to conduct Test 2 to assess for evidence of parallelism. The null hypothesis to be tested is specified as H_0: 430 nm and 665 nm lines are parallel. We know $n_A = 5 = n_B$.

From the Regression SS information and using equation (5.21), we have

$$SEQSS(X_1 X_2|X_1, \ X_2) = 478.9039 - 472.1574 = 6.7465.$$

This, together with MSRes $= 0.6888$, means the test statistic (5.20) becomes

$$F = 6.7465/0.6888 = 9.79$$

based on degrees of freedom $df_1 = 1$ and $df_2 = 5 + 5 - 4 = 6$ with 5% critical value of $F_{0.05,1,6} = 5.99$ (Table A.3). As test statistic exceeds critical value, we reject H_0 indicating that, statistically, the 430 and 665 lines appear not to be parallel ($p < 0.05$), as was hinted at within the data plot. We therefore have evidence that wavelength used can affect peak area measurement, suggesting that the procedure needs refinement before it can be accepted as valid for all potential wavelengths at which measurement could take place.

An alternative to the F statistic (5.20) is provided by the t statistic

$$t = b_3/se(b_3) \qquad (5.23)$$

corresponding to a simple t test of the β_3 coefficient in the multiple non-linear model. This test statistic is usually provided by default in statistical software regression output. The described comparison procedure can also be used, though in a modified form, to compare non-linear relationships and relationships from more than two data sets, such as may occur in assay experimentation.

Exercise 5.3

Blood plasma samples from malaria patients were subjected to colori-metric assay. Absorbance measurements, shown below, were obtained from the processed samples by two methods: using a DU-7 spectro-photometer set at 560 nm and using a photometer fitted with a 565 nm filter. Assess these data for any similarity between them and comment on the results obtained.

Sulfadoxine ($\mu g\ ml^{-1}$)	5	10	25	50	100
Spectrophotometer	0.018	0.031	0.072	0.128	0.224
Filter photometer	0.024	0.071	0.179	0.341	0.708

Source: M.D. Green, D.L. Mount and G.D. Todd, *Analyst (Cambridge)*, 1995, **120**, 2623.

7 MODEL BUILDING

The most important aspect of regression modelling is selection of the model to fit to a set of experimental data. When dealing with only two variables, it is simplest to plot the response against the experimental variable and consider the pattern, if any, that the points fall into. Several possibilities other than linear could occur as illustrated in Figure 5.8. Each plot shown is monotonic as it is either increasing, or decreasing, though with different trends.

The plot patterns shown Figures 5.8A to 5.8D, respectively, are referred to as *direct accelerated, direct decelerated, inverse decelerated*, and *inverse accelerated*. Figure 5.8A corresponds to response data exhibiting exponential growth such as may occur when assessing enzyme activity of genes over time. The second plot, Figure 5.8B, illustrates response data where the rate of change declines to zero, a data pattern prevalent in ion-electrode systems (the Nernst equation). Figure 5.8C corresponds to response data exhibiting exponential decay, *e.g.* radioactive decay, while Figure 5.8D displays a pattern typical of that associated with the change in mRNA in genes over time.

Relationships which could be used to model data conforming to the

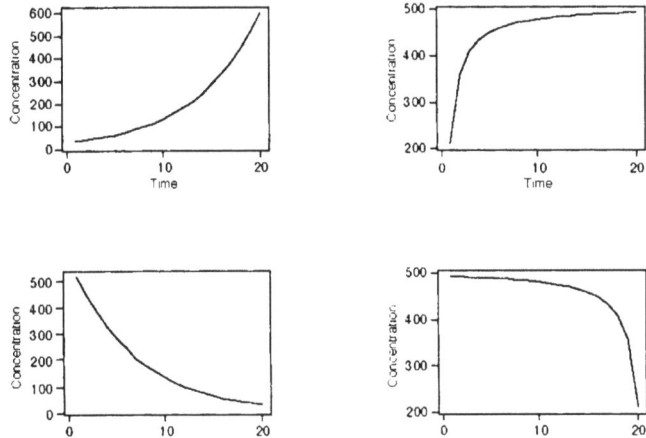

Figure 5.8 *Illustrations of non-linear data plots*

general patterns shown in Figure 5.8 are summarised in Table 5.6. This summary shows that several different models could fit each data pattern and that similar models could fit data exhibiting very different patterns. In such cases, model choice may be difficult though only one is generally fitted and fully assessed.

Table 5.6 *Potential models to fit to data exhibiting patterns illustrated in Figure 5.8*

Figure	Potential models
5.8A (direct accelerated)	$1/(\alpha + \beta X)$, $e^{\alpha + \beta X}$, $\alpha + \beta_1 X + \beta_2 X^2$
5.8B (direct decelerated)	$\alpha + \beta \log X$, $\alpha + \beta/X$, $\alpha(1 - e^{-\beta X})$
5.8C (inverse decelerated)	$1/(\alpha + \beta X)$, $e^{\alpha + \beta X}$, $\alpha + \beta_1 X + \beta_2 X^2$
5.8D (inverse accelerated)	$\alpha + \beta/X$, $Y^2 = \alpha + \beta X$

7.1 Non-linear Modelling

Curve fitting of non-linear functions occurs in many branches of chemistry such as spectroscopy, chromatography, and signal manipulation where it is necessary to estimate pertinent signal parameters (maximum amplitude, area). In such instances, a linear model may not be an appropriate description of the relationship between the variables. For example, the relationship between the measured voltage of ion-electrode systems and analyte concentration is often non-linear in pattern. Many non-linear relationships are *intrinsically linear* in that they can be re-expressed in linear form through the use of a suitable data transformation such as logarithms. Transforming data in this way enables the model to be re-expressed in a linear format allowing OLS estimation to be used to estimate model parameters.

Examples of non-linear models of this type include exponential, reciprocal, and power law with Table 5.7 showing the transformation and transformed linear format for each of these commonly occurring non-linear models. In each case, it can be seen that the transformation results in a linearised model but one expressed in terms of transformed data. It is this model we would fit by means of OLS estimation. Applicability of the proposed model can be assessed as for linear modelling, paying particular attention to checking practical validity of the fitted model. The power law, for example, is often used to relate reproducibility and repeatability to analyte concentration in collaborative trials.

Table 5.7 *Transformations and transformed model for common non-linear models*

Model	Model format	Transformation	Transformed linear model
exponential	$Y = e^{\alpha+\beta X}$	natural logarithm, $\log_e$	$\log_e Y = \alpha + \beta X$
reciprocal	$Y = 1/(\alpha + \beta X)$	reciprocal	$1/Y = \alpha + \beta X$
power	$Y = \alpha X^{\beta}$	logarithm to base 10, $\log_{10}$	$\log_{10} Y = \log_{10}\alpha + \beta\log_{10}X$

Not all non-linear models can be handled by simple transformation and linear least squares as no simple data transformation can linearise the proposed model. One such case is the relationship between the extent of matrix effects (I/I_0) and interferent concentration (c) in inductively coupled plasma atomic emission spectroscopy (ICPAES) which takes the form,

$$I/I_0 = 1 - \alpha(1 - e^{-\beta c}) \tag{5.24}$$

This model is *intrinsically non-linear* as no simple transformation can linearise the model. Parameter estimation, therefore, requires non-linear, or optimisation, methods.

7.2 Polynomial Modelling

Polynomial regression models, such as quadratic and cubic, occur often when the response measured is curvi-linear with the data plot producing a typical second order, non-monotonic shape, illustrated in Figure 5.9. Quadratic relationships occur often in calibration to describe the relationship between response and analyte concentration. Figures 5.8A and 5.8C also illustrate other forms of data plot for which quadratic models may be appropriate.

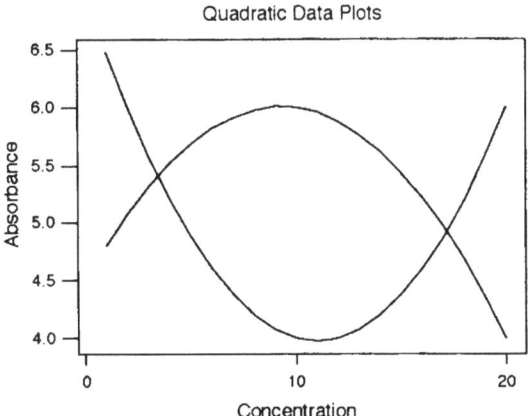

Figure 5.9 *Illustrations of quadratic dta plots*

A quadratic model is expressed as

$$Y = \alpha + \beta_1 X + \beta_2 X^2 + \epsilon$$

requiring estimation of three model parameters α, β_1, and β_2. OLS estimation can be used in this case because such models are basically linear in form even though one of the X terms corresponds to a power of X. Assessment of model fit would follow a similar structure to that explained for the simple linear model. In some applications of polynomial modelling, the independent variable(s) may be in coded form corresponding to deviations from the mean. This re-scaling can be beneficial if reducing the effect of multi-collinearity (high correlation between the X terms) is necessary.

8 MULTIPLE LINEAR REGRESSION

Simple regression modelling of a response as a function of only one experimental variable may not be appropriate in many modelling experiments. Often, a regression model including more than one experimental variable should be considered as the measured response is more likely to be affected by many variables than just a single one. Such a model is referred to as a *multiple regression model*. For example, the yield of a chemical process may depend on reaction time, temperature, pressure, and concentration of reactants, while in X-ray fluorescence, the concentration of an ingredient in a propellant slurry may be related to the X-ray intensity ratio of the slurry ingredients. In both of these cases, it is possible that the response is a function of not just one variable but many, all of which could be exerting a strong influence on the chemical response. Multiple regression analysis is highly useful in experimental situations where the experimenter can easily control the experimental variables and where investigation of a number of experimental variables simultaneously would be beneficial.

Both linear and non-linear multiple regression modelling can be considered. Multiple linear regression (MLR) modelling is based on expressing a response as a linear function of many variables, *e.g.*

$$Y = \alpha + \beta_1 X_1 + \beta_2 X_2 + \beta_3 X_3 + \epsilon$$

Such a model is given the general title of a *first order multiple model* because all the terms in X appear independently and to power one. This model is often used as an approximating function for the true relationship between a response and many controlled variables. A model of the form

$$Y = \alpha + \beta_1 X_1 + \beta_2 X_2 + \beta_3 X_1^2 + \beta_4 X_2^2 + \beta_5 X_1 X_2 + \epsilon$$

by contrast, would be called a *multiple non-linear*, or *second order multiple interaction model*. The nomenclature used reflects that some terms are quadratic (power, second order) and others contain more than one X term ($X_1 X_2$, interaction term).

When there are many experimental variables, combinations of them are experimented on with the response recorded at each combination of settings, as illustrated in Table 5.8. Experiments of this type can generate large amounts of data with the resultant regression computations matrix based. With these large data sets, it is best to use statistical software to provide the information on which model assessment can be based.

Table 5.8 *Typical data layout for a multiple regression experiment*

X_1	X_2	X_3	.	.	X_k	Y
x_{11}	x_{21}	x_{31}	.	.	x_{k1}	y_1
x_{12}	x_{22}	x_{32}	.	.	x_{k2}	y_1
x_{13}	x_{23}	x_{33}	.	.	x_{k3}	y_1
.	.	.	.	.	.	.
.	.	.	.	.	.	.
x_{1n}	x_{2n}	x_{3n}	.	.	x_{kn}	y_n

n, number of data points; k, number of experimental variables

The *general model* for MLR is specified as

$$Y = \alpha + \beta_1 X_1 + \beta_2 X_2 + \ldots + \beta_k X_k + \epsilon$$

where X_1, X_2,...., X_k represent the k experimental variables tested, α is the constant term, and ε the error. The β_i terms correspond to the *partial regression coefficients* and represent the expected change in response per unit change in the ith experimental variable, assuming all the remaining variables are held constant. Generally, we also *assume* the X variables are fixed and measured without error, constant response variance for each variable, and the recorded response exhibits normality characteristics. These assumptions mirror those underpinning linear regression though violations do not seriously affect the statistical viability of the model.

Parameter estimation in MLR again rests on the use of OLS estimation. To find the parameter estimates a, b_1, b_2, . . . , b_k of the coefficients in a multiple linear model, we construct the sum of squared deviations (residual mean square) of responses from model predictions (errors, residuals) as

$$\sum_{i=1}^{n} e_i^2 = \sum_{i=1}^{n} [y - (a + b_1 x_i + b_2 x_i + \ldots + b_k x_i)]^2 \tag{5.24}$$

The OLS estimation procedure involves minimising equation (5.24) using partial differentiation to provide the normal equations

$$na + b_1 \sum x_1 + b_2 \sum x_2 + \ldots + b_k \sum x_k = \sum y$$
$$a \sum x_1 + b_1 \sum x_1^2 + b_2 \sum x_1 x_2 + \ldots + b_k \sum x_1 x_k = \sum x_1 y$$
$$a \sum x_2 + b_1 \sum x_1 x_2 + b_2 \sum x_2^2 + \ldots + b_k \sum x_2 x_k = \sum x_2 y$$
$$\ldots$$
$$a \sum x_k + b_1 \sum x_1 x_k + b_2 \sum x_2 x_k + \ldots + b_k \sum x_k^2 = \sum x_k y$$

Solving these equations is not practicable especially if the number of experimental variables k is large. Hence, the use of matrix methods and the need to utilise statistical software when considering MLR.

MLR modelling rests on two aspects of validity, statistical and practical, just as in simple regression. Ideally, any fitted model must satisfy both criteria but in MLR, there are other considerations that must be examined to provide full model assessment. These include multi-collinearity and analysis for 'best' equation. Excel has multiple regression estimation facilities by default but unfortunately, these additional assessment elements are not available by default so we will use Minitab to illustrate MLR modelling. The Minitab dialog window for regression is presented in Figure 5.10 with the usual highlighting of data, checking, *etc.* necessary before implementing the estimation procedure.

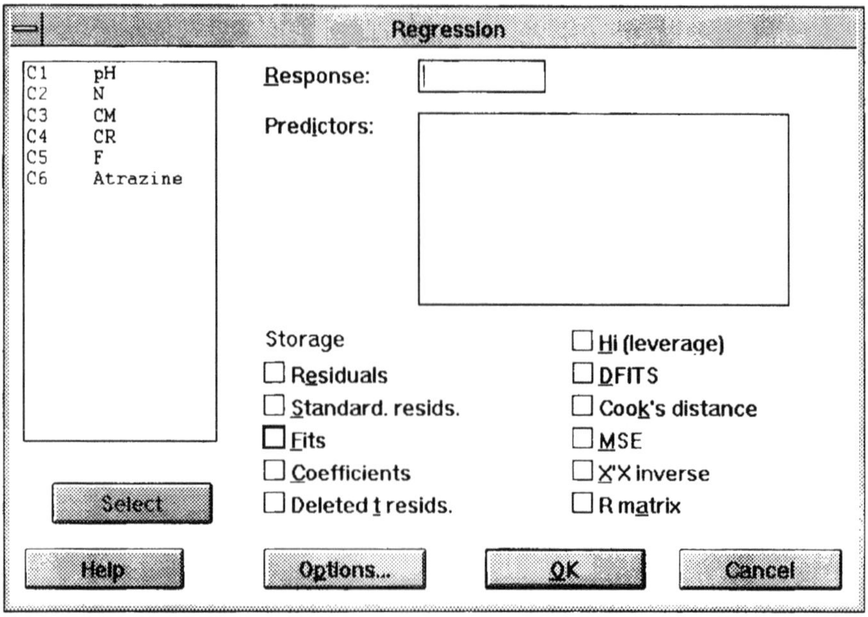

Figure 5.10 *Regression dialog window in Minitab*

The *ANOVA principle* of splitting response variability into two distinct elements: variation due to the regression model fitted (SSRegn) and residual variation (SSRes) is also applied in MLR modelling. Sum of squares (SS) and mean square (MS) terms again play the main roles with the main SS term, SSTotal, expressed as

$$SSTotal(SST) = SSRegression(SSRegn) + SSResidual(SSRes)$$

just as in simple regression. The calculation aspects and general ANOVA table for the multiple linear model are presented in Table 5.9.

Table 5.9 *Generalised ANOVA table for multiple linear model of a response Y as a function of k experimental variables*

Source	df	SS	MS
Regression	k	SSRegn= $b'X'Y = (\sum y)^2/n$	MSRegn = SSRegn/df
Residual	$n-k-1$	SSRes (by subtraction)	MSRes = SSRes/df
Total	$n-1$	SST= $\sum y^2 - (\sum y)^2/n$	

n, number of data points; k, number of experimental variables; df, degrees of freedom; SS, sum of squares; MS, mean square; Y, nx1 column vector of responses; X, nx(k + 1) matrix of the X observations (first column being n 1s); b, (k + 1)x1 column vector of unknown regression coefficients

Statistical validity can be assessed through the regression F test

$$F = MSRegn/MSRes \qquad (5.25)$$

with degrees of freedom $df_1 = k$ (Regression df) and $df_2 = n-k-1$ (Residual df). Additionally, we should also assess the adjusted coefficient of determination (R^2_{adj}) of the fitted multiple equation to help provide further confirmation of statistical validity of model fit. In general, we are looking for the p value to be less than 0.05 and R^2_{adj} to exceed 80%, as before.

The next step in model assessment, after the statistical checks, involves checking for the evidence of *multi-collinearity*, defined as strong association between the experimental variables. Presence of collinearity can affect the accuracy of coefficient estimates and the general applicability of the fitted model. Assessment is based on examining the correlations between all pairs of experimental variables where correlation measures the degree of linear association. Correlations in excess of 0.8 to 0.9, numerically, suggest that collinearity may exist hinting at a link between the variables being compared. Detection of such an effect implies that one of the variables could be discarded within the proposed model without serious loss of prediction power.

Example 5.8

An experiment was set up to assess whether the ion-interaction HPLC retention time of atrazine could be modelled as a function of five characteristics. The characteristics considered were pH of the mobile phase, ion-interaction reagent alkyl chainlength N, organic modifier concentration CM, ion-interaction reagent concentration CR, and flow rate F. The data matrix and responses are listed in Table B.2. It is thought that atrazine retention time may be related to these five characteristics through a multiple linear regression model.

Data entry in Minitab
Data are entered into columns C1 to C6 in the Minitab Data window (see Figure 1.4), C1 to C5 corresponding to the five experimental variables and C6 to the recorded atrazine recovery measurements.

Model
Since MLR has been proposed, then we can say that the model we wish to fit is

$$\text{retention time of atrazine} = \alpha + \beta_1 \text{pH} + \beta_2 N + \beta_3 CM + \beta_4 CR + \beta_5 F + \epsilon$$

The experimental variables are assumed fixed and measured without error, and retention time is assumed normally distributed.

Parameter estimation
Use of Minitab's Regression procedures, as described in Output 5.13, provides the fitted equation and related statistical information. The 'Residuals' and 'Fits' selection is again used to provide data for later practical and diagnostic analysis (see Example 5.10).

The fitted multiple linear equation is specified as

$$\text{retention time} = 224.5 - 3.192\text{pH} - 2.9973N - 3.2081CM - 1.0422CR - 28.909F$$

The coefficient estimates for pH, N, and CM are of similar magnitude and sign indicative that unit changes to the levels of these variables have similar effect on retention time. The largest coefficient is associated with the flow rate variable, F, suggesting that unit change in this variable can affect retention time substantially.

Statistical validity
Once we have a numerical form for the fitted model, the next stage in the analysis must be to check its statistical validity through formal statistical test and assessment of R^2_{adj}.

The null hypothesis to be tested is H_0: multiple linear model not appropriate with alternative reflecting valid model fit. The F test statistic (5.25) is given as 24.96 in the regression ANOVA table in Output 5.13 with associated p value of 0.000 ($p < 0.0005$). The small p value implies rejection of H_0 meaning that the fitted multiple linear model appears to be statistically valid ($p < 0.01$).

The adjusted coefficient of determination, equation (5.10), is estimated as 81.6% ['R-sq (adj)'] which exceeds the target of 80%, though

Output 5.13 *MLR output for Example 5.8*

Select **Stat** ▷ **Regression** ▷ **Regression** ▷ for *Response*, select **Atrazine** and click **Select** ▷ for *Predictors*, highlight **pH to F** and click **Select** ▷ for *Storage*, select **Residuals** and select **Fits** ▷ click **OK**.

Regression Analysis

The regression equation is
Atrazine = 244 − 3.19 pH − 3.00 N − 3.21 CM − 1.04 CR − 28.9 F

Predictor	Coef	Stdev	t-ratio	p
Constant	244.50	19.48	12.55	0.000
pH	− 3.192	1.053	− 3.03	0.006
N	− 2.9973	0.7897	− 3.80	0.001
CM	− 3.2081	0.3310	− 9.69	0.000
CR	− 1.0422	0.3798	− 2.74	0.012
F	− 28.909	5.984	− 4.83	0.000

s = 6.450   R-sq = 85.0%   R-sq(adj) = 81.6%

Analysis of Variance

SOURCE	DF	SS	MS	F	p
Regression	5	5191.2	1038.2	24.96	0.000
Error	22	915.2	41.6		
Total	27	6106.3			

SOURCE	DF	SEQ SS
pH	1	129.9
N	1	302.7
CM	1	3539.5
CR	1	248.4
F	1	970.8

only just, providing further evidence that the fitted multiple linear model has acceptable statistical validity. However, the closeness of R^2_{adj} to target may affect predictive ability as it suggests model only accounts for around 82% of the variability in retention time response. Around 18% of this variability is unaccounted for.

Multi-collinearity
The correlations required for this aspect of model checking are provided in Output 5.14. They indicate that retention time of atrazine appears most strongly associated, in a linear sense, to CM with a strong negative

correlation shown suggesting low concentration of methanol is associated with high retention time and high concentration with low retention time. The correlations between the experimental variables are very low indicating no evidence of collinearity between them.

Output 5.14 *Correlations between atrazine and experimental characteristics for Example 5.8*

Select **Stat** ▷ **Basic Statistics** ▷ **Correlation** ▷ for *Variables*, highlight **Atrazine, pH to F** and click **Select** ▷ click **OK**.

Correlations (Pearson)

	Atrazine	pH	N	CM	CR
pH	−0.146				
N	−0.213	−0.061			
CM	−0.734	−0.064	−0.064		
CR	−0.131	−0.054	−0.054	−0.057	
F	−0.308	−0.058	−0.058	−0.061	−0.051

MLR involves fitting a linear function of many experimental variables to a set of measured responses. In many cases, it could be that a response may depend on only a subset of these variables so a reduction in their number may be possible without serious loss of predictive ability. Assessment of the feasibility of reducing the number of experimental variables involves examining, statistically, if reduction is feasible and whether a 'best' equation in terms of fewer variables is a possibility. This check utilises the screening processes *best subsets regression* or *stepwise regression* though such procedures are only appropriate provided the original proposed model is statistically valid ($p < 0.05$, $R^2_{adj} > 80\%$). Any 'best' equation suggested by these procedures can never account for more variation in the response than the original fitted model, *i.e.* cannot increase the coefficient of determination R^2. However, the adjusted coefficient R^2_{adj}, equation (5.10), can increase with fewer variables but never such that it exceeds the original R^2 value.

Stepwise regression methods correspond to procedures for sequentially introducing or removing variables one at a time and testing at each step whether the retained variables still produce an effective regression model. This routine can be implemented through either *forward selection* (introduction of variables one at a time) or *backward elimination* (removal of variables one at a time). Addition, or removal, of variables is based on a partial F test with test statistic exceeding critical value providing justification for inclusion of a variable. The stepwise algorithm continues until no further variables can be added or deleted. It is at this point that the 'best' statistical model is reached. Best subsets, on the other hand, examines as many multi-variable models as possible, not just those that are statistically appropriate.

Both assessment processes can provide many potential models, so criteria for model comparison and decision on which is 'best' are required. Such criteria can be based on three model characteristics:

1. the adjusted coefficient of determination R^2_{adj}, *highest best*;
2. the residual mean square estimate of the error variance σ^2, *lowest best*;
3. derivation of Mallow's C_p statistic for measuring model specification, statistic given by

$$C_p = (SSRes_p/s^2_{y/x}) - n + 2p \qquad (5.26)$$

where $SSRes_p$ is the residual SS for the p variable subset model and $s^2_{y/x}$ is the estimate of regression variance for the model containing all variables, p = number of X terms + 1, C_p *equal to or lower than p best*.

For a model $Y = f(X_1, X_2, X_3, X_4, X_5, X_6)$, $p = 6 + 1 = 7$ as there are six X terms in the model so any 'best' six term model must produce a C_p statistic of 7 or less. The model $Y = f(X_1, X_2, X_3)$ has $p = 3 + 1 = 4$ as there are three X terms in the model so any 'best' three term model specification requires $C_p \leqslant 4$.

Stepwise methods primarily use criteria 1 and 2 while best subsets utilises all three. The 'best' model highlighted by both of these procedures will generally be the same. The advantage of the best subsets procedure is that information on a number of possible 'best' models, comprising different combinations of the variables, is provided. The stepwise procedure only provides details of 'best' models based on statistically valid variable addition or removal which are generally fewer in number.

R^2_{adj} and the residual mean square estimate are inversely related and are reflective of the level of variability in the response explained by the fitted regression model. 'Best' models will tend to have high R^2_{adj}, low residual mean square estimate, and C_p close to or less than p. Ill fitting models will tend to have low R^2_{adj}, high residual mean square estimate, and C_p much larger than p. As with most inferential data analysis, compromise between what is ideal and what is practicable may be necessary in choosing a 'best' model. A recent paper by Gilmour[9] has

---

[9] S.G. Gilmour, *The Statistician*, 1996, **45**, 49.

suggested that an adjusted C_p, representing a modification of equation (5.26), should be considered in preference to the quoted formula. Plots of C_p against p can also be considered to aid model selection.[10]

Example 5.9

Using the atrazine data of Example 5.8 and associated multiple linear model, we can assess whether retention time can be modelled using fewer experimental variables. Only the best subsets approach will be illustrated.

Best subsets output
Within Minitab, the menu commands **Stat** ▷ **Regression** ▷ **Best Subsets** provide access to Minitab's best subsets procedure. The output produced by this procedure for the atrazine data is shown in Output 5.15. The output provides information on R^2_{adj} ('Adj. R-sq'), C_p ('C-p'), and the square root of residual mean square estimate ('s') for the best five models relating retention time to the experimental variables for combinations of one to five of these variables.

'Best' equation
The output highlights three models that could be considered 'best' based on their residual mean square estimates. These models are summarised in Table 5.10. Considering these results, we can see that the first two suggestions have lower R^2_{adj} than appropriate ($< 80\%$) and inappropriate C_p values. Only the full model comprising all experimental variables provides acceptable values for all three features (highest R^2_{adj}, $C_p = 6$, lowest s). This looks to be the 'best' MLR model for the collected data. Checking for best equation, unfortunately, has not led to a reduction in the number of experimental variables making up the prediction model.

Table 5.10 *Summary of 'best' models for atrazine data of Example 5.9*

Experimental variables included	R^2_{adj}	C_p	s	Comments
pH, N, CM, F ($p = 5$)	76.4	11.5	7.3081	omit CR
N, CM, CR, F ($p = 5$)	75.1	13.2	7.5107	omit pH
pH, N, CM, CR, F ($p = 6$)	81.6	6.0	6.4497	include all variables

Application of stepwise regression, through both forward selection

---

[10] J.O. Rawlings, 'Applied Regression Analysis: A Research Tool', Wadsworth & Brooks/Cole, Belmont, California, 1988, pp. 183–186.

Output 5.15 *Best subsets output for Example 5.9*

Select **Stat** ▷ **Regression** ▷ **Best Subsets** ▷ for *Response*, select **Atrazine** and click **Select** ▷ for *Free predictors*, highlight **pH to F** and click **Select**. Select **Options** ▷ select the **Models of each size to print** box and enter **5** ▷ click **OK** ▷ click **OK**.

Best Subsets Regression

Response is Atrazine

Vars	R-sq	Adj. R-sq	C-p	s	p H	N	C M	C R	F
1	53.8	52.0	43.8	10.415			X		
1	9.5	6.0	108.8	14.578					X
1	4.6	0.9	116.1	14.972		X			
1	2.1	0.0	119.7	15.161	X				
1	1.7	0.0	120.3	15.193				X	
2	66.3	63.6	27.5	9.0718			X		X
2	60.6	57.5	35.8	9.8087		X	X		
2	57.5	54.1	40.3	10.184	X		X		
2	56.8	53.3	41.4	10.273			X	X	
2	14.9	8.1	103.0	14.420		X			X
3	74.3	71.1	17.7	8.0821		X	X		X
3	70.9	67.3	22.6	8.5974	X		X		X
3	70.0	66.3	24.0	8.7350			X	X	X
3	65.0	60.7	31.3	9.4303	X	X	X		
3	64.2	59.7	32.6	9.5503		X	X	X	
4	79.9	76.4	11.5	7.3081	X	X	X		X
4	78.8	75.1	13.2	7.5107		X	X	X	X
4	75.2	70.9	18.4	8.1144	X		X	X	X
4	69.1	63.7	27.3	9.0553	X	X	X	X	
4	21.0	7.3	97.9	14.479	X	X		X	X
5	85.0	81.6	6.0	6.4497	X	X	X	X	X

and backward elimination, resulted in the same 'best' model as the best subsets approach, as would be expected since both assess the likelihood of variable reduction while attempting to maintain model validity.

The final steps in MLR analysis comprise the practical and diagnostic checking of the 'best' model. The former is necessary, as before, to ensure that the model predictions are of acceptable accuracy for practicality of model to be confirmed. The latter can again investigate whether additional X terms may be needed and checks for any assumption violations (response non-normality, unequal variability). This final check should, ideally, be carried out on all possible 'best' models as predictive accuracy may vary with model due to combination of experimental variables present.

Example 5.10

As we have established that the full multiple regression model for the retention time data of Example 5.8 is the only one worth considering, we want to complete the analysis by checking practical validity and carrying out diagnostic checking of the residuals.

Practical validity
This can be achieved, as previously, by examining accuracy of model predictions. Output 5.16 provides a Minitab print of recorded and predicted retention times while Output 5.17 provides a graphical presentation of these figures. Some predictions are reasonable, *e.g.* Atrazine 23.86 and predicted 24.035, and others not so, *e.g.* Atrazine 27.53 and predicted 40.3719. The magnitude of the predictions, however, are at least of the correct order though differences appear to outweigh successes.

The plot in Output 5.17, which would be ideal if all points lay on the $Y = X$ line, highlights prediction to be acceptable at only a few points, with most tending to show either over-prediction (above $Y = X$) or under-prediction (below $Y = X$), the latter particularly so for higher atrazine levels. No detectable trend can be seen, however, in the pattern of over- and under-prediction. Overall, we have to conclude that the predictions are not of sufficient accuracy for the fitted multiple linear model to be accepted as valid on a practical basis.

Diagnostic checking
The final part of the analysis, diagnostic checking, is carried out to ensure that the variables fitted are not exhibiting undetected trends or patterns which may hint at model re-specification. The residual plots associated with the full fitted model are not shown but all showed some form of patterning. The variable plots hinted at column length differences with the CM plot also showing upward trend and the F plot, a pattern similar to the greater than symbol. The fits plot also had a distinct pattern indicative of predictions being reasonable till mid-recorded retention time after which, prediction accuracy wavered markedly. The normal plot had a positive trend but one which was more curved than linear suggesting non-normality for the atrazine response.

In summary, statistical validity of the full multiple linear model

Output 5.16 *Observed and predicted atrazine measurements for Example 5.10*

Select **File** ▷ **Display Data** ▷ for *Columns, constants, and matrices to display*, select **Atrazine** and click **Select**, and select **FITS1** and click **Select** ▷ click **OK**.

Data Display

Row	Atrazine	FITS1
1	16.34	11.7495
2	29.00	32.0554
3	57.50	55.8199
4	23.86	24.0351
5	39.49	37.9066
6	31.28	34.6608
7	22.84	16.6281
8	27.53	29.4500
9	27.53	40.3719
10	29.41	36.8434
11	25.67	26.0607
12	16.81	15.4015
13	24.69	21.9372
14	32.85	34.2546
15	65.33	60.6985
16	34.57	27.8867
17	67.69	58.2852
18	35.08	38.0500
19	22.12	38.1934
20	34.17	40.0756
21	16.67	14.1628
22	71.12	61.9372
23	44.30	42.1736
24	31.73	35.4933
25	23.69	20.2802
26	25.72	27.2674
27	21.42	17.8668
28	27.98	26.8446

proposed for the retention time data of Example 5.8 appears acceptable though no reduction in number of variables appears feasible. Practical validity and diagnostic checking call into question the practicality of this model in the form presented. Perhaps new variables or further terms based on the five experimental variables, *e.g.* interaction or power terms, need to be included to improve model applicability. In fact, inclusion of the interactions $N \times CR$ and $pH \times CR$ raise R^2_{adj} to 87.2%, an improvement of 5.6%. However, predictive ability does not improve though best subsets suggests pH could be removed from the seven variable model.

Output 5.17 *Plot of observed and predicted atrazine measurements for Example 5.10*

Select **Graph** ▷ **Plot** ▷ for *Graph 1 Y*, select **FITS1** and click **Select** ▷ for *Graph 1 X*, select **Atrazine** and click **Select**.

Select **Frame** ▷ **Axis** ▷ select the **Label 1** box and enter **Recorded Atrazine** ▷ select the **Label 2** box and enter **Model Predictions** ▷ click **OK**.

Select **Annotation** ▷ **Line** ▷ for **Points 1**, enter **15 15 62 62** ▷ click **OK** ▷ click **OK**.

The labels produced within the plot use further sub-menus (**Annotation** ▷ **Text**).

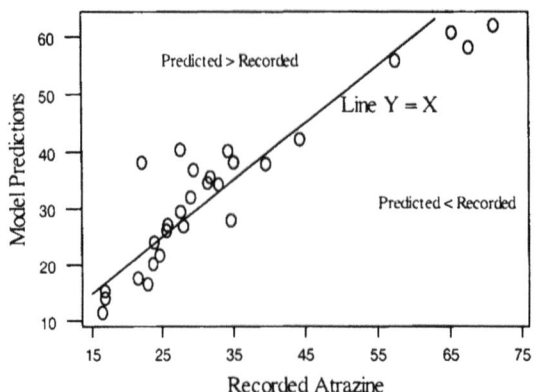

Perhaps, the atrazine response data cannot be modelled using the given experimental variables and may require inclusion of new variables before an acceptable model can be produced.

The analysis elements outlined for MLR represent the most appropriate mechanisms for assessing validity of a multiple model. They are by no means comprehensive but are sufficient for most practical purposes. Other procedures, such as tests of the regression coefficients, do exist to provide alternative mechanisms for analysing, in particular, the statistical validity of a fitted multiple regression model. Inclusion of interaction terms to the atrazine modelling problem was based on the change in R^2_{adj} caused by their inclusion. Such an approach has to be treated cautiously unless the inclusion can be justified on chemical grounds, because inclusion of all potential variable interactions will inevitably improve model validity but at the expense of making the model too cumbersome.

Exercise 5.4

An experiment was conducted into whether the fluorescence intensity of an aluminium complex could be related to five experimental factors:

volume of Solochrome Violet RS (SVRS, ml), pH, heating time (HT, seconds), cooling time (CT, seconds), and delay time (DT, seconds). Heating time refers to the heating of the aluminium solution in a microwave oven at 360W for the specified time period, while cooling time refers to the time the solution is left to cool. The delay time factor corresponds to the waiting time before measuring the fluorescence intensity set at 590 nm on a Perkin Elmer LS50 Luminescence Spectrometer. The collected data are presented in Table B.3. Using MLR, assess whether a multiple linear model for fluorescence can be developed.

9 FURTHER ASPECTS OF MULTIPLE REGRESSION MODELLING

9.1 Model Building

In experimental modelling, it is important to be able to select the relevant experimental variables on which to base a multiple regression model. Previous experience, or underlying chemical considerations, may help to indicate appropriate variables to consider. It may also be useful to screen the candidate variables to obtain a 'best' subset for model building though the model must contain enough independent variables for it to perform satisfactorily both statistically, and practically.

Plots of response against each experimental variable may indicate the type of simple relationship between these variables, *e.g.* linear, quadratic. Plots of response against an experimental variable taking account of the values of another experimental variable, similar to the interaction plot associated with factorial designs, can be used to indicate if interaction is present, *i.e.* whether a term such as $X_1 X_2$ should be included in model specification. Unfortunately, there is no best procedure for variable selection when developing multiple regression models with compromise necessary, together with use of judgement and experience, before an acceptable regression model can be adequately formulated.

9.2 Multiple Non-linear Modelling

In multi-variable investigations, the form of the relationship between a response and the experimental variables may be non-linear, *i.e.* model contains power and interaction terms. Typical non-linear models include the first order interaction model

$$Y = \alpha + \beta_1 X_1 + \beta_2 X_2 + \beta_3 X_3 + \beta_4 X_1 X_2 + \beta_5 X_1 X_3 + \epsilon$$

and the second order interaction model

$$Y = \alpha + \beta_1 X_1 + \beta_2 X_2 + \beta_3 X_1^2 + \beta_4 X_2^2 + \beta_5 X_1 X_2 + \epsilon$$

The term $X_1 X_2$ in each of these models is defined as an interaction term describing the non-uniform effect on the response of variable X_1 across the values of variable X_2. Assumptions are as for MLR with analysis based on the same principles. For multi-collinearity, however, high correlations between some of the components can be a natural consequence of the terms in the model, *e.g.* X_1 and $X_1 X_2$ may well be highly correlated because of presence of X_1 in both terms, so such evidence has to be treated cautiously.

9.3 Comparison of Multiple Regression Models

In multi-variable experiments, there may be cases where interest lies not in modelling *per se* but in comparing two multiple regression models, just as in the comparison of two linear equations discussed in Section 6. In multiple modelling, this generally refers to a comparison of two multiple models: *model 1* (the full model) and *model 2* (reduced model). The former represents the full proposed multiple model while the latter may represent a subset of the proposed model containing fewer X terms. Both models are fitted to the recorded responses with the comparison procedure utilising some of the resulting statistical information. Box 5.8 outlines the operation of this test procedure.

Box 5.8 *Test for comparing multiple regression models*

Hypotheses. The hypotheses for this comparison are specified as H_0: no difference between the two models against H_1: difference between the two models suggesting inclusion of the extra X variable(s) is important to the applicability of the full model.

Test statistic. The test statistic is based on using the residual sum of squares (SSRes) components for each fitted model. These are $SSRes_1$ and $SSRes_2$ with respective degrees of freedom df_1 and df_2 where subscript 1 refers to the full model and 2 the reduced model. We then evaluate the mean square terms $MS_{drop} = (SSRes_2 - SSRes_1)/(df_2 - df_1)$ and $MSRes_1 = SSRes_1/df_1$. Taking the ratio of these two terms provides the test statistic

$$F = MSdrop/MSRes \tag{5.27}$$

based on degrees of freedom $(df_2 - df_1, df_1)$.
Decision rule. As previous illustrations of the F test statistic.

As with most applications of F tests, small values of F will be indicative of acceptance of the null hypothesis and large values with rejection. For this comparison, acceptance of H_0 implies that inclusion of the specified X terms does not improve model applicability and so a reduced model excluding these terms could be worth considering.

10 WEIGHTED LEAST SQUARES

This application of least squares estimation can occur when replicate responses are measured at each X setting, *e.g.* triplicate absorbance readings at each concentration of chemical tested. In such a case, the variability in absorbance could change as concentration changes and when modelling such data, we must try to account for this proportioning effect as the variance about the regression line may not be constant for the experimental variable (invalid variance assumption). The method of *weighted least squares* (WLS)[11] provides the procedure for model fitting in such a case based on the same concepts and principles as OLS estimation except that weights w_i are brought into the calculations.

For WLS within linear modelling, the residual mean square expression (5.1) is re-expressed as

$$S_w = \sum_{i=1}^{n} e_i^2 \sum_{i=1}^{n} w_i \big(y = (a + bx_i)\big)^2 \qquad (5.28)$$

incorporating the specified weights. Minimising equation (5.28) through partial differentiation provides the WLS estimate of slope as

$$b = \frac{\sum w_i x_i y_i - \frac{\left(\sum w_i x_i\right)\left(\sum w_i y_i\right)}{n}}{\sum w_i x_i^2 - \frac{\left(\sum w_i x_i\right)^2}{n}} \qquad (5.29)$$

and intercept as

$$a = \frac{\sum w_i y_i - b \sum w_i x_i}{n} \qquad (5.30)$$

Often, the weightings w_i may simply be the reciprocal of the variances of the replicate observations at each X setting, *i.e.* $w_i = 1/\text{variance}(Y_i)$. Comparable statistical and assessment principles would again be

---

[11] Reference 7, pp. 124–128.

applied to discern the practicality of fitting a WLS model to experimental data.

11 SMOOTHING

Least squares regression may not always be appropriate when modelling a chemical process, particularly if no single relationship may fit through all the data points as they do not exhibit a smooth enough pattern. In such a case, it may be necessary to consider smoothing techniques where several linked relationships are fitted to data using different formats for different ranges of the experimental variable. *Spline functions*[12] is a particular illustration of this method with cubic splines most often used. These refer to fitting a series of cubic equations joined at specific points. Such curve fitting techniques have found application in such as radioimmunoassays and atomic emission spectroscopy.

[12] Reference 7, p. 137.

CHAPTER 6

Non-parametric Inferential Data Analysis

1 INTRODUCTION

In Chapters 2, 3, and 4, exploratory and inferential data analysis techniques for one-, two-, and multi-sample experimentation, based on the use of parametric inference procedures for the inferential data analysis aspect, were introduced. Parametric procedures are based on the assumption of normality for the quantitative measured response. In chemical experimentation, however, this may not always be appropriate as the response may be on a scale which does not conform to normality, *e.g.* percent recovery in a recovery experiment, or it may be of qualitative type, *e.g.* ranks or categorical data. To cater for these cases, we need to have alternative means of carrying out inferential data analysis. *Non-parametric*, or *distribution-free*, methods provide such alternatives and this chapter will discuss these methods in relation to inferential data analysis.

Non-parametric tests are based on minimal assumptions concerning the nature of the measured response and are mostly based on simple ranking principles which are unaffected by the nature of the response data. Most are computationally simpler than the comparable parametric procedure and are, therefore, easier to calculate and implement. A drawback of ranking, however, is that the magnitude of the observations is lost and so important information on magnitude differences within the data is not properly utilised. This leads to such routines generally being less powerful than their parametric equivalents. In other words, if the data can be handled by a parametric procedure, it is better to do so, as use of an alternative non-parametric procedure results in the loss of much of the information contained within the collected data.

2 THE PRINCIPLE OF RANKING OF EXPERIMENTAL DATA

Many non-parametric inference procedures are based on the simple concept of *ranking* of experimental data and using this re-specification as the basis of the related inference procedure. To rank data, we express them in ascending order of magnitude and then provide each data value with a rank (1, 2, 3, *etc.*) related to its 'position' in the list. When tied values occur, *i.e.* a particular measurement occurs more than once, we assign the measured value a rank equivalent to the average of the position values the tied observations cover.

Consider the data set 35, 31, 37, 40, 37, 41, 43, 41, 39, and 41. To rank these observations, we must first express them in ascending order of magnitude in an ordered list. The 'position' of each recorded observation in this list is then used to assign it an appropriate rank, as illustrated in Table 6.1. The ranks of all individually occurring observations correspond exactly to their position value. The two exceptions, 37 and 41, represent tied observations. The value 37 appears twice, in positions 3 and 4. The rank for observation 37 is produced by averaging its position numbers to give $(3 + 4)/2 = 3\frac{1}{2}$. The other tied observation, 41, which occurs three times, is similarly treated with its rank evaluated as $(7 + 8 + 9)/3 = 8$. Ranking is, therefore, a simple data re-coding principle which can be easily achieved without resorting to software.

Table 6.1 *Illustration of data ranking*

Ordered list	31	35	37	37	39	40	41	41	41	43
Position	1	2	3	4	5	6	7	8	9	10
Rank	1	2	$3\frac{1}{2}$	$3\frac{1}{2}$	5	6	8	8	8	10

However, for larger data sets and given the need to produce a comprehensive summary of data for analysis purposes, use of software becomes imperative. Excel has a simple RANK command which, at present, ranks individually occurring observations correctly but, unfortunately, does not handle ranking of tied observations properly. Use of this command returns either the lowest or highest rank among tied observations. For the data illustrated above, this could mean observation 37 being assigned a rank 3 (lowest position between 3 and 4) compared with a correct rank of $3\frac{1}{2}$ and observation 41, a rank of 7 (lowest position between 7, 8, and 9) compared with a correct rank of 8. Clearly, this ranking facility cannot be used directly. A simple 'manual' workaround based on it is necessary. Such a workaround will now be explained.

Consider the set of observations mentioned above. Such data would be entered into cells A1:A10 of the Excel spreadsheet. Using column B for the ranks, array-enter in cell B1 the expression,

$$= \text{RANK}(A1, A\$1 : A\$10, 1) + (\text{SUM}(1^*(A\$1 = A1 : A\$10)) - 1)/2 \quad (6.1)$$

where array-entry is achieved by holding down the Ctrl and Shift keys while pressing the Enter key. This formula must then be copied and pasted into cells B2:B10. Expression (6.1) is based on Excel's RANK command returning the lowest position value for tied observations. For a return of highest position value, simply replace plus (+) in expression (6.1) with minus (−). All illustrations of the ranking workaround in this chapter will use expression (6.1) as presented.

The first part of expression (6.1) represents Excel's RANK command for ranking data based on all the observations in the data set (A$1:A$10). The term 1 specifies ranking in ascending order of magnitude, *i.e.* smallest to largest. The SUM part in expression (6.1) represents a simple correction for ties. It checks the data, through (A1=A$1:A$10), for the number of observations equal to the value being ranked. Subtraction of one and then division by 2 enables the appropriate correction to Excel's RANK value to be determined.

To describe how expression (6.1) operates, consider the previous data set 35, 31, 37, 40, 37, 41, 43, 41, 39, and 41, the ranks for which are summarised in Table 6.1. For observation 31, RANK will provide 1 with the SUM expression returning $(1*1-2)/2$, *i.e.* 0, as there is only one observation of value 31 in the data set. Hence, this observation would be assigned a rank of 1, as obtained previously. The same reasoning can be applied to the ranking of the other individually occurring observations.

The rank for tied observations can be determined similarly. For observation 37, which appears twice, RANK will provide 3 and the SUM expression will be $(1*2-1)/2$, *i.e.* $\frac{1}{2}$, as there are two observations of 37 in the data. This will mean a rank of $3\frac{1}{2}$ being provided for this observation. For the other tied observation 41, which appears three times, RANK will provide 7 and SUM will return 1, *i.e.* $(1*3-1)/2$, resulting in the observation being assigned a rank of 8. Ranks produced by this workaround will therefore exactly match the correct ranks for all observations in a data set, whether tied or not.

3 INFERENCE METHODS FOR TWO SAMPLE EXPERIMENTS

Two sample experimentation and associated parametric inference procedures were discussed in Chapter 2, Section 7. For data not conforming to the normality assumption underpinning these procedures, we would require to use alternative non-parametric procedures to produce the objective (inferential) analysis of the recorded data. Non-parametric test and estimation procedures, for such experimentation, are based on using the median as the measure of location, the principle of ranking, and simple calculational components.

3.1 Hypothesis Test for Difference in Median Responses

For two sample inference, a non-parametric alternative to the two sample t tests, equation (2.13) or (2.15), is the *Mann–Whitney test*, also called *Wilcoxon's rank sum test*. This test is based on comparing the medians using all the data and simple ranking to obtain the test statistic. The essential steps associated with the Mann–Whitney test of inference are provided in Box 6.1. Using the median as the base parameter of the

---

Box 6.1 *Mann–Whitney test*

Assumption. Two independent random samples of sizes m and n, and measurement scale of data is at least ordinal.

Hypotheses. The hypotheses associated with this procedure mirror those of the two sample t test (see Box 2.7) except that medians are used in the parameterised form of the hypotheses. Generally, the null hypothesis H_0 reflects no difference between the medians with three forms of alternative H_1 possible:

  1. H_1: difference between the two medians ($M_1 \neq M_2$),
  2. H_1: median of population 1 lower ($M_1 < M_2$),
  3. H_1: median of population 1 higher ($M_1 > M_2$).

Exploratory data analysis (EDA). Again, simple plots and summaries of the two data sets should be investigated to assess the information contained within the data concerning the experimental objective(s).

Test statistic. To determine the test statistic, we combine the two sets of data into a large group and rank the observations in order of magnitude, noting the data set source of each rank value. The Mann–Whitney test statistic, based on this grouping and ranking, is expressed as

$$T = S - m(m+1)/2 \qquad (6.2)$$

where S is the sum of the ranks of the sample 1 observations and m is the number of sample 1 observations.

Decision rule. Testing at the $100\alpha\%$ significance level can be operated using either test statistic or p value approach if appropriate.

test statistic approach: this operates in a similar way to the parametric t test. Lower critical values $T_{\alpha,m,n}$ are displayed in Table A.7 while upper critical values are determined as

$$T_{1-\alpha,m,n} = mn - T_{\alpha,m,n} \qquad (6.3)$$

  1. two-sided alternative (two-tailed test), $T_{\alpha/2,m,n} \leq T \leq T_{1-\alpha/2,m,n} \Rightarrow$ accept H_0,
  2. one-sided alternative H_1: < (one-tailed test), $T \geq T_{\alpha,m,n} \Rightarrow$ accept H_0,
  3. one-sided alternative H_1: > (one-tailed test), $T \leq T_{1-\alpha,m,n} \Rightarrow$ accept H_0.

p value approach: p value > significance level $\Rightarrow$ accept H_0.

---

inference procedure does not affect the conclusion reached since mean and median of data are often very similar.

When ties occur in the ranks, the Mann–Whitney test statistic (6.2) can be adjusted to suit. This adjustment has only a negligible effect on the value of the test statistic unless there are large numbers of tied observations. When software output specifies a p value, it should be noted that the value quoted is generally estimated using a normal approximation procedure so is, therefore, only an approximation to the true p value.

Example 6.1

A comparison was made of the percent recovery of simazine using solid phase extraction empore discs for two concentrations of spiked samples. The purpose of the investigation was to assess whether the concentration of spiking affects recovery. The recovery data are presented in Table 6.2.

Table 6.2 *Recovery data for Example 6.1*

Concentration 10 ng l^{-1}	77	75	82	73	82	75	79	83	81	71
Concentration 100 ng l^{-1}	69	74	76	68	69	73	71	68	71	67

Assumptions
Assume the two sets of recovery data are independent.

Hypotheses
Construction of hypotheses for this problem is relatively straightforward based on the experimental objective of assessing whether recovery rates differ with concentration of the spiked samples. This means the null hypothesis will refer to no difference, H_0: no difference in median recovery rate $(M_{10} = M_{100})$, and the alternative will be two-sided, H_1: difference in the median recovery rates $(M_{10} \neq M_{100})$.

Exploratory data analysis
The data plot in Output 6.1 shows that recovery rates do appear to differ with concentration. At 10 ng l^{-1}, recovery is generally higher and marginally more variable. At 100 ng l^{-1}, recovery is distinctly lower by comparison and appears less variable. In both cases, recovery is not above 85%. Given that 100% recovery would be ideal, the measurements quoted would appear to suggest that the extraction method tested, irrespective of sample concentration, is not as reliable as might be hoped.

Output 6.2 contains summary statistics associated with the recovery

Output 6.1 *Data plot of recovery data for Example 6.1*

Data entered in cells A1:B11, labels in cell 1 of each column. Menu commands as Output 2.2.

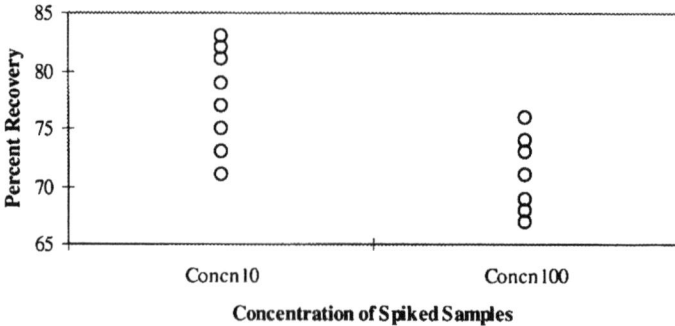

Plot of Recovery Data for Two Concentrations of Spiked Samples

Output 6.2 *Summary statistics for recovery data of Example 6.1*

Data entered in cells A1:B11, labels in cell 1 of each column. Menu commands as Output 2.4.

	Concn10	*Concn100*
Mean	77.8	70.6
Standard Error	1.332	0.933
Median	78	70
Standard Deviation	4.211	2.951
Sample Variance	17.733	8.711
Kurtosis	−1.388	−0.63
Skewness	−0.272	0.652
Range	12	9
Minimum	71	67
Maximum	83	76
Sum	778	706
Count	10	10
Confidence Level (95%)	3.012	2.111

data. They provide further evidence of recovery differences as the mean at 10 ng l^{-1} is greater than that associated with 100 ng l^{-1} by over 7% and the median by 8%. The *RSDs* are 5.4% and 4.2%, respectively, supporting the conclusion that there appears a marginal difference in the consistency of reported recovery rates.

Test statistic

Using the array-entry principle for ranking in Excel (see Section 6.2), ranks for the 10 ng l^{-1} concentration data can be derived as shown in Output 6.3. S within the test statistic (6.2) is shown to be 146.5 ('S') and with $m = n = 10$, the Mann–Whitney test statistic (6.2) becomes

$$T = 146.5 - 10(10 + 1)/2 = 91.5$$

as given by '*T*'. The lower 5% ($\alpha = 0.05$) two-tail critical value from Table A.7 is determined as $T_{0.025,10,10} = 24$ based on the alternative being two-sided. The upper critical [equation (6.3)] is evaluated as

$$T_{0.975,10,10} = 10.10 - 24 = 76$$

Since *T* lies outwith the range covered by the critical values of 24 and

Output 6.3 *Mann–Whitney test output for Example 6.1*

Data entered in cells A1:B11, labels in cell 1 of each column and column D.

In cell *C2*, array-enter (hold down Ctrl + Shift as Enter is pressed) the formula =RANK(A2,A\$1:B\$11,1)+(SUM(1*(A2=A\$1:B\$11))−1)/2. Copy and paste this formula to cells C3:C11 to use expression (6.1) to generate the data ranks.

In cell **E2**, enter =COUNT(A2:A11). In cell **E3**, enter =SUM(C2:C11) to generate the S term of test statistic (6.2). In cell **E4**, enter =E3−E2*(E2+1)/2 to generate the Mann–Whitney test statistic (6.2).

Concn10	Concn100	Ranks-10	Test Stat	
77	69	15	m	10
75	74	12.5	S	146.5
82	76	18.5	T	**91.5**
73	68	9.5		
82	69	18.5		
75	73	12.5		
79	71	16		
83	68	20		
81	71	17		
71	67	7		

76, we must reject H_0 and conclude that there appears to be sufficient evidence to suggest that the recovery rates of simazine differ according the concentration of simazine present in the sample ($p < 0.05$).

Initial impressions of the simazine recovery data suggested a difference with the concentration of the spiked sample. This was backed-up by the inferential analysis which found statistical evidence of difference ($p < 0.05$). Such a result suggests that recovery of simazine may depend on the concentration present and for such a recovery experiment, this is not a good result as recovery should ideally be independent of concentration present. The difference in the precision of the rates reported also gives cause for concern. This, together with the low quoted recovery rates, would appear to suggest that the tested recovery procedure may not be adequate for its intended purpose.

3.2 Confidence Interval for Difference in Median Responses

In two sample non-parametric inference, it is also possible to construct a confidence interval estimate similar to equation (2.17), but this time in respect of the difference between the two medians, to enable the magnitude of the difference effect to be estimated. The associated calculation procedure is cumbersome and is based on evaluating the differences between the ordered observations across each sample and using these figures to determine lower and upper bounds of the confidence interval.[1]

3.3 Hypothesis Tests for Variability

Non-parametric alternatives to the variance ratio F test (2.18) for testing variability in two sample experiments also exist. Two such tests are the *Ansari–Bradley test*[2] and the *Moses test*, both based on data ranking. The former merges the two data sets into one, as in the Mann–Whitney test, and ranks the smallest and largest with the same ranks in a pairing procedure until all measurements are assigned a rank. The sum of the ranks of the measurements in the first sample becomes the associated test statistic.

Exercise 6.1

A comparison was made between the recovered weights of ibuprofen using two standards: an internal one and an ibuprofen one. The

[1] W.W. Daniel, 'Applied Nonparametric Statistics', 2nd Edn., PWS-Kent Publishing Company, Boston, Massachusetts, 1990, pp. 97–101.
[2] Reference 1, pp. 103–107.

recovered weight, in mg, for two groups of eight specimens by each method are presented below. Do the methods differ?

| *Internal standard* | 192.00 | 193.50 | 190.04 | 186.07 | 185.92 | 191.64 | 185.43 | 187.21 |
| *Ibuprofen standard* | 187.80 | 194.30 | 194.84 | 189.77 | 188.53 | 196.85 | 177.91 | 183.21 |

4 INFERENCE METHODS FOR PAIRED SAMPLE EXPERIMENTS

Parametric inference and estimation procedures associated with paired sample experimentation were discussed in Chapter 2, Section 8. Non-parametric inference alternatives also exist for the analysis of such data if assuming normality for such is inappropriate. Again, tests and estimation procedures are based on testing medians using ranks as the appropriate base data.

4.1 Hypothesis Test for Median Difference in Responses

An equivalent non-parametric procedure to the paired comparison *t* test (2.20) is the *Wilcoxon signed-ranks test* which enables the median difference in the paired observations to be compared against a target of zero difference, reflecting no detectable difference between the treatments tested. The procedure is based on ranking the numerical differences for each matched pairing, ignoring zero differences, and using the ranks as the basis of test statistic determination. Box 6.2 summarises the steps associated with the implementation of the Wilcoxon signed-ranks procedure. The decision on acceptance or rejection of the null hypothesis is comparable to the mechanism associated with the Mann–Whitney test (see Box 6.1). It will be, in the main, the same decision that would be arrived at had the paired comparison *t* test (2.20) been used, the only difference being the way the experimental data contribute to the determination of the test statistic.

For the hypotheses, the general rule is that the target median difference M_D is set equal to 0 as we are generally testing for no difference though specified levels of this difference can be accommodated easily. Software mainly produce T_+ but often the test statistic may be expressed as T_-, the sum of the ranks of the negative differences. T_- can be evaluated from T_+ by the expression

$$T_- = n(n+1)/2 - T_+ \tag{6.6}$$

When using T_- as the test statistic, the decision rule for alternative

Box 6.2 *Wilcoxon signed-ranks test*

Difference. Analysis of paired sample experimental data is based on assessing a difference D between each pair of measurements. Generally, D = treatment 1 − treatment 2 is used though order of determination is unimportant.

Assumptions. Data are random, sampled population is symmetric, and measurement scale of data is at least interval.

Hypotheses. Null hypothesis specification again reflects no difference, H_0: no difference between treatments (median difference $M_D = 0$). Three forms of alternative hypothesis are again possible:

1. H_1: difference between treatments (median difference $M_D \neq 0$),
2. H_1: treatment 1 lower (median difference $M_D < 0$),
3. H_1: treatment 1 higher (median difference $M_D > 0$).

Exploratory data analysis (EDA). Use data plots and difference summaries to obtain an initial picture of the data.

Test statistic. To determine the test statistic, we determine the differences D and rank them in order of numerical magnitude, ignoring zero differences. The Wilcoxon signed-ranks test statistic, based on these ranks, is expressed as

$$T_+ = \textit{sum of the ranks of the positive differences} \qquad (6.4)$$

Decision rule. Deciding on accepting or rejecting the null hypothesis at the $100\alpha\%$ significance level is based on similar rules to those shown for the Mann–Whitney test (see Box 6.1).

test statistic approach: this decision mechanism depends on the number of non-zero differences, n, and the nature and significance level of the test. It is based on tabulated lower critical values, $T_{\alpha,n}$, displayed in Table A.8 with upper critical values determined as

$$T_{1-\alpha,n} = n(n+1)/2 - T_{\alpha,n} \qquad (6.5)$$

1. two-sided alternative (two-tailed test), $T_{\alpha/2,n} \leq T_+ \leq T_{1-\alpha/2,n} \Rightarrow$ accept H_0,
2. one-sided alternative H_1: < (one-tailed test), $T_+ \geq T_{\alpha,n} \Rightarrow$ accept H_0,
3. one-sided alternative H_1: > (one-tailed test), $T_+ \leq T_{1-\alpha,n} \Rightarrow$ accept H_0.

p value approach: as previous illustrations.

hypotheses 2 and 3 requires modification to $T_- \leq T_{1-\alpha,n} \Rightarrow$ accept H_0 and $T_- \geq T_{\alpha,n} \Rightarrow$ accept H_0, respectively. p Value estimation in software is again approximate, based as it is on the use of a normal approximation procedure.

Example 6.2

As part of a study into UV and electrochemical (EC) detection, samples of benzocane, a local anaesthetic, were analysed by reversed phase HPLC with mobile phase methanol : 1% ammonium acetate : 1% potassium nitrate (80:15:5). The retention time in minutes, for ten samples split in half and assigned to each detection system randomly, are presented in Table 6.3. Do the detection systems produce different retention times?

Table 6.3 *Retention time data for Example 6.2*

Sample	1	2	3	4	5	6	7	8	9	10
UV	3.1	3.2	3.3	3.1	3.2	3.3	3.3	3.1	3.2	3.1
EC	3.2	3.4	3.3	3.1	3.3	3.5	3.4	3.2	3.2	3.2

Difference
For this example, the required difference will be determined as $D =$ UV $-$ EC based simply on the order of data presentation.

Hypotheses
The posed question is concerned with assessing for evidence of a difference in retention times, a general directional difference reflective of a two-sided alternative hypothesis. The requisite hypotheses are therefore H_0: retention times similar for both detection methods (UV $=$ EC, median difference $M_D = 0$) and H_1: retention times different (UV $\neq$ EC, median difference $M_D \neq 0$).

Exploratory data analysis
The data plot in Output 6.4 provides distinct evidence that the detection systems provide different retention times in most cases. For all samples bar three, EC produces a higher retention time. In addition, the measured retention time appears to be more variable for the EC detection system (more variation in the points).

Calculation of retention time differences is straightforward in Excel as Output 6.5 illustrates. The differences D take only three numeric values. All are negative bar three and the median difference is -0.1.

Output 6.4 *Data plot of retention time data of Example 6.2*

Data entered into cells A1:C11, labels in cell 1 of each column. Menu commands as Output 2.14.

Plot of Retention Times Against Sample

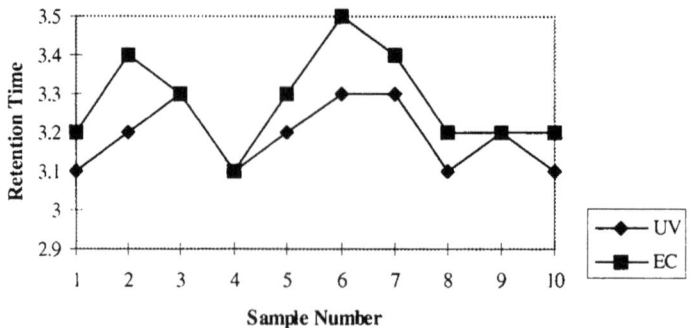

Output 6.5 *Retention time difference summaries for Example 6.2*

Data entered in cells A1:C11, labels in cell 1 of each column. Menu commands as Output 2.15.

Sample	UV	EC	D=UV−EC		D=UV−EC
1	3.1	3.2	−0.1	Mean	−0.09
2	3.2	3.4	−0.2	Standard Error	0.0233
3	3.3	3.3	0	Median	−0.1
4	3.1	3.1	0	Standard Deviation	0.0738
5	3.2	3.3	−0.1	Sample Variance	0.0054
6	3.3	3.5	−0.2	Kurtosis	−0.73
7	3.3	3.4	−0.1	Skewness	−0.17
8	3.1	3.2	−0.1	Range	0.2
9	3.2	3.2	0	Minimum	−0.2
10	3.1	3.2	−0.1	Maximum	0
				Sum	−0.9
				Count	10
				Confidence Level (95.0%)	0.052784

Output 6.6 *Data plot of retention time differences*

Difference data are in cells D1:D11, label in cell D1. Menu commands as Output 2.16.

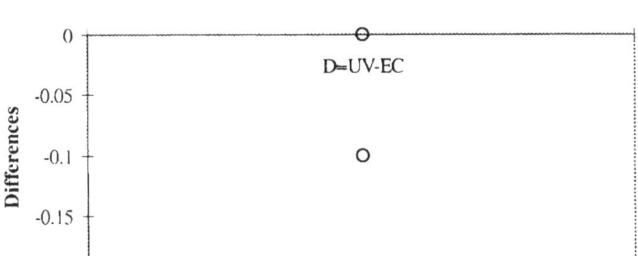

Plot of Retention Time Differences

This points to EC generally producing higher retention times and system difference being apparent.

The plot of retention time differences in Output 6.6 amply illustrates the differences interpretation. Three groupings of differences emerge and as two of these are related to negative differences, the evidence appears to be pointing to EC producing longer retention times.

Test statistic
Excel derivation of the test statistic (6.4) is shown in Output 6.7 using the array-entry ranking principle summarised in Section 6.2 modified to account for the presence of zero differences (the COUNTIF command for column I). The test statistic T_+ is specified to be 0 ('T+'). The number of non-zero differences, n, is indicated to be 7 ('n') based on use of the COUNT and COUNTIF commands.

Based on the alternative hypothesis being two-sided and using a 5% ($\alpha = 0.05$) significance level, Table A.8 provides the lower critical $T_{0.025,7} = 3$. Using equation (6.5), the upper critical is evaluated as

$$T_{0.975,7} = 7(7 + 1)/2 - 3 = 25$$

Since the test statistic of 0 lies outwith the range specified by the critical values, the null hypothesis is rejected and we can conclude that there appears sufficient statistical evidence to imply a difference in retention time according to detection system employed ($p < 0.05$). Such a conclusion mirrors the initial impressions though the consistency of the measurements could give cause for concern.

Output 6.7 *Wilcoxon signed-ranks output for Example 6.2*

Difference data are in cells D1 to D11, labels in row 1 of each column and column J.

In cell **G2**, enter =ABS(D2). Copy and paste this formula into cells G3:G11 to generate the required numerical differences.

In cell **H2**, array-enter (hold down Ctrl + Shift as Enter is pressed) the formula =RANK(G2,G\$2:G\$11,1)+(SUM(1*(G2=G\$2:G\$11))−1)/2. Copy and paste this formula into cells H3:H11 to use expression (6.1) to produce the requisite ranks of the absolute differences.

In cell **I2**, enter = IF(D2<>0,H2−COUNTIF(D\$2:D\$11,"=0")," "). Copy and paste this formula into cells H3:H11 to re-specify the ranks accounting for the presence of zero differences.

In cell **K2**, enter COUNT(D2:D11)−COUNTIF(D2:D11,"=0") to generate the number of non-zero differences *n*. In cell **K3**, enter. =SUMIF (D2:D11,">0",I2:I11) to generate the Wilcoxon signed-ranks test statistic (6.4).

D=UV−EC	AbsD	RnksAbD	RnksNzD	Test Stat	
−0.1	0.1	6	3	n	7
−0.2	0.2	9.5	6.5	T+	0
0	0	2			
0	0	2			
−0.1	0.1	6	3		
−0.2	0.2	9.5	6.5		
−0.1	0.1	6	3		
−0.1	0.1	6	3		
0	0	2			
−0.1	0.1	6	3		

4.2 Confidence Interval for Median Difference in Responses

A non-parametric confidence interval for the median difference, comparable to that for mean difference (2.21), can also be evaluated[3] if an estimate of this difference would be beneficial as part of the data analysis. The computational procedure is cumbersome based, as it is, on evaluating all possible pairwise observation averages, ranking these in order of magnitude, and determining the lower and upper bounds from the ordered array for the required confidence level of the interval.

---

[3] Reference 1, pp. 155–162.

Exercise 6.2

Data were collected on the nitrate levels (NO_3^- mg kg^{-1}) in dry milk
by two analytical procedures: a spectrophotometric method based on
the quantitative nitration of 2-*sec*-butylphenol in concentrated sulphuric acid and the French official reduction spectrophotometric
reference method (AFNor). The measured data presented below refer to
the means of three replicate measurements on each piece of paired
sample material tested. Do the methods provide measurement data of
equal accuracy?

Sample	1	2	3	4	5	6	7	8	9	10
Butylphenol	17.77	10.68	11.21	15.39	15.25	20.18	23.68	14.19	29.51	13.15
AFNor	16.92	10.33	11.90	15.97	15.58	19.81	23.18	13.37	29.26	13.48

Source: V.P. Bintoro, D. Cantin-Esnault and J. Alary, *Analyst (Cambridge)*, 1995, **120**,
2747.

5 INFERENCE FOR A CRD BASED EXPERIMENT

As with one- and two-sample experimentation, data collected from
experiments based on the CRD structure (see Chapter 3, Section 2) may
not conform to the underpinning assumptions of the associated parametric ANOVA methods (normality and equal variability). Use of a
non-parametric alternative to the ANOVA *F* test (3.1) and SNK follow-
up (3.2) may therefore be necessary in order to be able to carry out
relevant inferential data analysis.

5.1 The Kruskal–Wallis Test of Treatment Differences

The non-parametric alternative to the treatment *F* test (3.1) associated
with a CRD-based experiment is referred to as the *Kruskal–Wallis test*
and represents a simple extension of the two sample Mann–Whitney
test to the case of *k* samples. It is also referred to as *analysis of variance
by ranks*. Application of this test generally occurs if normality for the
measured response cannot be safely assumed or if the response data
consist of ranks. Box 6.3 contains an outline of the analysis for such a
case based on the Kruskal–Wallis test as the main test of treatment
effect.

Box 6.3 *Kruskal–Wallis test for treatment differences*

Assumptions. Measured data constitute k random samples and underlying populations are identical except for location differences in at least one.

Hypotheses. The hypotheses related to the Kruskal–Wallis test are essentially the same as those presented for the comparable parametric inference (see Chapter 3, Section 2.5) namely, H_0: no difference between treatments (median treatment effects same) and H_1: treatments differ (median treatment effects different).

Exploratory data analysis (EDA). Exploratory analysis of the data is again advisable to gain initial insight into what they indicate in respect of the experimental objective.

Test statistic. To determine the test statistic, the first step involves pooling the observations into a large group and ranking the pooled data in ascending order, smallest to largest. The ranks associated with each of the k treatments are then summed to generate the rank totals C_1 to C_k. From these, we use

$$H = \frac{12}{N(N+1)} \sum_{j=1}^{k} \frac{C_j^2}{n} - 3(N+1) \qquad (6.7)$$

to calculate the Kruskal–Wallis test statistic where N is the total number of measurements, k is the number of treatments, and n is the number of observations associated with each treatment.

Decision rule. This test statistic approximately follows a chi-square distribution with $(k-1)$ degrees of freedom enabling the χ^2 table (Table A.2) to be used for critical value determination. As with most inference procedures, low values of the test statistic (H < critical) or high p values (p > significance level) will be indicative of acceptance of H_0 and the conclusion that the treatments tested do not appear to differ on a statistical basis.

As previously, the specified hypotheses reflect only general hypotheses and do not provide any information on how a treatment difference, if detected, is reflected within the measured responses. Follow-up will again be necessary to enhance the conclusion. In the test statistic calculation, significant differences in the total ranks provide evidence of treatment differences as the position of the measurements associated with each treatment is differing. The test statistic (6.7) can be modified easily to accommodate an experiment based on unequal numbers of observations per treatment through replacing N by $\sum n_j$ and n by n_j, $j = 1, 2, \ldots, k$.

If there is a substantial number of tied observations, test statistic (6.7) can be adjusted by including a correction factor based on the extent of tied observations present. Inclusion of this factor generally inflates the test statistic but, when H is large enough to signify significant treatment differences, adjustment is not necessary as the conclusion reached is unaffected. Test statistic (6.7) could also be computed by carrying out ANOVA on the data ranks to provide

$$H = SSTreatment/MSTotal \tag{6.8}$$

where MSTotal = SSTotal/Total df. The value produced for equation (6.8) corresponds to test statistic (6.7) adjusted for ties.

Example 6.3

Before being used in production, chemical raw material is often checked for purity. For chemical powders, for instance, powder samples are heated and the residual ash content, representing the impurities, weighed. The purity, expressed as a percentage, is then determined by

$$purity = [1 - (ash\ weight/initial\ weight)]^* 100\%$$

A pharmaceutical company wishes to compare the purities of a chemical powder that it receives from its four regular suppliers using a CRD structure. Six batches of the powder were randomly selected from the deliveries of each supplier and the percent purity of each determined as shown in Table 6.4. The investigator was not prepared to assume normality for the purity measurements. Does percent purity differ with supplier?

Table 6.4 *Purity measurements for Example 6.3*

Supplier	A	B	C	D
	99.3	99.8	98.2	99.6
	99.4	97.4	97.2	98.7
	98.8	98.9	97.4	99.2
	99.4	99.0	98.3	99.2
	98.9	98.6	97.7	99.4
	99.1	98.8	97.5	99.7

Source: E.R. Dougherty, 'Probability and Statistics for the Engineering, Computing and Physical Sciences', Prentice Hall, Englewood Cliffs, New Jersey, 1990: reproduced with the permission of Prentice Hall, Inc.

Hypotheses
The null hypothesis for this supplier comparison is H_0: no difference between the suppliers (all provide similar median purity measurements).

Exploratory data analysis
An Excel generated plot of the purity data with respect to supplier is presented in Output 6.8. The plot shows that three of the suppliers, A, B, and D, have similar purity measurements but that supplier C appears very much lower than the rest. The variability in the reported measurements is similar for all suppliers except B which has much greater variation, mainly due to one sample with a much lower purity measurement.

Output 6.8 *Data plot of purity measurements for Example 6.3*

Purity data are in cells A1:D7, labels in cell 1 of each column. Menu commands as Output 2.2.

Purity Measurements against Supplier

The summaries in Output 6.9 confirm these initial observations. The medians for A, B, and D are similar but that for C differs markedly indicative of supplier C being most different. The respective *RSD*s are 0.26%, 0.79%, 0.45%, and 0.36% showing that the measurements are very consistent in themselves but this variability differs between the suppliers with supplier B worst by far.

Output 6.9 *Purity measurement summaries for Example 6.3*

Purity data are in cells A1:D7, labels in cell 1 of each column. Menu commands as Output 2.4.

	Suppl A	*Suppl B*	*Suppl C*	*Suppl D*
Mean	99.15	98.75	97.72	99.3
Standard Error	0.1057	0.3181	0.1815	0.1461
Median	99.2	98.85	97.6	99.3
Standard Deviation	0.2588	0.7791	0.4446	0.3578
Sample Variance	0.067	0.607	0.198	0.128
Kurtosis	-2.00	2.38	-1.70	0.74
Skewness	-0.42	-0.81	0.45	-0.83
Range	0.6	2.4	1.1	1
Minimum	98.8	97.4	97.2	98.7
Maximum	99.4	99.8	98.3	99.7
Sum	594.9	592.5	586.3	595.8
Count	6	6	6	6
Confidence Level (95%)	0.272	0.818	0.467	0.375

Test statistic

Generation of the test statistic in Excel, as Output 6.10 illustrates, is relatively straightforward using the ranking modification process explained in Section 6.2. The rank totals C_i are provided as 96, 71.5, 25.5, and 107 ('Sum Rnks'). With $N = 24$ ('N') and $n = 6$ ('n'), the test statistic (6.7) is calculated to be 13.09 ('H').

The corresponding critical value, at the 5% significance level from Table A.2, is $\chi^2_{0.05,3} = 7.81$. As test statistic exceeds critical value, we reject the null hypothesis and conclude that there appears to be a statistically significant difference in the purity measurements associated with the suppliers ($p < 0.05$). This confirms the initial impressions of supplier difference but does not, however, provide an indication of how this difference is occurring. A follow-up check is therefore necessary to pinpoint how the suppliers are differing on a statistical basis.

5.2 Multiple Comparison Associated with the Kruskal–Wallis Test

The Kruskal–Wallis test, as with the ANOVA F test (3.1), only provides a statistical test of general treatment differences. To pinpoint how the treatments differ, we need to employ a non-parametric follow-up. Pairwise Mann–Whitney tests could be considered but testing all

Output 6.10 *Kruskal–Wallis test statistic derivation for Example 6.3*

Purity data in cells A1:D7, labels in cell 1 of each column and in cells D9:D11 and D13:D15.

In cell **E2**, array-enter (hold down Ctrl + Shift as Enter is pressed) the formula **=RANK(A2,$A$1:$D$7,1)+(SUM(1*(A2=$A$1:$D$7))–1)/2**. Copy and paste this formula into cells E3:E7, F2:F7, G2:G7, and H2:H7 to use expression (6.1) to provide the ranks of each data value.

In cell **E9**, enter **=COUNT(E2:E7)**. Copy and paste this formula into cells F9, G9, and H9 to produce *n*.

In cell **E10**, enter **=SUM(E2:E7)**. Copy and paste this formula into cells F10, G10, and H10 to provide the rank totals C_j for each treatment.

In cell **E11**, enter **=AVERAGE(E2:E7)**. Copy and paste this formula into cells F11, G11, and H11 to generate the average ranks for each treatment.

In cell **E14**, enter **=COUNT(A1:D7)**. In cell **E15**, enter **=(12/ (E14*(E14+1)))*SUMSQ(E10:H10)/E9 – 3*(E14+1)** to generate the Kruskal–Wallis test statistic (6.7).

Suppl A	Suppl B	Suppl C	Suppl D	Rnks A	Ranks B	Rnks C	Rnks D
99.3	99.8	98.2	99.6	18	24	6	22
99.4	97.4	97.2	98.7	20	2.5	1	9
98.8	98.9	97.4	99.2	10.5	12.5	2.5	16.5
99.4	99	98.3	99.2	20	14	7	16.5
98.9	98.6	97.7	99.4	12.5	8	5	20
99.1	98.8	97.5	99.7	15	10.5	4	23
			n	6	6	6	6
			Sum Rnks	96	71.5	25.5	107
			Ave Rnks	16	11.92	4.25	17.83
			Test Stat				
			N	24			
			H	**13.09**			

possible treatment pairings this way could increase the probability of a Type I error, *i.e.* raise the probability of rejecting a true null hypothesis. It is best to use a tailor-made non-parametric multiple comparison for the same reasons as were discussed when describing the background to ANOVA based multiple comparisons (see Chapter 3, Section 3.2).

Such a procedure is available and is called *Dunn's procedure*[4] which, like the SNK procedure, provides a mechanism for treatment comparison which ensures that the experimentwise error rate (overall

[4] O.J. Dunn, *Technometrics*, 1964, **6**, 241.

significance level) for all pairwise tests is constrained to $100\alpha\%$. Currently, this follow-up is not available in Excel and so it requires manual derivation. An outline of the calculation and decision elements of this non-parametric follow-up is provided in Box 6.4.

Box 6.4 *Dunn's non-parametric multiple comparison*

Hypotheses. The null hypothesis to be tested in this multiple comparison is H_0: no difference between compared treatments (medians similar). The alternative specifies difference between the compared treatments.

Test statistics. The numerical difference between the average ranks for each pair of treatments become the test statistics for the pairwise comparisons within this follow-up procedure.

Critical value. The critical value necessary for comparison with the difference test statistics is expressed as

$$W_{ij} = z_{\alpha/[k(k-1)]}\sqrt{\frac{N(N+1)}{12}\left(\frac{2}{n}\right)} \qquad (6.9)$$

where $100\alpha\%$ is the experimentwise error rate (significance level), k is the number of treatments, z refers to the $100\alpha/[k(k-1)]\%$ critical z value from Table A.4, N is the total number of data measurements, and n is the number of observations per treatment.

Decision rule. If the numerical difference in average ranks for treatments i and j ($|\bar{R}_i - \bar{R}_j|$) is less than the specified critical value (6.9), we accept H_0 and declare that the compared treatments appear not to differ significantly.

In most CRD based experiments, the number of measurements associated with the compared treatments should, ideally, be the same. If the number of observations in the compared treatments differs, then the term $2/n$ in equation (6.9) can be modified to $(1/n_i + 1/n_j)$ to accommodate the differing numbers of observations associated with treatments i and j. Adjustment of the critical value (6.9), if there are large numbers of tied observations, can be considered but the effect on the value produced is only minor. Choice of experimentwise error rate, $100\alpha\%$, is based partly on k, the number of treatments tested, and should be larger for large k. The general rule is to select $100\alpha\%$ larger than that customarily used in standard inference, *i.e.* use 10% ($\alpha = 0.1$), 15% ($\alpha = 0.15$), or even 20% ($\alpha = 0.2$).

Example 6.4

From Example 6.3, we know that the purity measurements differ statistically with respect to supplier but we do not know how this difference occurs. We need to implement a non-parametric multiple comparison to pinpoint how the suppliers are differing. We will do so using Dunn's procedure based on an experimentwise error rate of 10% ($\alpha = 0.1$). From the design structure, we know $n = 6$, $k = 4$, and $N = 24$.

Hypotheses
The null hypothesis we are testing in all pairwise comparisons is H_0: no difference in median purity between the suppliers compared.

Test statistics
The average ranks for the purity measurements associated with each supplier are available in Output 6.10 (row 'Ave Rnks') and are summarised in Table 6.5. This table also provides the test statistics, $|\bar{R}_i - \bar{R}_j|$, for each pairwise supplier comparison.

Table 6.5 *Dunn's procedure test statistics for Example 6.4*

			C	Supplier B	A	D
		Average rank	*4.25*	*11.92*	*16*	*17.83*
	C	*4.25*	–	7.67	11.75 *	13.58 *
Supplier	B	*11.92*		–	4.08	5.91
	A	*16*			–	1.83
	D	*17.83*				–

*, difference statistically significant at 10% level

Critical value
With $\alpha = 0.1$ and $k = 4$, we have $\alpha/[k(k-1)] = 0.0083$ with Table A.4 providing $z_{0.0083}$ as 2.3954. Given that $n = 6$ for all suppliers and $N = 24$, the requisite critical value (6.9) for all supplier comparisons is,

$$W_{ij} = (2.3954)\sqrt{\frac{24(24+1)}{12}\left(\frac{2}{6}\right)} = 9.78$$

From the average rank differences in Table 6.5, we can see that the difference between suppliers C and A, and suppliers C and D are statistically significant based on the associated difference exceeding the critical value of 9.78. A symbol '*' (significant at the 10% level) is inserted in Table 6.5 at the corresponding difference to indicate the detection of a statistically significant difference ($p < 0.1$). No other supplier differences are deemed statistically significant. The major differences appear, therefore, to be between supplier C and the rest as hinted at in the exploratory analysis discussed in Example 6.2.

5.3 Linear Contrasts

Frequently, we may also require to test a planned comparison (contrast) of the treatments as illustrated in Chapter 3, Section 3.3 for ANOVA-based procedures. An equivalent non-parametric contrast exists and is expressed in terms of the k mean ranges as

$$L = c_1 \bar{R}_1 + c_2 \bar{R}_2 + \ldots + c + k \bar{R}_k$$

based on the constraint that the constants c_j add to 0, *i.e.* $\sum c_j = 0$.[5] A $100(1 - \alpha)\%$ confidence interval estimate for a non-parametric contrast, similar to the contrast confidence interval (3.3), is expressed as

$$\text{estimate of } L \pm \chi^2_{\alpha, k-1} \sqrt{\frac{N(N+1)}{12} \left(\sum_j \frac{c_j^2}{n} \right)} \qquad (6.10)$$

based on the previous definitions of k, N, c_j, and n.

Exercise 6.3

The data presented below refer to the percent recovery of propazine from spiked solutions containing 100 ng l^{-1} propazine. Solid phase extraction using GC thermionic specific detection was used with varying levels of pH adjustment. Does level of pH adjustment affect propazine recovery?

pH adjustment	none	pH 5	pH > 8
	89.5	86.5	82.1
	91.3	88.2	79.3
	85.1	84.1	85.4
	92.4	87.3	84.1
	87.6	89.2	81.2
	90.8	90.2	83.7

[5] J.H. Zar, 'Biostatistical Analysis', 3rd Edn., Prentice Hall International, Upper Saddle River, New Jersey, 1996, p. 229.

6 INFERENCE FOR AN RBD BASED EXPERIMENT

As with CRD-based experimentation, it may be necessary in an RBD-based experiment to replace ANOVA as the inferential tool if normality for the response cannot be safely assumed. As such, we would need to utilise non-parametric approaches to replace the ANOVA F test (3.8) and SNK multiple comparison (3.2) as the necessary inferential base of the data analysis.

6.1 The Friedman Test of Treatment Differences

For an RBD-based experiment, the non-parametric alternative to the treatment F test (3.8) is provided by the *Friedman test*. The test is again based on data ranking with ranks replacing the recorded measurements in the test statistic calculation. A summary of the analysis components based on the Friedman test as the test of treatment difference is presented in Box 6.5.

Box 6.5 *Friedman test for treatment differences*

Assumptions. Data constitute k random samples of size n and there is no interaction between blocks and treatments.

Hypotheses. The hypotheses related to this test are as those of the Kruskal–Wallis test (see Box 6.3).

Exploratory data analysis (EDA). Exploratory analysis of the data is again advisable to gain initial insight into what they indicate in respect of the experimental objective.

Test statistic. To determine the test statistic, the first step requires the observations within each block to be ranked in ascending order of magnitude, smallest to largest. The resultant ranks associated with each of the k treatments are then summed to generate the rank totals C_1 to C_k. From these, we use

$$S = \frac{12}{kn(k+1)} \sum_{j=1}^{k} C_j^2 - 3n(k+1) \qquad (6.11)$$

to calculate the Friedman test statistic where k is the number of treatments and n is the number of blocks.

Decision rule. This test statistic, as with the Kruskal–Wallis test statistic (6.7), approximately follows a chi-square distribution with $(k-1)$ degrees of freedom meaning Table A.2 can be used to provide the necessary critical values. Again, low values of the test statistic ($S <$ critical) or high p values (p value > significance level) will imply acceptance of the null hypothesis H_0 and the conclusion that the treatments do not appear to differ on a statistical basis.

Again, test statistic (6.11) only provides a test of general difference with a follow-up necessary if significant treatment differences are detected. As with other non-parametric procedures, test statistic (6.11) can be adjusted to account for tied observations though adjustment is unnecessary if the value of S is sufficient to reject the null hypothesis and suggest existence of significant treatment differences.

Example 6.5

An experiment was carried out to investigate the reaction time of a chemical process using five different ingredients. It is thought ingredients may differ in their effect according to batch of raw material tested, so the experimenter blocked on the batch factor and conducted the experiment using an RBD structure. The reaction time data collected are presented in Table 6.6. Do the ingredients used affect reaction time?

Table 6.6 *Reaction time data for Example 6.5*

		Ingredient				
		A	B	C	D	E
	1	8	7	7	1	3
	2	7	8	11	3	2
Batch	*3*	9	4	10	5	1
	4	10	6	8	6	6
	5	8	3	8	2	4

Source: D.C. Montgomery, 'Design and Analysis of Experiments', 4th Edn., Wiley, New York, 1997: reproduced with the permission of John Wiley & Sons, Inc.

Hypotheses
The null hypothesis to be tested is H_0: no difference in median reaction time with ingredient.

Exploratory data analysis
From the data plot presented in Output 6.11, we can see that three groupings appear to be emerging, these being (A and C), B, and (D and E). A and C have longest reaction times and D and E shortest, suggesting a wide difference between these two groups. The range of results looks relatively similar for all ingredients.

The summary statistics in Output 6.12 produce the same picture in respect of groupings of the medians. The *RSD*s for the five ingredients (13.6%, 37%, 18.7%, 61%, 60.1%) differ markedly suggesting significant variation in the consistency of the measured reaction times. Initially, it would appear that ingredient differences exist and such differences could be statistically important.

Output 6.11 *Data plot of reaction time measurements for Example 6.5*

Reaction time data are in cells A1:F6, labels in cell 1 of each column and block information in column A. Menu commands as Output 2.2.

Plot of Reaction Time Against Ingredient

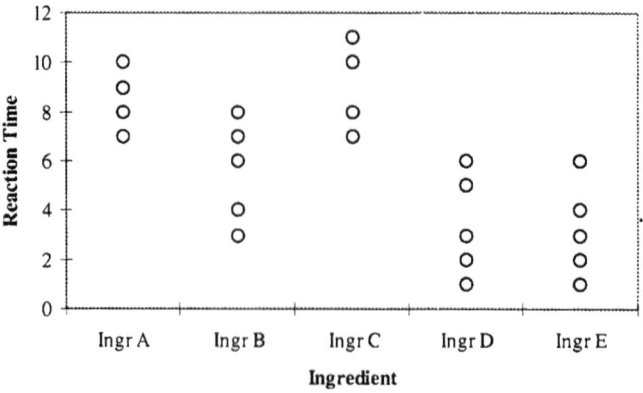

Output 6.12 *Reaction time summaries for Example 6.5*

Reaction time data are in cells A1:F6, labels in cell 1 of each column and block information in column A. Menu commands as Output 2.4.

	Ingr A	Ingr B	Ingr C	Ingr D	Ingr E
Mean	8.4	5.6	8.8	3.4	3.2
Standard Error	0.510	0.927	0.735	0.927	0.860
Median	8	6	8	3	3
Standard Deviation	1.140	2.074	1.643	2.074	1.924
Sample Variance	1.3	4.3	2.7	4.3	3.7
Kurtosis	−0.18	−1.96	−1.69	−1.96	−0.02
Skewness	0.40	−0.24	0.52	0.24	0.59
Range	3	5	4	5	5
Minimum	7	3	7	1	1
Maximum	10	8	11	6	6
Sum	42	28	44	17	16
Count	5	5	5	5	5
Confidence Level (95%)	1.416	2.575	2.040	2.575	2.388

Output 6.13 *Friedman test statistic derivation for Example 6.5*

Reaction time data are in cells A1 to F6, labels in cell 1 of each column and in cells F8 to F13, and block information in column A.

In cell **G2**, array-enter (hold down Ctrl + Shift as Enter is pressed) the formula **=RANK(B2,$B2:$F2,1)+(SUM(1*(B2=$B2:$F2))−1)/2**. Copy and paste this formula into cells H2:K2, G3:K3, G4:K4, G5:K5, and G6:K6 to use expression (6.1) to rank the data by block for each treatment.

In cell **G8**, enter **=SUM(G2:G6)**. Copy and paste this formula into cells H8, I8, J8, and K8 to generate the rank totals C_j for each treatment.

In cell **G11**, enter **=COUNTA(B1:F1)** to generate k. In cell **G12**, enter **=COUNT(B2:B6)** to generate n. In cell **G13**, enter **=(12/(G11*G12*(G11+1)))* SUMSQ(G8:K8)−3*G12*(G11+1)** to generate the Friedman test statistic (6.11).

Batch	Ingr A	Ingr B	Ingr C	Ingr D	Ingr E	Rnks A	Rnks B	Rnks C	Rnks D	Rnks E
1	8	7	7	1	3	5	3.5	3.5	1	2
2	7	8	11	3	2	3	4	5	2	1
3	9	4	10	5	1	4	2	5	3	1
4	10	6	8	6	6	5	2	4	2	2
5	8	3	8	2	4	4.5	2	4.5	1	3
				Sum Rnks		21.5	13.5	22	9	9
				Test Stat						
				k		5				
				n		5				
				S		**13.24**				

Test statistic

Output 6.13 shows Excel generation of the test statistic (6.11) using the ranking modification process explained in Section 2. The total ranks C_j are provided as 21.5, 13.5, 22, 9, and 9 ('Sum Rnks') and with $k = 5$ ('k') and $n = 5$ ('n'), the test statistic (6.11) is calculated to be 13.24 ('S').

As $k = 5$, this test statistic must be compared against a critical value of $\chi^2_{0.05,4}$ (5% significance level) which, from Table A.2, is 9.49. Rejection of the null hypothesis is indicated as test statistic exceeds critical value. Therefore, it appears there is sufficient evidence within the reaction time data to indicate a difference between the ingredients ($p < 0.05$). This only provides information on general ingredient difference so we must again consider a follow-up to pinpoint the specific differences between the ingredients.

6.2 Multiple Comparison Associated with Friedman's Test

Common to all previous illustrations of design structure experimentation, the main test of inference can only provide information on general
treatment differences. Use of a follow-up is necessary to help understood more of how the treatment differences are occurring. The non-
parametric multiple comparison for use in association with the
Friedman test, described in Box 6.6, differs marginally in its operation
from Dunn's procedure (see Section 5.2) though the underlying philosophy and mechanism of application are similar. This form of non-
parametric follow-up[6] is not currently available in Excel and so requires
manual derivation.

Box 6.6 *Multiple comparison for use with Friedman test*

Hypotheses. The null hypothesis to be tested in this multiple comparison is
as previously, *i.e.* H_0: no difference between compared treatments
(medians similar).
Test statistics. The numerical difference between the rank totals for each
pair of treatments become the test statistics for the pairwise comparisons
of this follow-up procedure.
Critical value. The critical value for comparison with each test statistic is
specified as

$$W = z_{\alpha/[k(k-1)]} \sqrt{\frac{nk(k+1)}{6}} \qquad (6.12)$$

where n is the number of blocks in the design structure, and $100\alpha\%$, k,
and z are as defined for Dunn's procedure (see Box 6.4).
Decision rule. If the numerical difference in rank totals for treatment i and
j ($|R_i - R_j|$) is less than the critical value (6.12), we accept H_0 and declare
that the compared treatments appear not to be statistically different.

As with use of Dunn's procedure, choice of experimentwise error
rate, $100\alpha\%$, is important and largely depends on k. Again, we select a
value of $100\alpha\%$ larger than generally used in inferential data analysis
(10%, 15%, 20%) with larger values chosen for larger k.

Example 6.6

Example 6.5 showed that there was a significant difference in reaction
time with respect to ingredient. This specifies that ingredients differ but

---

[6] Reference 1, pp. 274–275.

does not indicate how. To pinpoint this, we will carry out the follow-up explained above based on an experimentwise error rate of 15% ($\alpha = 0.15$), choice of higher value due to the large number of ingredients being compared. We know that $k = 5$ and $n = 5$ from the form of experiment conducted.

Hypotheses
The null hypothesis to be tested in all pairwise comparisons is H_0: no difference in median reaction time with ingredient.

Test statistics
The rank totals for reaction time for each ingredient are displayed in Output 6.13 in the row 'Sum Rnks' and are summarised in Table 6.7. From these totals, we determine the test statistics, $|R_i - R_j|$, for each pairwise ingredient comparisons. These are also shown in Table 6.7.

Table 6.7 *Test statistics for multiple comparison of Example 6.6*

				Ingredient		
		Rank total	D, E 9	B 13.5	A 21.5	C 22
Ingredient	D, E	9	–	4.5	12.5*	13.0*
	B	13.5		–	8.0	8.5
	A	21.5			–	0.5
	C	22				–

*, difference statistically significant at 15% level

Critical value
With $\alpha = 0.15$ and $k = 5$, we have $\alpha/[k(k-1)] = 0.0075$. From Table A.4, we find $z_{0.0075}$ to be 2.4324. Given that $n = 5$, the necessary critical value (6.12) for the ingredient comparisons can be evaluated as

$$W = (2,4324)\sqrt{\frac{5(5)(5+1)}{6}} = 12.162$$

From the rank total differences in Table 6.7, we can deduce that ingredients D and E differ significantly from A and C (test statistic > critical value, $p < 0.15$). No other differences appear statistically important. This result conforms to our initial impressions of the ingredient differences in respect of the two groups mentioned in Example 6.5.

Exercise 6.4

A sample of fertiliser, taken from each of six production batches, is divided into three equal parts and analysed for percent potash by three different analysts. Analyst A has not used the associated analytical procedure before. Is there a bias between analysts?

Analyst		A	B	C
	1	15.1	14.7	15.2
	2	15.3	15.6	15.5
Batch	*3*	14.9	15.4	15.3
	4	15.5	15.6	15.6
	5	15.4	14.8	15.2
	6	14.8	15.1	15.2

Source: C.J. Brookes, I.G. Betteley and S.M. Loxston, 'Fundamentals of Mathematics and Statistics for Students of Chemistry and Allied Subjects', Wiley, Chichester, 1979: reproduced with the permission of John Wiley & Sons, Ltd.

7 INFERENCE ASSOCIATED WITH A TWO FACTOR FACTORIAL DESIGN

As with one factor designs, assuming normality for data collected through a factorial design structure may not be appropriate mitigating against the use of the ANOVA-based F tests (4.1) to (4.3) for assessing the statistical importance of each factorial effect. Access to a non-parametric alternative would therefore be useful to enable such data to be objectively assessed. Such a procedure exists[7] based on an extension of the Kruskal–Wallis procedure using ranks of the cell measurements for each treatment combination. The other analysis components illustrated in Chapter 4 for such designs, with the exception of diagnostic checking, can still be used as the base of the data analysis, the only change being the form of inferential procedure used to assess general factorial effect.

For a two factor factorial design with a levels of factor A, b levels of factor B, and n replications of each treatment combination, we again have three possible tests: factor A, factor B, and interaction $A \times B$. The non-parametric equivalents to test statistics (4.1), (4.2), and (4.3) are as follows, with order of testing as indicated in Chapter 4, Section 2.4, *i.e.* interaction test *first*:

---

[7] C. Barnard, F. Gilbert and P. McGregor, 'Asking Questions in Biology', Longman, Harlow, Essex, 1993, pp. 130–135.

$$\text{factor } A: \quad H_A = \frac{12}{N(N+1)} \sum_{i=1}^{a} \frac{R_{i.}^2}{bn} - 3(N+1) \tag{6.13}$$

$$\text{factor } A: \quad H_B = \frac{12}{N(N+1)} \sum_{j=1}^{b} \frac{R_{.j}^2}{an} - 3(N+1) \tag{6.14}$$

$$\text{interaction } A \times B: \quad H_{A \times B} = \frac{12}{N(N+1)} \sum_{i=1}^{a} \sum_{j=1}^{b} \frac{R_{ij}^2}{n} - 3(N+1) - H_A - H_B \tag{6.15}$$

where $N = abn$ is the total number of observations, R_{ij} is the sum of the ranks for the combination of factor A level i and factor B level j, $R_{i.} = \sum_{j=1}^{b} R_{ij}$ is the sum of the ranks for the observations for factor A level i, and $R_{.j} = \sum_{i=1}^{a} R_{ij}$ is the sum of the ranks for the observations for factor B level j. All tests follow approximate χ^2 distributions with degrees of freedom $(a-1)$ for factor A, $(b-1)$ for factor B, and $(a-1)(b-1)$ for the interaction $A \times B$. Again, these tests represent only general tests of factor and interaction effects with appropriate follow-up necessary to enhance the conclusion in respect of interaction or factor–level differences, whichever is appropriate.

8 LINEAR REGRESSION

In Chapter 5, linear regression modelling was discussed based on the use of least squares estimation of model parameters. Often, it may be that the statistical assumptions underpinning use of this procedure (normality of response, equal variability) are not valid for a particular set of data and so use of an alternative, and less assumption dependent, procedure may be necessary. Two such non-parametric methods are *Thiel's method* and the *Brown–Mood method*.[8]

The former involves, for linear modelling, evaluating all possible sample slopes

$$S_{ij} = (Y_j - Y_i)/(X_j - X_i), i < j \tag{6.16}$$

for each possible (X, Y) pairing and using the median of these estimates as the slope estimate. Using this estimate, a series of intercept estimates are produced with, again, the median estimate chosen as the estimate of intercept. Statistical validity of a non-parametric linear regression

[8] Reference 1, pp. 427–441.

equation can be checked through the use of non-parametric inference procedures related to the parameter estimation procedure used.

9 TESTING THE NORMALITY OF CHEMICAL EXPERIMENTAL DATA

Often, when dealing with chemical data, it may be necessary to check whether such data can be assumed to be normally distributed. This can be particularly important if we need to justify the use of parametric inference for data assessment. For example, we could be collecting data on the concentration of a toxic chemical in water samples or the yield of a chemical reaction. In either case, it may be necessary to check that the responses collected are approximately normally distributed in order to provide justification for using parametric inference to analyse the collected data objectively. Simple data plots and formal inference tests can be considered.

9.1 Normal Probability Plot

A *normal probability plot* provides a simple visual means of assessing the normality of recorded experimental data. The plot is a simple X–Y plot of the normal scores of the recorded data against the ordered data measurements where normal scores refer to data we would expect to obtain if a sample of the same size was selected from a standard normal distribution (normal, mean 0, standard deviation 1). If the plot exhibits a reasonable positive linear trend, then normality of the measured data can be assumed. Non-linear patterns will tend to suggest non-normality, *e.g.* skew type data, indicating a need to consider transforming the data to enable normality to be more readily acceptable (see Chapter 3, Section 9). Fitting a line through the points in the plot and taking the inverse of the slope can provide an estimate of the data standard deviation.

In Excel, there is no default mechanism for producing a normal plot of data. However, use of an approximation method[9] and Excel's inherent commands can enable such a plot to be produced. The first step in this procedure involves ranking the data in order of magnitude and using the ranks to determine an approximate standard normal probability

$$p_i = (i - 3/8)/(n + 1/4) \tag{6.17}$$

[9] T.A. Ryan and B.L. Joiner, Technical Report Minitab Inc., 1990, 1.3–1.4.

for the *i*th ordered observation in a sample of *n* measurements. Using this approximation, we can estimate the corresponding normal scores (p_ith percentage point of the standard normal distribution) as

$$normal\ score = 4.91[p_i^{0.14} - (1 - p_i)^{0.14}] \tag{6.18}$$

A plot of ordered observations against normal scores (6.18) can now be produced and assessed as Example 6.7 will demonstrate.

Example 6.7

As part of a quality assurance study, the data presented in Table 6.8 were collected. They refer to the measurement, in g, of $FeSO_4$ in iron tablets using the official BPC method. Can these data be assumed normally distributed?

Table 6.8 *Recorded $FeSO_4$ measurements for Example 6.7*

0.1685	0.1845	0.1823	0.1713	0.1705	0.1802	0.1705	0.1778	0.1853	0.1811
0.1762	0.1835	0.1732	0.1812	0.1769	0.1821	0.1769	0.1821	0.1816	0.1864
0.1749	0.1749	0.1728	0.1779	0.1790	0.1740	0.1790	0.1777	0.1777	0.1747
0.1790	0.1777	0.1735	0.1747	0.1790	0.1735	0.1824	0.1832	0.1773	0.1812

Summaries
The presented data have mean 0.1779, median 0.1778, and standard deviation 0.0044. The similarity of mean and median indicate potential symmetry of data suggesting that normality of $FeSO_4$ measurements may be acceptable.

Normal plot
Output 6.14 contains an Excel generated normal plot for the collected $FeSO_4$ data containing also a reference line corresponding to the *z* scores, expression (2.6), for the normal distribution which the data are being compared against (normal, mean 0.1779, standard deviation 0.0044). There appears to be a degree of clustering in mid-range with only a few low and high measurements. Despite these patterns, the trend in the points looks reasonably linear compared to the *z* score line providing evidence that it would appear reasonably safe to assume normality for the recorded $FeSO_4$ data.

In Minitab, the menu commands **Graph** ▷ **Normal Plot** provide access to the %Normplot macro for production of a normal plot. The *y* axis of the plot corresponds to the logarithm of the calculated normal

Output 6.14 *Normal probability plot of FeSO₄ data of Example 6.7*

$FeSO_4$ data are in cells A1:A41, labels in cell 1 of each column. First sort the data in order of magnitude using **Data ▷ Sort**.

In cell **B2**, array-enter (hold down Ctrl + Shift as Enter is pressed) the formula **=RANK(A2,A\$2:A\$41,1)+(SUM(1*(A2=A\$2:A\$41))−1)/2**. Copy and paste this formula into cells B3:B41 to generate the ranks of the ordered data through use of expression (6.1).

In cell **C2**, enter **=(B2-0.375)/(COUNT(A\$2:A\$41)+0.25)**. Copy and paste this formula into cells C3:C41 to generate the approximate standard normal probabilities (6.17) for the ordered data.

In cell **D2**, enter **=4.91*(C2^0.14−(1−C2)^0.14)**. Copy and paste this formula into cells D3:D41 to generate the normal scores (6.18) of the ordered data.

In cell **E2**, enter **=(A2−0.1779)/0.0044**. Copy and paste this formula into cells E3:E41 to generate the z scores, expression (2.4), for the normal distribution (mean 0.1779, standard deviation 0.004) the $FeSO_4$ data are to be compared against.

Plot the normal scores (D1:D41) and z scores (E1:E41) against the ordered $FeSO_4$ data (A1:A41) using the menu commands of Output 5.1 choosing Chart type 2 in Step 3 of ChartWizard and plotting only the line through the z scores.

Normal Plot of FeSO4 Data

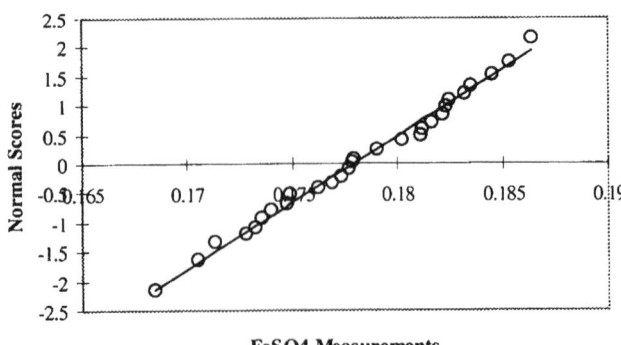

FeSO4 Measurements

probabilities and the x axis to the ordered observations. The plot resembles that which would be produced when using normal probability paper. A linear reference line, that estimates the cumulative normal distribution function for the population from which the data are drawn, is also provided to aid the decision making.[10]

---

[10] Minitab Reference Manual Release 10Xtra, Minitab Inc., 1995, pp. 19.44–19.46.

A normal probability plot is one graphical tool for checking data normality. Other plots such as *stem-and-leaf, histogram, quantile–quantile plot (Q–Q)*, and *rankit plot* could equally be used. Stem-and-leaf and histogram are most appropriate for large data sets with data exhibiting a symmetrical pattern, comparable to that shown in Figure 2.6, suggesting normality for response data. For *Q–Q* and rankit plots, linearity of trend again provides the evidence base for normality, as in the normal plot.

Exercise 6.5

A chemical engineer collected the following data on the percentage of toxic impurities removed by a waste treatment process. Is it valid to assume these data to be normally distributed?

40	36	49	47	55	35	42	51	49	60
42	38	53	51	62	48	39	53	52	63
50	37	52	50	59	51	40	50	51	61

Source: J.S. Milton and J.C. Arnold, 'Introduction to Probability and Statistics', 2nd Edn., McGraw-Hill, New York, 1990: reproduced with the permission of The McGraw-Hill Companies.

9.2 Statistical Tests for Assessing Normality

The use of a normal probability plot provides a simple and visual means of assessing normality of experimental data. However, such an assessment is subjective and it is also often useful to include a more objective analysis through adoption of a formal statistical test of normality. Several such tests exist including *Lilliefors test*, the *Ryan–Joiner test*, and the *Anderson–Darling test*. All are based on testing a null hypothesis that the experimental data conform to normality.

Lilliefors test[11] is generally applied when the mean μ and/or variance σ^2 of the normal distribution which the data are to be tested against are unknown. It represents a modification of the *Kolmogorov–Smirnov goodness-of-fit* procedure based on tables of critical values specially constructed for the task. It is based on comparing cumulative distributions for the sample data and a fitted normal with small deviations implying normality acceptable (small test statistic) and large deviations normality unacceptable.

The Ryan–Joiner test[12] is similar to the *Shapiro–Wilks test* of

[11] L.H. Lilliefors, *J. Am. Stat. Assoc.*, 1967, **62**, 399.
[12] Reference 9, pp. 1.5–1.7.

normality. It is based on the principle underlying the normal probability plot that linearity is indicative of normality and that an objective measure of linearity can be found in the correlation coefficient, values near one numerically tending to imply linearity and thus, normality of response data. The formal decision on acceptance or rejection of normality is based on a modified decision rule from those outlined in this and previous chapters. Essentially, if the estimate of correlation exceeds the associated critical value, we accept normality.

The Anderson–Darling test[13] is based on ordering the data and using the z scores of the normal distribution which the data are being compared against. Small values of the test statistic are commensurate with normality being acceptable.

All three of these normality test procedures are available in Minitab through the **Graph ▷ Normal Plot** menu commands. However, many software-based procedures only allow normality testing of data using the sample mean and sample standard deviation as estimates of the unknown mean μ and unknown standard deviation σ for the normal distribution against which the data are to be tested. Often we may wish to assess normality of data with respect to pre-set mean and/or standard deviation values. In such cases, we may require to develop a procedure using software commands to assess data normality correctly as the in-built procedures may not allow over-riding of defaults.

---

[13] M. Meloun, J. Militky and M. Forina, 'Chemometrics for Analytical Chemistry Volume 1: PC-Aided Statistical Analysis', Ellis Horwood, Chichester, 1994, pp. 80–81.

CHAPTER 7

Two-level Factorial Designs

1 INTRODUCTION

Factorial experimentation is a powerful technique for scientific and technological research. However, when many factors require to be assessed, classical factorial designs, based on multiple replication of many treatment combinations, can be inefficient and costly to implement.[1] *Two-level factorial designs* represent a particular type of factorial structure which are of great benefit in multi-factor experimentation through the reduction in factor levels tested, and through their ability to estimate the effects of all factors and their interactions independently. Such structures provide good estimation precision enabling information on a chemical phenomenon to be obtained easily and efficiently. Two-level designs are extremely useful for exploratory purposes (screening) within product and process development, and can also play a useful role within process optimisation, *e.g.* optimisation of chemical instruments.

These factorial designs are generally used to study the effect on a response of k factors, each set at two specified levels, based on the premise of linearity of response across the factor levels. The levels tested are often denoted as 'low' and 'high' to represent extreme levels for each factor, *e.g.* the absence or presence of a catalyst, a temperature of 40 °C or 50 °C, and a pH of 5.5 or 8.5. Each level of every factor is tested with each level of every other factor resulting in a total of 2^k treatment combinations and giving rise to all treatment combinations being tested. Further reduction in experimental effort is attained by only carrying out one experiment for each treatment combination (single replicate) though replication may be necessary if repeatability and reproducibility elements are also to be considered.

To illustrate the operation of such a design, consider a two-level factorial design in three factors A, B, and C. In this experiment, there are $2 \times 2 \times 2 = 2^3 = 8$ treatment combinations giving rise to the design

---

[1] L. Davies, 'Efficiency in Research, Development and Production: The Statistical Analysis of Chemical Experiments', The Royal Society of Chemistry, Cambridge, 1993.

matrix of Table 7.1 where a single replicate response is measured for each treatment combination. The minus sign $(-)$ denotes that the factor has been set at the 'low' level and the plus $(+)$ sign that the factor has been set at the 'high' level. It can be seen in Table 7.1 that all combinations of the factors and their levels are included in the experimental structure meaning that the full experimental region has been covered. In addition, each factor–level, or combination of levels, are experimented on an equal number of times. In this illustration, this corresponds to four times $(4-$ and $4+$ in each column).

Table 7.1 2^3 *factorial design matrix*

Combination	A	B	C	Response
(1)	−	−	−	x
a	+	−	−	x
b	−	+	−	x
ab	+	+	−	x
c	−	−	+	x
ac	+	−	+	x
bc	−	+	+	x
abc	+	+	+	x

x, *denotes a response measurement*

The 'Combination' column in Table 7.1 specifies the eight treatment combinations tested using a standard form of notation for two-level designs. Such combinations would be tested in a random order, if feasible though there may be a necessity to run the two-level factorial experiment on the basis of trend-free run orders which negate the need for randomising run order.[2,3] The notation is based on denoting the higher levels of the factors A, B, and C by the lowercase letters *a*, *b*, and *c*, respectively, while lower levels are denoted by the absence of the letter referring to the factor. Thus, the treatment combination *(1)* denotes all factors at low level while combination *a* denotes factor A high level and B and C both at low level. Combination *b* denotes factor A low level, B high and C low while the combination *ab* corresponds to factors A and B at high level and C low level. The other four treatment combinations, *c*, *ac*, *bc*, and *abc*, can be similarly defined.

All full two-level factorial designs are based on a group of properties which highlight the appropriateness of such designs. These properties are

---

[2] C.-S. Cheung and M. Jacroux, *J. Am. Statist. Ass.*, 1988, **83**, 1152.
[3] P.C. Wang and H.W. Jan, *The Statistician*, 1995, **44**, 379.

- every treatment combination is included and so the design points provide full coverage of the experimental region,
- the design is also perfectly balanced as every factor level, and factor-level combination, appears an equal number of times,
- all response measurements can be used to estimate the factor effects providing estimates of high precision,
- all main effects and interaction effects can be estimated independently.

2 CONTRASTS AND EFFECT ESTIMATION

Analysis of data from two-level factorial designs is based on the specification of *contrasts* and the *estimation of factor effects*. Contrasts essentially represent measures of the difference in the level of response caused by changing factor levels, whilst factor effect estimation involves determining the average effect on the response resulting from a change in the level of the factor or combination of levels. Both aspects will now be described.

2.1 Contrasts

Contrasts can be constructed for all factor effects. A *main effect contrast* is constructed as the difference between the responses at the high factor level and the responses at the low factor level, *i.e.* (high level − low level). The *contrast for a two factor interaction* is constructed as the difference between the responses when each factor is set at the same level and the responses when each factor is set at opposing levels. Such general definitions can be easily extended to higher order interactions.

To explain contrasts further, we will again use the 2^3 design in three factors A, B, and C. Let (1), *a*, *b*, *ab*, *c*, *ac*, *bc*, and *abc* represent the measured response of the eight treatment combinations. For factor A, the contrast would be

$$contrast_A = [a + ab + ac + abc - (1) - b - c - bc] \qquad (7.1)$$

where the responses *a*, *ab*, *ac*, and *abc* represent those for factor A at the high level and (1), *b*, *c*, and *bc* those for A at the low level. Contrasts for B and C can be similarly defined.

For the interaction A × B, the contrast can be summarised as the difference between the responses for factors A and B tested at the same level and the responses for A and B at opposite levels, *i.e.* (same − opposite). This leads to the A × B contrast being defined as ($A_L B_L$ + $A_H B_H - A_L B_H - A_H B_L$) where L and H refer to 'low' and 'high', respectively. Based on the standard notation, this contrast is expressed as

$$contrast_{A \times B} = [(1) + ab + c + abc - a - b - ac - bc] \qquad (7.2)$$

based on (1) and c representing $A_L B_L$, ab and abc representing $A_H B_H$, b and bc representing $A_L B_H$, and a and ac representing $A_H B_L$. This contrast is independent of the third factor C as it is essentially (low C + high C − low C − high C) so cancelling out the effect of C. The contrasts for the interaction A × C and B × C can be similarly defined.

The contrast for the three factor interaction A × B × C is a measure of the differences in a related two factor interaction at the two levels of the third factor. One possible way of finding this contrast is to take the difference in the interaction A × B at the two levels of C. This results in,

$$contrast_{A \times B \times C} = AB_{highC} - AB_{lowC} \qquad (7.3)$$

From the A × B contrast definition, we have

$$AB_{highC} = AB_{same} \text{ at high } C - AB_{opp} \text{ at high } C$$

while,

$$AB_{lowC} = AB_{same} \text{ at low } C - AB_{opp} \text{ at low } C$$

In the usual notation, these become

$$AB_{highC} = c + abc - ac - bc$$

and.

$$AB_{lowC} = (1) + ab - a - b$$

Combining these using equation (7.3) provides the contrast expression,

$$contrast_{A \times B \times C} = [a + b + c + abc - (1) - ab - ac - bc] \qquad (7.4)$$

These effect contrasts for a 2^3 design involving three factors can be succinctly summarised in tabular form as Table 7.2 shows.

As each contrast is expressed in terms of the treatment combinations tested, their contrast coefficients can be easily summarised in a simple table, Table 7.3. The coefficients in each column of this table sum to 0 for each factorial effect and multiplication of any pair of columns

Table 7.2 *Contrasts for a three factor two-level experiment*

Effect	Contrast
A	$a + ab + ac + abc - (1) - b - c - bc$
B	$b + ab + bc + abc - (1) - a - c - ac$
AB	$(1) + ab + c + abc - a - b - ac - bc$
C	$c + ac + bc + abc - (1) - a - b - ab$
AC	$(1) + b + ac + abc - a - ab - c - bc$
BC	$(1) + a + bc + abc - b - ab - c - ac$
ABC	$a + b + c + abc - (1) - ab - ac - bc$

Table 7.3 *Signs of contrasts for a three factor two-level factorial experiment*

Treatment combination	Mean	Factorial effect A	B	C	AB	AC	BC	ABC
(1)	+	−	−	−	+	+	+	−
a	+	+	−	−	−	−	+	+
b	+	−	+	−	−	+	−	+
ab	+	+	+	−	+	−	−	−
c	+	−	−	+	+	−	−	+
ac	+	+	−	+	−	+	−	−
bc	+	−	+	+	−	−	+	−
abc	+	+	+	+	+	+	+	+

provides another column in the table, *e.g.* multiplying columns A and B, row by row, produces column AB. Contrasts exhibiting these properties are called *orthogonal contrasts* resulting in each factorial effect being independently estimable. On the basis of this property, two-level designs are often referred to as *orthogonal designs* and, as each factor level and combination of levels appear the same number of times in the design structure, they also exhibit a *balancing property*.

2.2 Effect of Each Factor

Analysis of results from two-level designs is based on estimating and analysing the effect of each factor and factor interaction. An effect estimate is essentially an average measurement reflecting how changing levels affect average response. Using the contrasts specified in Table 7.2, effect estimates are generally expressed as

$$(effect\ contrast)/(2^{k-1}) \tag{7.5}$$

as each factor-level, or combination of factor-levels, is experimented on 2^{k-1} times. Estimating factor effects can be carried out using Yates's method[4] but, as with all previous chapters, software estimation will be the only approach considered.

Effect estimates can be either positive or negative with appropriate interpretation depending on the objective of the experimentation. For maximising a response, positive estimates would imply high level, or same level for interaction effects, with negative estimates suggesting low level best, or opposite levels for interaction effects. For response minimisation, estimate values result in opposite interpretation in respect of effects tested (positive $\Rightarrow$ low level or opposite levels for interaction effects, negative $\Rightarrow$ high level or same levels for interaction effects). In two-level experimentation, it is hoped that effects will break down into two distinct categories, important (active) and unimportant (inactive) effects. This is defined as *effect sparsity* where it is assumed only a few effects are active and have any real effects on the response. The former would reflect effects strongly influencing the response and the latter, those exerting little, or no influence, on the response measured.

Effect estimation also depends on the factor levels chosen for experimentation so choice of levels is important as it can influence the chemical importance of the associated controllable factors. However, it is generally not feasible to specify a *'target'* estimate for discerning important effects so interpretation is based solely on comparing estimates within themselves, generally on a multiplicative difference basis.

Most statistical software can handle two-level designs relatively easily using regression based approaches while the calculational procedures of such designs can be easily programmed into spreadsheets such as Excel. In Minitab, procedures for analysis of two-level designs are accessed through the menu commands **Stat ▷ DOE ▷ Fit Factorial Model**, the dialog window for which is shown in Figure 7.1. As usual, data highlighting and filling and checking of boxes is necessary before implementing the procedures.

Example 7.1

When separating phenols by reverse phase HPLC, several factors can influence the separation. A study was set up to examine which of three factors are important to the optimisation of the separation of eleven priority pollutant phenols on an isocratic HPLC (solvent composition cannot be altered during the chromatographic run). The chemical factors chosen to be studied were the concentration of acetic acid (AA), the proportion of methanol (methanol:water, MW), and the concentration of citric acid (CA).

---

[4] E. Morgan, 'Chemometrics: Experimental Design', ACOL Series, Wiley, Chichester, 1991, pp. 118–119.

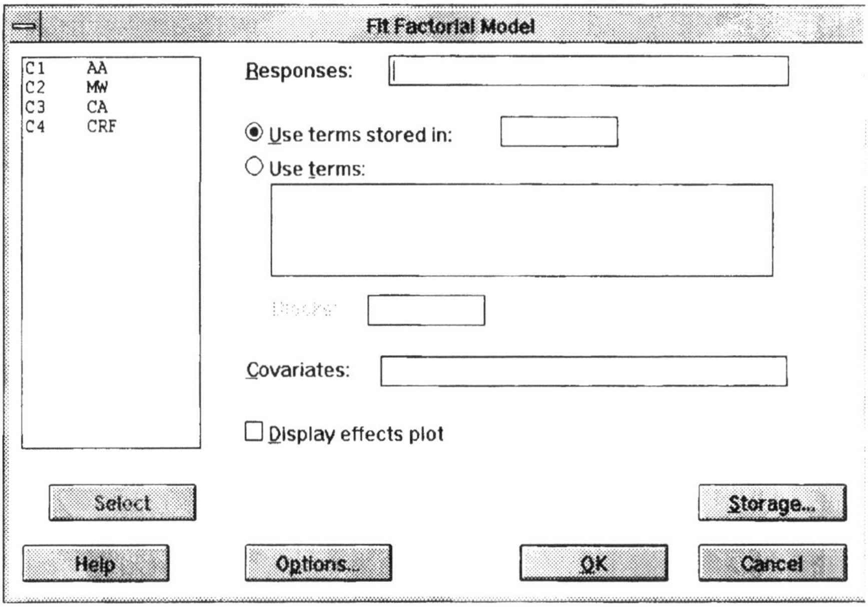

Figure 7.1 *Fit factorial model dialog window in Minitab*

Factors AA and CA are added to the mobile phase as they can reduce the degree of peak tailing, a severe phenomenon in the HPLC of phenols. The response chosen to be monitored was the chromatographic response function (CRF), a summation term of the individual resolutions between pairs of peaks. A high CRF will be obtained if the peaks are separated at the base-line and the degree of peak tailing is small. The design matrix and CRF responses obtained are presented in Table 7.4 based on each experimental run being conducted in a randomised order.

Data entry in Minitab
Data entry in Minitab involves entering the design matrix and response measurements into columns C1 to C4 of the Minitab spreadsheet (see Figure 1.4). For the design matrix, this entails entering in each of the first three columns a sequence of codes -1 ('low' level) and 1 ('high' level) to correspond to the two factor levels tested for each factor, based on the ordering shown in Table 7.4. Output 7.1 contains the Minitab output of effect estimates for the recorded CRF data based on using the Fit Factorial Model estimation procedure (see Figure 7.1). The output contains effect estimates (column 'Effect') and an incomplete ANOVA table, the latter due to single replication not providing sufficient data for error estimation (insufficient degrees of freedom).

Table 7.4 *Design matrix and CRF response data for Example 7.1*

Combination	AA (mol dm^{-3})	Factors MW (%)	CA (g dm^{-3})	Recorded CRF
(1)	0.004	70	2	10.0
a	0.01	70	2	9.5
b	0.004	80	2	11.0
ab	0.01	80	2	10.7
c	0.004	70	6	9.3
ac	0.01	70	6	8.8
bc	0.004	80	6	11.9
abc	0.01	80	6	11.7

Source: Reproduced from 'Chemometrics: Experimental Design' (E. Morgan) with the permission of the University of Greenwich.

Output 7.1 *Effect estimates for CRF measurements of Example 7.1*

Select **Stat** ▷ **DOE** ▷ **Fit Factorial Model** ▷ for *Responses*, select **CRF** and click **Select** ▷ select **Use terms**, select the empty box, and enter **(AA MW CA)3**. Select **Storage** ▷ for *Storage*, click **Effects** ▷click **OK** ▷click **OK**.

Fractional Factorial Fit

Estimated Effects and Coefficients for CRF

Term	Effect	Coef
Constant		10.3625
AA	−0.3750	−0.1875
MW	1.9250	0.9625
CA	0.1250	0.0625
AA*MW	0.1250	0.0625
AA*CA	0.0250	0.0125
MW*CA	0.8250	0.4125
AA*MW*CA	0.0250	0.0125

Analysis of Variance for CRF

Source	DF	Seq SS	Adj SS	Adj MS	F	P
Main Effects	3	7.72375	7.72375	2.57458	**	
2-Way Interactions	3	1.39375	1.39375	0.46458	**	
3-Way Interactions	1	0.00125	0.00125	0.00125	**	
Residual Error	0	0.00000	0.00000	0.00000		
Total	7	9.11875				

Effect estimate

To illustrate effect estimate derivation, consider factor AA. We know $k = 3$ so $2^{k-1} = 4$ represents the number of times each level of AA was experimented on. From the design matrix in Table 7.4, the contrast expression (7.1) and the effect estimate expression (7.5) produce the estimated effect of AA as,

$$[a + ab + ac + abc - (1) - b - c - bc]/4$$
$$= [9.5 + 10.7 + 8.8 + 11.7 - 10.0 - 11.0 - 9.3 - 11.9]/4 = -0.375$$

as printed in Output 7.1 (row 'AA', column 'Effect'). For the interaction AA × MW, the effect estimate using Table 7.4, expression (7.2), and expression (7.5), is

$$[(1) + ab + c + abc - a - b - ac - bc]/4$$
$$= [10.0 + 10.7 + 9.3 + 11.7 - 9.5 - 11.0 - 8.8 - 11.9]/4 = 0.125$$

as specified in Output 7.1 (row 'AA*MW', column 'Effect').

From the estimates obtained, we can see that two effects, MW and the interaction MWxCA, stand out from the rest numerically suggesting important effects. The estimate for AA also differs from the others but to a lesser extent, though it is three times the CA effect estimate.

3 INITIAL ANALYSIS COMPONENTS

Numerical assessment of effect estimates represents the first step in the data analysis for two-level designs. If the number of factors is large, however, this approach becomes impractical. Graphical presentations of the effect estimates then become the most appropriate way of presenting and assessing the estimate information.

3.1 Exploratory Data Analysis

Exploratory data analysis (EDA), through simple plots and summaries for each factor, can again be useful to gain initial insight into how the factors tested are affecting the response. Obviously, such an approach cannot provide interaction information but can provide a useful starting point for any analysis.

3.2 Effect Estimate Plots

This graphical form of analysis is particularly appropriate for two-level designs based on single replicates as it enables visual presentations of the effect estimates to be displayed to help ascertain which effects are important and which are unimportant. Several forms of graphical presentation can be adopted for this purpose including normal plot, half-normal plot, active contrast plot, pareto chart, and dotplot. Often, it is advisable to use at least two of these plots to assess effect estimates though all should produce comparable results in respect of important and unimportant effects.

The first, and most often used, is the *normal plot*. This corresponds to a plot of the normal scores of each effect estimate (as defined in Chapter 6, Section 9.1) against the estimates themselves. An 'ideal' normal plot should exhibit a distinct split between unimportant and important effects. Unimportant effects should cluster around zero and have corresponding effect estimate values such that a simple, almost vertical, straight line can be drawn through them linking them together. Important effects, by contrast, should appear at the bottom left (large negative) or top right (large positive) of the plot in positions very different from those occupied by the unimportant effects.

Figure 7.2 is based on a four factor experiment and illustrates an 'ideal' normal plot showing distinct separation of the important effects from those deemed unimportant. Only four effects appear important, C in the bottom left corner and D, A × C, and A top right. The remaining effects appear unimportant as they· cluster around 0 and can be approximately joined by a simple near vertical line. From such a plot, we would have evidence for the existence of effect sparsity specifying distinct effect groupings according to the importance of their influence on the measured response.

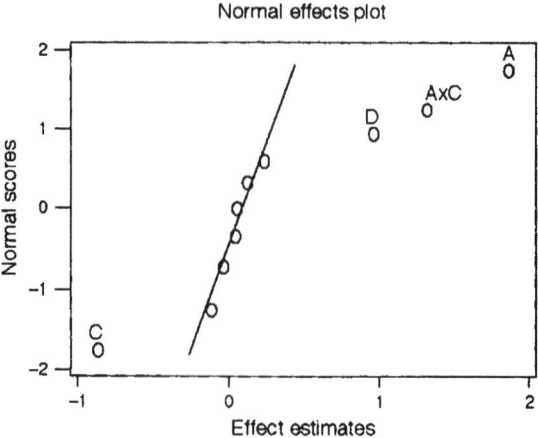

Figure 7.2 *Normal plot of effect estimates*

Example 7.2

Example 7.1 illustrated effect estimation for the CRF experiment outlined. A normal plot of these effect estimates is presented in Output 7.2 based on using a column of labels for plot derivation.

Output 7.2 *Normal plot of effect estimates for CRF measurements for Example 7.2*

Select **Calc** ▷ **Functions** ▷ select the **Input column** box, select **EFFE1** and click **Select** ▷ for *Results in*, enter **NSCORES** ▷ select **Normal scores** ▷ click **OK**.
Select **Graph** ▷ **Plot** ▷ for *Graph 1 Y*, select **NSCORES** and click **Select** ▷ for *Graph 1 X*, select **EFFE1** and click **Select**. Select **Frame** ▷ **Axis** ▷ select the **Label 1** box and enter **Effect estimates** in box ▷ select the **Label 2** box and enter **Normal scores** in box ▷ click **OK**. Select **Annotation** ▷ **Title** ▷ for *Title 1*, enter **CRF Effects plot** in box ▷ click **OK**. Remainder of plot produced using **Annotation** ▷ **Text** for the point labelling and **Annotation** ▷ **Line** for the line through unimportant effects.

CRF Effects plot

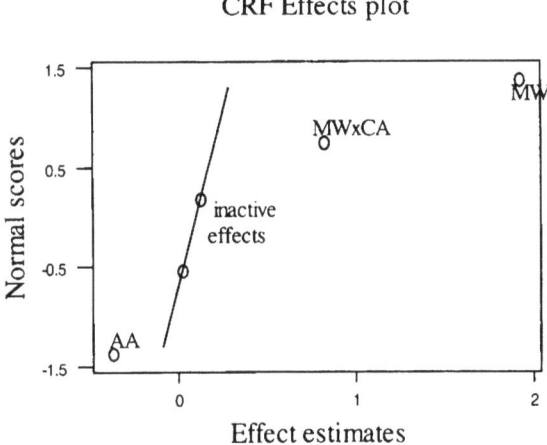

Normal plot of effect estimates
We can see that the normal plot in Output 7.2 exhibits an 'ideal' pattern with distinct separation of effects shown. The same three effects, as indicated in Example 7.1, stand out, these being MW, the interaction MW × CA, and AA, the latter the least of the three. MW and MW × CA have large positive estimates while AA is large negative.

For MW, the positive estimate suggests that as maximum CRF is optimal, a high proportion of methanol may be best, *i.e.* MW 80%. The positive estimate for the interaction MW × CA suggests that the same level for both factors may be best for maximum CRF, *i.e.* both low or both high. This suggests that either of the combinations (MW 70%, CA 2 g dm^{-3}) or (MW 80%, CA 6 g dm^{-3}) may be best. AA is the only important effect providing a negative estimate suggestive that low level of AA may be best, *i.e.* AA 0.004 mol dm^{-3}.

The *half-normal plot*, developed by Cuthbert Daniel,[5] uses the absolute values of the effect estimates to provide a picture of the relative size of effects. The plot is similar to the normal plot except that the lower left of the graph (*negative* estimates) is flipped over with important effects ideally showing up as points appearing in the top right corner, deviating markedly from the straight line through the unimportant effects. Guardrail values for half-normal plots enabling statistical decisions to be made as to which effects are statistically significant in their influence on the response and which not also exist. Often, however, analysis based on this technique can produce a misleading picture of effect importance.[6,7]

The *active contrast plot*[8] is based on plotting the probability of obtaining the estimated factor effects under the assumption of effect sparsity (only a few effects are active) and is displayed in a form similar to a horizontal histogram. Construction of the plot depends on the prior probability of an effect being important, as defined by the experimenter and a scale factor describing how much larger effect estimates for 'real' factors are compared with unimportant factors.

A *pareto chart*, illustrated in Figure 7.3 for a four factor experiment, is similar in form to the active contrast plot and is particularly useful when factor effects exhibit effect sparsity. Important effects will lie in the top right corner of the plot, as illustrated, with a distinct 'elbow' effect apparent at the jump from unimportant to important effects.

A fifth graphical presentation of effect estimates is that of a *dotplot*, similar to those used in previous chapters for exploratory data analysis. In this case, a dotplot of effect estimates simply represents a projection of the points in the normal plot on to the *x*-axis. Important effects will lie at the ends of the plot while unimportant effects will tend to cluster around the zero point.

Example 7.3

Output 7.3 contains a Minitab generated dotplot of effect estimates for the CRF experiment of Example 7.1.

[5] C. Daniel, *Technometrics*, 1959, **1**, 311.
[6] C.K. Bayne and I.B. Rubin, 'Practical Experimental Design and Optimisation Methods for Chemists', VCH Publishers Inc., Deerfield Beach, Florida, 1986, pp. 96–103.
[7] G.A.R. Taylor, *The Statistician*, 1994, **43**, 529.
[8] P.D. Haaland, 'Experimental Design in Biotechnology', Marcel Dekker Inc., New York, 1989, pp. 66–68.

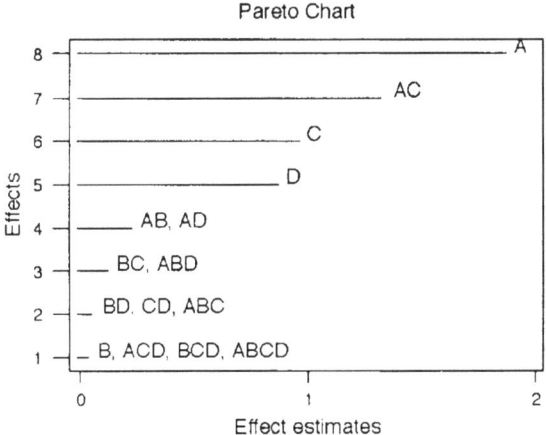

Figure 7.3 *Pareto chart of effect estimates*

Output 7.3 *Dotplot of effect estimates for CRF measurements for Example 7.3*

Select **Graph** ▷ **Character Graphs** ▷ **Dotplot** ▷ for *Variables* select **EFFE1** and click **Select** ▷ click **OK**.

Character Dotplot

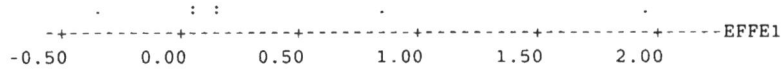

Dotplot of effect estimates
The plot presented conforms to the hoped for pattern just as the normal plot did in Example 7.2. Four effects, the unimportant effects, cluster around the zero point while three effects differ markedly from these representing the important (active) effects. These effects are, reading from left to right, AA, MW × CA, and MW.

3.3 Data Plots and Summaries

Plot and summary analyses of important effect estimates provide the final part of the initial data analysis for two-level designs. They provide a means whereby we can understand the active effects and explain how they affect the measured response.

For important main effects, we plot the average response, in conjunction with response range, for each level while for two factor interactions, interaction plots and summaries should be used. Ideally, we want to use these mechanisms to decide on an optimal combination of factors corresponding to the 'best' results obtained, maximum or minimum response with good consistency. As in all experimentation, however, this goal may not always be attainable and it may be necessary to trade-off to decide on the 'best' results.

Example 7.4

Effect estimate analysis for the CRF experiment in Example 7.1 indicated that effects MW, MW × CA, and AA appeared to be the most important. To understand how these effects affect CRF, we should examine plots and summaries, and describe the chemical implications of them. Such summary information is displayed in Output 7.4 for the main effects and Output 7.5 for the highlighted interaction.

Main effects
The main effects plot indicates that the major change in CRF occurs when the proportion of methanol (MW) increases from 70% to 80%. A slight decrease in CRF is highlighted as concentration of acetic acid (AA) increases while no real change appears to occur for changes to the concentration of citric acid (CA). The summary statistics highlight that changing methanol levels appears to increase CRF by about two units, on average. The effects of changes in the levels of the other factors on mean CRF are less marked.

Variability differences, through assessment of the range (maximum − minimum), only affect citric acid with the high level (6 g dm^{-3}) giving rise to more variable CRF measurements than the low level (2 g dm^{-3}). Overall, it would appear that, for maximum CRF, AA 0.004 mol dm^{-3}, MW 80%, and CA 6 g dm^{-3} could be the best treatment combination, though 2 g dm^{-3} for CA may be better in order to achieve greater consistency of CRF response measurement.

Methanol and citric acid interaction
From the plot in Output 7.5, we can see that, as the methanol proportion increases, a different pattern emerges for the two levels of citric acid. For a low concentration of citric acid (2 g dm^{-3}), CRF marginally increases while for high concentration (6 g dm^{-3}), the increase in CRF is more marked suggesting that higher CRF measurements may be feasible by using a high proportion of methanol and high concentration of citric acid. The summaries presented indicate that maximum CRF appears to occur at combination (MW 80%, CA 6 g dm^{-3}) with a low range of 0.2 (11.9 − 11.7). No other combination

Output 7.4 *Main effect plots and summaries of CRF measurements for Example 7.4*

Select **Stat** ▷ **ANOVA** ▷ **Main Effects Plot** ▷ for *Factors*, highlight **AA MW and CA**, and click **Select** ▷ for *Raw response data in*, click the box, select **CRF** and click **Select** ▷click **OK**.

Main Effects Plot - Means for CRF

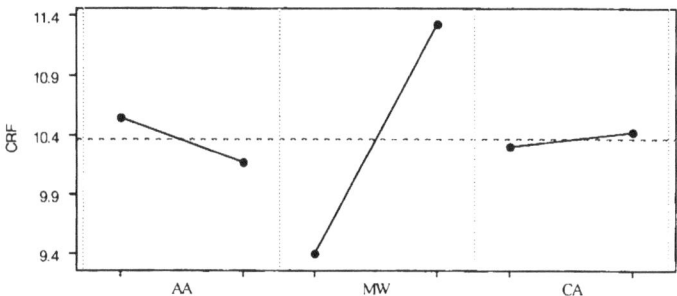

Select **Stat** ▷ **Basic Statistics** ▷ **Descriptive Statistics** ▷ for *Variables*, select **CRF** and click **Select** ▷ select **By variable**, click the empty box, select **AA**, and click **Select** ▷ for *Display options*, select **Tabular form** ▷ click **OK**.

Descriptive Statistics

Variable	AA	N	Mean	Median	TrMean	StDev	SEMean
CRF	−1	4	10.550	10.500	10.550	1.139	0.569
	1	4	10.175	10.100	10.175	1.284	0.642
Variable	AA	Min	Max	Q1	Q3		
CRF	−1	9.300	11.900	9.475	11.675		
	1	8.800	11.700	8.975	11.450		

As above except use MW as By variable.
Descriptive Statistics

Variable	MW	N	Mean	Median	TrMean	StDev	SEMean
CRF	70	4	9.400	9.400	9.400	0.497	0.248
	80	4	11.325	11.350	11.325	0.568	0.284
Variable	MW	Min	Max	Q1	Q3		
CRF	70	8.800	10.000	8.925	9.875		
	80	10.700	11.900	10.775	11.850		

As above except use CA as By variable.
Descriptive Statistics

Variable	CA	N	Mean	Median	TrMean	StDev	SEMean
CRF	2	4	10.300	10.350	10.300	0.678	0.339
	6	4	10.425	10.500	10.425	1.603	0.801
Variable	CA	Min	Max	Q1	Q3		
CRF	2	9.500	11.000	9.625	10.925		
	6	8.800	11.900	8.925	11.850		

Output 7.5 *Methanol and citric acid interaction plot for Example 7.4*

Select **Stat** ▷ **ANOVA** ▷ **Interactions Plot** ▷ for *Factors*, select **CA** and click
Select, and select **MW** and click **Select** ▷ for *Raw response data in*, click the
box, select **CRF** and click **Select** ▷ for *Title*, enter **Methanol x Citric Acid
Interaction Plot** ▷ click **OK**.

Methanol x Citric Acid Interaction Plot

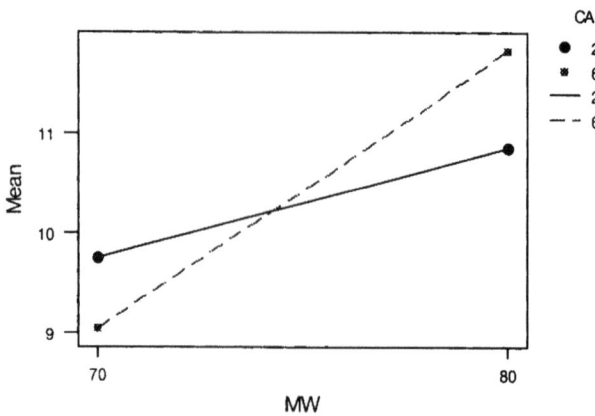

Select **Stat** ▷ **Tables** ▷ **Cross Tabulation** ▷ for *Classification variables*,
highlight **MW and CA**, and click **Select**. Select **Summaries** ▷ for *Associated
variables*, select **CRF** and click **Select** ▷ for *Display*, select **Means**, **Minimums**
and **Maximums** ▷ click **OK** ▷ click **OK**.

Tabulated Statistics

ROWS: MW   COLUMNS: CA

	2	6	ALL
70	9.750	9.050	9.400
	9.500	8.800	8.800
	10.000	9.300	10.000
80	10.850	11.800	11.325
	10.700	11.700	10.700
	11.000	11.900	11.900
ALL	10.300	10.425	10.363
	9.500	8.800	8.800
	11.000	11.900	11.900

CELL CONTENTS –
                    CRF:MEAN
                    MINIMUM
                    MAXIMUM

comes near to this one in either mean or consistency. The original data in Table 7.4 highlight that this combination did provide the best results with AA set at $0.004 \, mol \, dm^{-3}$.

Summary of findings

In two-level experiments, it is generally useful to summarise the findings in a simple table from which an optimal treatment combination may emerge. Table 7.5 provides such a summary indicating that the optimal combination, based on the experiment undertaken, may be AA 0.004 mol dm^{-3}, MW 80%, and CA 6 g dm^{-3}. Further experimentation around this combination may prove fruitful in respect of procedure optimisation based on the conclusion that the combination of MW and CA appears important though AA has a small main effect also.

Table 7.5 *Summary of analysis of CRF experiment for Example 7.4*

Source	AA ($mol \, dm^{-3}$)	Factors MW (%)	CA ($g \, dm^{-3}$)
Main effects	0.004	80	6
MW × CA interaction		80	6
Possible optimal	0.004	80	6

Square and *cube plots* represent further data plotting mechanisms. The square plot is a simple square-shaped plot suitable for two factor interactions and could be used as an alternative to the interaction plot of combination means. A cube plot is essentially a cube-shaped plot suitable for three factor interactions. The vertex points for each plot type correspond to the responses/average response (interaction means), and/or range of responses, for the associated factor–level combination. The vertex which provides the 'best' results in respect of experimental objective would obviously suggest optimal treatment combination.

Exercise 7.1

The data below were collected from a recovery experiment investigating spectrophotometric determination of colourants in synthetic food mixtures. The mixtures comprised different proportions (mg l^{-1}) of the colourants E-123 (Amaranth), E-124 (Ponceau 4R), and E-120 (Carminic acid). The measurement data refer to percent recovery of Ponceau 4R as the mean of three determinations measured at 561.5 nm in the first derivative ratio spectrum using Amaranth (16 mg 1^{-1}) as divisor. Which factors and factor interactions influence recovery most?

E-123	E-124	E-120	% recovery
12	12	12	99.83
24	12	12	100.75
12	24	12	101.00
24	24	12	95.75
12	12	24	97.25
24	12	24	99.83
12	24	24	97.08
24	24	24	98.54

Source: Reprinted from J.J. Berzas Nevado, C. Guiberteau Cabanillas and A.M. Contento Salcedo, *Talanta*, 1995, **42**, 2043–2051 with kind permission of Elsevier Science-NL, Sara Burgerhartstraat 25, 1055 KV Amsterdam, The Netherlands.

4 STATISTICAL COMPONENTS OF ANALYSIS

The data analyses discussed so far have utilised only some of the analysis and interpretational mechanisms associated with two-level designs. Other, more statistically based, techniques could also be included particularly when software is being used. Three such procedures will now be discussed.

4.1 Statistical Assessment of Proposed Model

Based on effect estimates analysis, we can construct a response model which we believe may be able to explain the measured response in terms of the active effects highlighted. By so doing, the significance of such effects can be examined on a statistical basis as well as enabling some diagnostic checking to take place. Model construction involves pooling the unimportant effects to form the error component. This hinges on assuming, for practical purposes, that these effects are negligible in their influence on the measured response.

As with one factor and factorial designs, this involves setting up an ANOVA table for the model sources of variation deemed to be affecting the response. The sums of squares (SSs) for active effects are expressed as

$$SS(\mathit{Effect}) = (\mathit{Effect\ estimate})^2 2^{k-2} \qquad (7.6)$$

with each SS based on one degree of freedom. The total sum of squares (SSTotal) is given by

$$TotalSS = \sum_{j=1}^{2^k} X_j^2 - \frac{\left(\sum_{j=1}^{2^k} Xj \right)^2}{2^k} \tag{7.7}$$

with associated degrees of freedom $(2^k - 1)$ where X_j refers to the response measurement for the jth combination tested. The residual sum of squares (SSRes) is determined by subtraction in the usual manner.

As the sum of squares of each active effect is based on only one degree of freedom, then SS(Effect) = MS(Effect) for active effects included in the response model. Test statistic construction is as before, *i.e.*

$$F = MS(Effect)/MSRes \tag{7.8}$$

enabling the statistical significance of the specified effects to be tested. Construction of the ANOVA table can be carried out manually or by use of statistical software though software, such as Minitab, may require the proposed response model to be hierarchical, *i.e.* terms contributing to later terms must be included in the model even if unimportant. This difficulty, however, can be overcome by manual calculation or by using regression methods to fit a model of response as a function of the active effects, main effects and interaction effects where appropriate.

Example 7.5

From the CRF analysis carried out in Examples 7.1 to 7.4, we could propose a model of the form

$$CRF = MW + MW \times CA + AA + error$$

to explain the CRF measurements collected. The error term comprises those effects deemed unimportant from the earlier analysis, *i.e.* CA, $AA \times MW$, $AA \times CA$, and $AA \times MW \times CA$. This pooling of terms assumes that these effects are negligible in their influence on the CRF response. Table 7.6 contains the manually constructed ANOVA table based on using equations (7.6), (7.7), and (7.8), the latter for test statistic determination.

Analysis of proposed model
The critical values for test statistic comparison from Table A.3 are $F_{0.05,1,4} = 7.71$ and $F_{0.01,1,4} = 21.20$ (the 0.1% critical value is 74.14). Based on these, we can see why the MW effect and the interaction

Table 7.6 *ANOVA table for proposed model for CRF for Example 7.5*

Source	DF	SS	MS	F	p
MW	1	7.4112	7.4112	871.91	< 0.001
MW × CA	1	1.3612	1.3612	160.44	< 0.001
AA	1	0.2812	0.2812	33.08	< 0.01
Residual	4	0.0338	0.0085		
Total	7	9.1187			

MW × CA are classified as significant at the 0.1% significance level ($p < 0.001$) while the AA effect is only significant at a lower level ($p < 0.01$). We can conclude, however, that the three active effects specified are all highly significant, much as expected since these effects differed markedly from the others within the experiment in their influence on CRF.

4.2 Prediction

Once a potentially optimal treatment combination has been extracted from the data analysis, it may be useful to predict the response for this combination and compare it with the measured response at the same, or nearly equivalent, treatment combination. The prediction estimated by this process represents an estimate of what we believe would be the response if the factor-level combination suggested was run. Comparing these two figures enables a simple assessment of the viability of the suggested optimal to be carried out. If the prediction is better than the recorded experimental measurement with respect to the type of response necessary, then the suggested optimal combination may well be 'best' for the experiment considered. Prediction is determined as

$$\bar{X} + \frac{1}{2}\left[\sum(\text{optimal factor levels})^*(\text{corresponding effect estimates})\right] \quad (7.9)$$

where $\bar{X}$ is the overall mean of the response data. The summation in equation (7.9) simply means sum the product of factor level and effect estimate across the significant effects. For interaction effects, the optimal level to use refers to a simple multiple of the corresponding main effect levels.

Example 7.6

Example 7.5 indicated that all three active effects were statistically significant and Example 7.4 hinted at the combination AA 0.004 mol dm^{-3}, MW 80%, and CA 6 g dm^{-3} as being possibly best. We want to predict the CRF for this combination and compare the result against the measured experimental CRF response of 11.9 for that combination (see Table 7.4).

Prediction of optimal
The steps in the prediction calculation are shown in Table 7.7 based on the optimal levels of MW +, MW × CA + (same best), and AA −. The predicted CRF is shown to be 11.925. This prediction is marginally higher than the corresponding experimental measurement of 11.9 suggesting that the specified optimal combination may well be 'best' based on the factor levels tested in the original experiment.

Table 7.7 *CRF prediction for potential optimal combination for Example 7.6*

Significant effects	MW	MW × CA	AA
Optimal levels	+	+	−
Effect estimate	1.925	0.825	−0.375

mean CRF = 10.3625; prediction = $10.3625 + \frac{1}{2}[(+)(1.925) + (+)(0.825) + (-)(-0.375)]$
= 11.925

This concept of prediction is often extended to assess all predicted results to investigate model adequacy fully. Essentially, this corresponds to fitting a multiple regression model to the response data based on the highlighted effects and carrying out a practical validity check of the fitted model.[9] Ideally, if model is a good explanation of the response, measurements and predicted responses should match well. Inadequate matching would be indicative of inappropriate model pointing to missing explanatory factors as a likely cause.

4.3 Diagnostic Checking

Through prediction, we can obtain estimates of the error in the proposed response model (the residuals), enabling diagnostic checking of the proposed model to be considered. Such analysis is useful, in two-level designs, for assessing factor influences on response variability for both active and inactive factors. Often, inactive factors can have a strong influence on response variability even though no detectable effect on response measurement can be found. Additionally, a plot of residuals against run order can provide information on the randomness of the response data in a similar way to the procedures associated with

[9] Reference 3, pp. 51–52.

Statistical Process Control (SPC). A formal statistical test for assessing how factors affect response variability also exists.[10]

Discussion so far has centred on the analysis of a three factor experiment. In two-level experiments, the number of factors can be increased easily though the number of experiments will obviously double for each new factor introduced. A 2^4 experiment requires 16 experiments, double that of a 2^3, while a 2^5 experiment requires 32 experiments, double that of a 2^4 and four times that of a 2^3. Although there is an increase in experimentation, the increase in the number of factors enables more wide ranging conclusions on factor effects to be reached. The basis of the structure and analysis of such designs is the same as that illustrated for the 2^3 design structure. Contrasts for factor effects can be easily constructed using the general principle introduced in Section 2.1. For a 2^4 design structure, this results in the signs of contrasts displayed in Table 7.8 where, again, each effect column adds to 0 and each column can be derived from a pair of columns, just as occurred for the 2^3 design structure (Table 7.3). Example 7.7 will be used to illustrate the analysis for a 2^4 design as well as indicating how to deal with a problem which does not conform to the ideal.

Table 7.8 *Signs of contrasts for a four factor two-level factorial experiment*

Treatment Combn	Mean	A	B	C	D	AB	AC	AD	BC	BD	CD	ABC	ABD	ACD	BCD	ABCD
(1)	+	−	−	−	−	+	+	+	+	+	+	−	−	−	−	+
a	+	+	−	−	−	−	−	−	+	+	+	+	+	+	−	−
b	+	−	+	−	−	−	+	+	−	−	+	+	+	−	+	−
ab	+	+	+	−	−	+	−	−	−	−	+	−	−	+	+	+
c	+	−	−	+	−	+	−	+	−	+	−	+	−	+	+	−
ac	+	+	−	+	−	−	+	−	−	+	−	−	+	−	+	+
bc	+	−	+	+	−	−	−	+	+	−	−	−	+	+	−	+
abc	+	+	+	+	−	+	+	−	+	−	−	+	−	−	−	−
d	+	−	−	−	+	+	+	−	+	−	−	−	+	+	+	−
ad	+	+	−	−	+	−	−	+	+	−	−	+	−	−	+	+
bd	+	−	+	−	+	−	+	−	−	+	−	+	−	+	−	+
abd	+	+	+	−	+	+	−	+	−	+	−	−	+	−	−	−
cd	+	−	−	+	+	+	−	−	−	−	+	+	+	−	−	+
acd	+	+	−	+	+	−	+	+	−	−	+	−	−	+	−	−
bcd	+	−	+	+	+	−	−	−	+	−	+	−	−	−	+	−
abcd	+	+	+	+	+	+	+	+	+	+	+	+	+	+	+	+

[10] D.C. Montgomery, 'Design and Analysis of Experiments', 4th Edn., Wiley, New York, 1997, p. 395.

Example 7.7

An experiment was planned to investigate the determination of aluminium through complexation with Solochrome Violet RS (SVRS). Four factors were identified for experimentation, each planned to be set at only two levels. The factors chosen were SVRS volume (ml), pH, heating time (seconds), and delay time (seconds). Heating time refers to the heating of the aluminium solution in a microwave oven at 360W for the specified time period. The delay time factor corresponds to the waiting time before measuring the fluorescence intensity at 590 nm on a Perkin Elmer LS50 Luminescence Spectrometer. The design matrix and collected fluorescence intensity measurements are presented in Table B.4. Output 7.6 represents the Minitab derived effect estimates and normal plot for the presented data.

Effect estimates
Effect estimation shows that SVRS, the interaction SVRS × pH, and heating time (HeatTime) appear to be the most important effects. The remaining effects split into two groups, one based on numerical estimates of the order 10 to 20 and the other on numerical estimates below 10. Given the magnitude of these estimates, it is difficult to state that these effects are unimportant and so can be discarded from the remaining analysis.

Normal plot of effect estimates
The normal plot in Output 7.6 does not conform to the ideal pattern of two distinct groups of estimates, important effects and unimportant effects. SVRS (A), the interaction SVRS × pH (AB), and heating time (C) do, however, stand out as large negative estimates. These effects would obviously require further analysis. Which of the remaining effects to consider as important is a more difficult decision as all are clustered close together with none obviously different.

Data analysis
The advice, in this case, would be to analyse the main effects and possibly all two factor interactions even though certain three factor interactions provide similar effect estimates. By analysing all the main effects and two factor interactions, it will be feasible to obtain a good overall picture of how the tested factors affect fluorescence intensity of the aluminium complex. From such an analysis, an optimal combination of factors could result from which it may be possible to specify a model for the fluorescence response measurement and suggest avenues for future experimentation.

Output 7.6 *Effect estimates and normal plot for fluorescence measurements of Example 7.7*

Select **Stat** ▷ **DOE** ▷ **Fit Factorial Model** ▷ for *Responses*, select **Flscnce** and click **Select** ▷ select **Use terms**, select the empty box, and enter **(SVRS pH HeatTime DelTime)4**.

Select **Storage** ▷ for *Storage*, select **Effects** ▷ click **OK**. Select **Display effects plot** ▷ click **OK** ▷ click **OK**.

Fractional Factorial Fit
Estimated Effects and Coefficients for Flsscnce

Term	Effect	Coef
Constant		121.50
SVRS	−80.00	−40.00
pH	7.00	3.50
HeatTime	−33.00	−16.50
DelTime	10.75	5.37
SVRS*pH	−52.00	−26.00
SVRS*HeatTime	5.00	2.50
SVRS*DelTime	−7.25	−3.62
pH*HeatTime	21.50	10.75
pH*DelTime	−14.25	−7.13
HeatTime*DelTime	17.75	8.88
SVRS*pH*HeatTime	−17.50	−8.75
SVRS*pH*DelTime	12.75	6.38
SVRS*HeatTime*DelTime	−17.25	−8.62
pH*HeatTime*DelTime	7.25	3.63
SVRS*pH*HeatTime*DelTime	−5.75	−2.88

Analysis of Variance for Flscnce

Source	DF	Seq SS	Adj SS	Adj MS	F	P
Main Effects	4	30614.3	30614.3	7653.6	**	
2-Way Interactions	6	15047.8	15047.8	2508.0	**	
3-Way Interactions	4	3275.8	3275.8	818.9	**	
4-Way Interactions	1	132.3	132.3	132.3	**	
Residual Error	0	0.0	0.0	0.0		
Total	15	49070.0				

```
 Nscores -                              *      BC
         -
     1.2+                             *          CD
         -                          *         ABD
         -                         *        D
         -                        *      BCD
         -                       **      AC,B
     0.0+                      *     ABCD
         -                   * *     BD,AD
         -                * ACD
         -                * ABC
         -            *   C
    -1.2+        *    AB
         -
         -   *   A
         -
            ----+---------+---------+---------+---------+--Effects
             -75      -50       -25        0        25
        A = SVRS     B = pH     C = HeatTime    D = DelTime
```

Exercise 7.2

An experiment was conducted into the factors affecting the recorded responses of an automatic chemical analyser used to determine the activity of a serum enzyme, alkaline phosphatase. The aim of the experiment was to determine which, if any, of the factors and their interactions influence the measured response (phosphatase enzyme activity). From past experience, it is thought that the factors influencing the phosphatase activity are zinc sulphate (A), magnesium sulphate (B), pH (C), disodium *p*-nitrophenyl phosphate (D), and 2-amino-2-methyl-propan-1-ol (E). The levels chosen for testing were

	Levels		
Factors	−	+	*Units*
A	40	80	μmol dm^{-3}
B	1.50	2.50	μmol dm^{-3}
C	10.00	10.70	dimensionless
D	10	20	mmol dm^{-3}
E	0.20	0.60	mol dm^{-3}

The design basis is therefore a 2^5 factorial design requiring 32 experimental runs. The design matrix and the response obtained (enzyme activity) are given Table B.5. Are any effects active? Can a model for enzyme activity be specified?

5 THE USE OF REPLICATION

Most instances where two-level designs are used, *e.g.* in screening experiments and in laboratory experimentation, only one observation is recorded for each treatment combination. It is possible to construct two-level designs such that replicate observations are obtained for each treatment combination. Replication provides data from which the experimental error can be estimated and enables more formal statistical analysis to be considered though effects analysis and data interpretation should still be utilised.[11]

Use of replication can also be beneficial if assessing which effects affect response variability is of more interest rather than what affects response measurement. Replicate measurements at each treatment combination can provide a variability estimate for that combination. It is this summary which would be used as the response to be analysed in a response variability study using the analysis mechanisms explained in this chapter.

---

[11] Reference 1, pp. 51–52.

Two-level factorial designs are based on the assumption of linearity of factor effects though the designs are generally robust to violations of this assumption. Replicating *centre points* in such a design enables curvature of effects to be estimated. Such an effect is useful to test for since optimal factor levels may occur in the interior of the experimental region rather than at the extremes. Centre points, consisting of replications of the combination of factors set at a mid-level (0), can be added to the design structure easily and do not affect the factor effect estimates derived from such designs. Increasing factor levels from two to three levels to use three-level design structures could also be considered to help assess curvature effects though this, inevitably, will result in increased experimentation.

6 FRACTIONAL FACTORIAL DESIGNS

Two-level and three-level factorial experiments can become quite demanding in terms of the number of experimental runs when k is large which can make such experiments too large to be practical. For example, a 2^6 design requires 64 experimental runs while a 2^8 design requires 256 experimental runs. In these designs, most of the degrees of freedom are associated with third and higher order interactions. Since these are difficult to interpret and are often negligible in their effect on a response, then such designs can have substantial degrees of freedom associated with negligible effects. Such effects can be pooled to provide a form of error estimation. Reducing the amount of experimentation by discounting higher order interactions is therefore worthy of consideration. *Fractional factorial designs*[12-14] cater for such approaches and are also useful when all treatment combinations cannot be applied under homogeneous conditions due to lack of resources, such as manpower, raw material, time, and cost, or because of physical constraints, such as access to instruments and storage of material. Orthogonal arrays (OA) are particular illustrations of fractional designs used often in optimisation of instruments.

A major use of fractional factorial designs is in *screening experiments* in which many controllable factors are considered in order to identify those factors (if any) that are active. Such experiments are usually performed in the early stages of a project (exploratory/investigative experiments) with the factors identified as important then investigated more fully in subsequent experiments. They provide for small, efficient designs though there is no 'best' design as choice

---

[12] Reference 4, pp. 151–188.
[13] Reference 10, pp. 372–422.
[14] Reference 1, pp. 89–105.

depends on the experimental requirements and constraints. In fact, it is often prudent to carry out a sequence of fractional factorial experiments, each succeeding experiment being influenced by the results of the preceding experiment. Fractional factorial designs are also used in *ruggedness studies* where small deviations to procedure elements are investigated to assess the resistance for the reported measurements to these deviations.

For example, suppose we wished to conduct an experiment involving six factors, each set at two levels ($2^6 = 64$ combinations), and that we are constrained to running at most 32 experimental runs. The full two-level factorial design, based as it is on 64 runs, is therefore a non-starter. Three possible fractional structures could be considered, as summarised in Table 7.9. The first is the half fraction (2^{6-1} design) which requires 32 runs, the second is a quarter fraction (2^{6-2} design) based on only 16 runs, and the third is the eighth fraction (2^{6-3} design) structure requiring least experimentation at 8 runs. The number of runs, and thereby treatment combinations to be tested, differ with design structure. Obviously, the 2^{6-3} design requires fewest runs but we must be wary of using this criterion for deciding on design to use. Consideration, within design choice, must also be given to how the designs produce effect estimates and how accurate they can be taken to be.

Table 7.9 *Possible six factor fractional factorial design structures*

Fraction	Number of runs	Resolution	Design
half	32	VI	2^{6-1}
quarter	16	IV	2^{6-2}
eighth	8	III	2^{6-3}

One way of aiding the choice of design to use is based on the concept of *resolution* and *aliasing*. These refer to the ability of the design to estimate factor and interaction effects independently. High resolution is generally best as this enables independent estimation of most effects to be attainable. A resolution IV design, such as the 2^{6-2} design in Table 7.9, is based on the premise that main effects are independent of two factor interactions but two factor interactions are aliased with each other, aliasing meaning the effects cannot be independently estimated. The 2^{6-3} design in Table 7.9 is classified as of resolution III which means main effects and two factor interactions cannot be independently estimated. The best design in terms of estimation precision is the half fraction (resolution VI) as it enables all main effects and two factor interactions, at the least, to be estimated independent of each other. The drawback of this choice, however, would be the number of runs required but reducing this affects design resolution and, therefore, estimation precision. Extensive tables of fractional designs for up to 11 factors are available.[15]

Fractional factorial experiments, therefore, are based on implementing one-half, one-quarter, or even one-eighth of the total factorial plan, *i.e.* a fraction of the possible treatment combinations. When deciding on the most appropriate fractional factorial design to use, it is best to choose one with high resolution consistent with the degree of fractionation required so providing less restrictive assumptions in respect of which interactions to be assumed negligible in their effect (better alias structure). Most dedicated statistical software such as Minitab, SAS, and GLIM have fractional factorial design construction and analysis facilities though there is a restriction on the number of available designs from which to choose.

Analysis methods for fractional factorial designs are based on those illustrated in this chapter namely, effect estimates, effect plots, data plots and summaries, proposed model, and diagnostic checking. Fuller details on fractional replication can be found in the texts by Bayne and Rubin,[16] Davies,[17] and Kuehl.[18]

7 RESPONSE SURFACE METHODS

Response surface methods (RSM) are powerful experimental design tools applicable to product/process development, improvement, and optimisation.[19,20] They represent a collection of experimental design and multiple regression based techniques that can be used to analyse problems where several factors may influence a response and where the goal is to optimise (maximise/minimise) system performance. The response model within such designs can be either first or second order, the latter being particularly useful if curvature in response is anticipated.

Response surface methods are based on sequential procedures using two-level factorial designs to identify important factors, with subsequent experiments on specified treatment combinations used to home in on a region of factor levels that appear most likely to produce optimum response. Based on these results, an attempt is made to characterise the

[15] Reference 10, pp. 683–699.
[16] Reference 6, pp. 85–96.
[17] Reference 1, pp. 89–116.
[18] R.O. Kuehl, 'Statistical Principles in Research Design and Analysis', Duxbury Press, Belmont, California, pp. 390–422.
[19] Reference 4, pp. 189–272.
[20] Reference 1, pp. 144–162.

response surface with a polynomial model leading to the vicinity of the optimum by rapid and efficient means.

RSM techniques were developed in the 1950s for the purpose of determining optimal operating conditions in the chemical process industry.[21] Later work concentrated on the analysis principles for RSM strategies together with aspects of design structure such as rotatability, robustness, and optimality. Chemists have used RSM techniques to help optimise experimental routines, while the food industry has made use of them to optimise the properties of foodstuffs and beverages. RSM strategies have also found application in the biological and clinical sciences within pollution studies and cancer research.

The relationship between the response variable Y and the factors X_1, $X_2, \ldots, X_k$ in RSM applications is expressed as

$$Y = f(X_1, X_2, \ldots, X_k) + \varepsilon$$

where ε represents the noise (error) observed in the response Y. The function $f(X_1, X_2, \ldots, X_k)$ is called the *response surface* which can be unimodal (one optimum) or multi-modal (local and global optimums) and which is represented graphically by means of contour plots. Such functions are approximations to the true response function as the form of functional relationship between response and factors is generally unknown. Using experimental results, we attempt to predict the direction of movement in order to achieve the optimal response.

Experimental design structures utilised in RSM studies include *Central Composite Designs*, *Face Centred Cube Designs*, and *Box–Behnken Designs*, all of which stem from two-level multi-factor factorial structures. Such designs are appropriate for estimation of main effects, two factor interaction effects, and the quadratic effect of experimental factors, the latter only if centre points have been included in the experimental structure implemented. Other optimisation procedures based on experimental design approaches include *Taguchi Designs*[22,23] and *Simplex Optimisation*.[24]

8 MIXTURE EXPERIMENTS

In mixture experiments, the experimental factors correspond to components or ingredients of a mixture resulting in these factors not being

[21] G.E.P. Box and K.B. Wilson, *J. Roy. Stat. Soc.*, *Series B*, 1951, **13**, 1.
[22] Reference 10, pp.622–641.
[23] R.H. Lochner and J.E. Matar, 'Designing for Quality: An Introduction to the Best of Taguchi and Western Methods of Experimental Design', Chapman and Hall, London, 1990.
[24] Reference 1, pp. 128–133.

independent. The response in such experiments may be a chemical solution, a food product, or construction materials. Such a response depends on the proportions of ingredients rather than their amounts with variations in the proportions affecting the end product. The relationship between the measured response and the ingredient proportions is therefore important and is the basis of the associated analysis.[25]

For k components, we denote the proportions as $x_1, x_2, \ldots, x_k$ where $0 \leq x_i \leq 1$ and $\sum x_i = 1$, *i.e.* 100%. For $k = 2$ components, the factor space will be $0 \leq x_1 \leq 1$, $0 \leq x_2 \leq 1$, and $x_1 + x_2 = 1$, *i.e.* factor levels to be tested will lie on the line $x_1 + x_2 = 1$. For $k = 3$, the factor space is a triangle whose vertices result in *pure blends*, *i.e.* mixtures containing 100% of a single component. The design basis of most mixture experiments is generally that of a simplex design such as the $\{k,m\}$ *simplex lattice* (fitting k components into a response surface polynomial of degree m) and the *simplex centroid*. Linear, quadratic, and cubic response models are mostly used in mixture experiments.

[25] Reference 18, pp. 448–462.

CHAPTER 8

Multivariate Analysis Methods in Chemistry

1 INTRODUCTION

Use of analytical procedures within chemistry can often produce large data sets comprising measurements on a number of variables over a large number of samples. Handling, interpretation and prediction techniques are therefore needed to help identify or display structure in the recorded chemical data, in order to extract the chemical information from it. Such techniques come within the branch of statistics referred to as *Multivariate Analysis Methods* (MVA) with mathematical statistics, calculus, geometry, and algebra providing the necessary mathematical underpinning. As most data sets assessed by MVA techniques are large, computer software is an important working tool and so the emphasis in this chapter will again be on showing how software can be used to aid interpretation of multivariate data. The most commonly used MVA methods in chemistry include *Principal Component Analysis* (PCA), *Principal Components Regression* (PCR), *Factor Analysis* (FA) in a variety of forms, *Statistical Discriminant Analysis* (SDA), *Cluster Analysis* through the *k*-nearest neighbour (*K*-NN) and soft independent modelling of class analogy (SIMCA) techniques, and recently, *Artificial Neural Networks* (ANN). Numerous texts have been written specifically describing MVA methods within chemistry[1-4] and in general terms.[5]

In chemical data handling, the application of MVA methods is often discussed under the heading of *Chemometrics* covering methods for multivariate exploratory data analysis, supervised learning, unsuper-

[1] M.J. Adams, 'Chemometrics in Analytical Spectroscopy', The Royal Society of Chemistry, Cambridge, 1995.

[2] R.G. Brereton, 'Multivariate Pattern Recognition in Chemometrics', Elsevier, Amsterdam, 1992.

[3] D.L. Massart, B.G.M. Vandeginste, S.N. Deming, Y. Michotte and L. Kaufman, 'Chemometrics: A Textbook', Elsevier, Amsterdam, 1988.

[4] M. Meloun, J. Militky and M. Forina, 'Chemometrics for Analytical Chemistry Volume 2', Ellis Horwood, Chichester, 1994.

[5] B.S. Everitt and G. Dunn, 'Applied Multivariate Data Analysis', Arnold, London, 1991.

vised learning, and pattern recognition techniques. The associated methods are based on the simultaneous analysis of multiple measurements of a large number of variables (characters or attributes) and, in many instances, are generalisations of classical univariate statistical methods. The main thrust of MVA methods is to investigate the underlying structure of data to help reveal points of similarity and dissimilarity as well as inter-relationships between the experimental units (sample material) and their response measurements. The purpose of MVA is to try to formulate answers to some or all of the following questions:

- Is there a structure within the collected observations which is chemically interpretable?
- Can the measurements be reduced or put in summary form displaying their chemical properties?
- Do the data obtained give any insight into the chemical mechanisms which helped generate them?
- How many response variables are necessary to describe the chemical properties of the data adequately?

MVA techniques have applications in many branches of chemistry such as spectroscopy, gas chromatography (GC), pyrolysis mass spectra (PyMS), nuclear magnetic resonance studies (NMR), environmental chemistry, geochemistry, and chemotaxonomy. For example, they can be used to obtain a preliminary idea of the relationship between chemical structure and odour for substances or objects. In kinetics, where spectral data may be collected, they can help to determine the number of absorbing components and the concentration and spectrum of each component. In the biological sciences, they occur in bacterial taxonomy, microbial systematics, biodiversity, modelling phenetic variation in organisms, and genetic finger printing.

The methods associated with MVA represent a 'mixed' bag of analysis procedures which can be used to investigate large data sets with respect to a number of diverse objectives. These include,

1. data reduction (structural simplification, dimensionality reduction);
2. sorting and grouping;
3. detection of the level of dependence among the measured variables;
4. construction of a viable prediction system (modelling);
5. hypotheses construction and testing.

In this chapter, we will discuss multivariate techniques appropriate only to objectives 1 and 2: *data reduction* and *sorting and grouping*.

In data reduction, the focus is on the recorded variables and the structure which they may exhibit. These correlated responses are replaced by uncorrelated components which contain most of the chemical information appropriate to the data. It is hoped that this replacement of original data by components will highlight the underlying chemical structure of the data. For sorting and grouping, sets of 'similar' samples are formed based on their similarity across a number of associated characteristics. Through this detection of similarity, rules for classifying unknown samples to distinctly separate groupings may be constructed. Multivariate methods, therefore, investigate the relationship between variables measured across different samples or defined groups of samples with similar characteristics.

Multivariate data sets comprise large amounts of data with matrix methods forming the mathematical and computational basis of most. Statistical software plays an important role in the application of MVA methods due, in large part, to the size and nature of the associated data set but also to the need to provide an easily implemented analysis method with a capability of presenting results simply and succinctly for analysis purposes. At present, Excel does not have MVA procedures by default so illustration of MVA methods will be based on Minitab which has a comprehensive array of in-built multivariate routines.

For *data reduction*, we generally collect the measurements on each of p variables across n samples giving rise to an $n \times p$ data matrix as illustrated in Table 8.1. The rows of the table specify the response measurements for each variable for each sample. Each column provides the values of a single response variable across all the samples. Such a data set could occur in the analysis of multicomponent mixtures by fluorescence spectroscopy. Each column of the data matrix would represent a spectrum and each row a particular wavelength where the number of wavelengths exceeds the number of spectra. Interest, in this case, would lie in determining the number of components contributing to the emission spectra and the chemical identity of each component (explanation of data structure). MVA methods for data reduction include *principal component analysis* (PCA, see Section 2) and *factor analysis* (see Section 4), both of which assess how closely related (correlated) variables are in order to try to reduce the number of variables describing the data.

For the *sorting and grouping* technique to be described in this chapter, data are collected on p response variables from random samples of sizes $n_1, n_2, ..., n_k$ associated with k distinct groups. In other words, data are collected according to known group structure prior to implementation of the associated analysis method. Data of this type, as illustrated in Table 8.2, could occur in a food analysis where several comparable

Table 8.1 *Typical data layout for a data reduction investigation*

| | | \multicolumn{6}{c}{Response variables} |||||||
		X_1	X_2	X_3	.	.	X_p
Samples	1	x_{11}	x_{12}	x_{13}	.	.	x_{1p}
	2	x_{21}	x_{22}	x_{23}	.	.	x_{2p}
	3	x_{31}	x_{32}	x_{33}	.	.	x_{3p}
	.	.	.	.	.	.	.
	.	.	.	.	.	.	.
	n	x_{n1}	x_{n2}	x_{n3}	.	.	x_{np}

p, number of response variables; n number of samples tested; x denotes a response measurement

products containing specified flavouring compounds are tested. The products would be the groupings and the concentration of the flavour compounds, the response variables with interest lying in assessing whether the products can be distinguished between on the basis of the concentration of the flavour compounds. The MVA method appropriate to this type of study is *statistical discriminant analysis* (SDA, see Section 5) which attempts to construct a mechanism for distinguishing between known groups of samples in terms of their common characteristics. A further sorting and grouping procedure is *cluster analysis* (see Section 7) which differs from SDA in that prior grouping of data is not

Table 8.2 *Typical data layout for a sorting and grouping investigation*

| | Samples | \multicolumn{6}{c}{Response variables} |||||||
		X_1	X_2	X_3	.	.	X_p
Group 1	1	x	x	x	.	.	x
	2	x	x	x	.	.	x
	.	.	.	.	.	.	.
	n_1	x	x	x	.	.	x
Group 2	1	x	x	x	.	.	x
	2	x	x	x	.	.	x
	.	.	.	.	.	.	.
	n_2	x	x	x	.	.	x
.	.	.	.	.	.	.	.
.	.	.	.	.	.	.	.
Group k	1	x	x	x	.	.	x
	2	x	x	x	.	.	x
	.	.	.	.	.	.	.
	n_k	x	x	x	.	.	x

p, number of response variables; n_i, number of samples tested in group i; k, number of groups sampled; x denotes a response measurement necessary with data laid out in a similar fashion to Table 8.1, the general starting point.

In many multivariate methods, the collected response data may need to be *standardised* prior to analysis so that it has zero mean and unit variance. The values of each variable would then not depend on possible arbitrary and conflicting units of measurement where large data measurements may dominate, and so produce a potentially distorted data interpretation. This re-scaling of data can be particularly useful with response variables measured on widely different numerical scales, *e.g.* one variable measured as %m/m (g per 100 g) and one on the pH scale. Moreover, certain software packages only operate on standardised data when implementing multivariate methods. Minitab, for instance, automatically standardises the experimental data before implementing its PCA routine.

The application of multivariate methods to multivariate data is based mostly on the determination of *derived*, or *latent*, *variables* expressed as linear combinations of the original response variables, *e.g.*

$$Z = 0.33X_1 + 0.72X_2 - 0.83X_3 + 0.05X_4 + 0.01X_5 - 0.77X_6 - 0.04X_7$$

similar to a multiple linear regression equation with no constant. The *coefficients* of each variable in the derived combination measure the relative importance of that variable to the aspect of the data it explains. Large coefficients, relative to other coefficients, suggest a large contribution while small coefficients near zero are indicative of a minor contribution.

The mathematics underpinning MVA methods is based on the solution of the *eigenvalue equation*

$$Sw = \lambda w \qquad (8.1)$$

where S is a $p \times p$ square data matrix, w is a column vector, and λ is a scalar. The solution to this equation involves solving for the λ terms, the *eigenvalues*, and the w vectors, the *eigenvectors*, which satisfy this equation. To find the λ terms, the eigenvalue equation is re-expressed as

$$(S - \lambda I)w = 0 \qquad (8.2)$$

where I is the $p \times p$ identity matrix and 0 is a column vector of 0s. The values of λ satisfying equation (8.2) also satisfy the determinant expression

$$|S - \lambda I| = 0 \qquad (8.3)$$

which is a polynomial in λ. Solving the polynomial expression (8.3) provides the eigenvalues. The vector w associated with each eigenvalue is derived by solving equation (8.2) for each eigenvalue solution.

The eigenvalues from equation (8.1) provide a measure of the importance of the 'derived variables' while the eigenvectors provide the coefficient estimates for each response variable within the 'derived variables'. Interpretation of these results is based on the magnitude of the eigenvalues such that the largest eigenvalue with associated eigenvector produces the most important 'derived variable', the next largest eigenvalue with associated eigenvector the next most important 'derived variable', and so on. Each eigenvalue, therefore, measures the relative importance of each 'derived variable' in respect of what it explains of the chemical patterns within the response data. It is generally hoped that the first few 'derived variables' are all that will be necessary to explain most of the data variation adequately where the total number of 'derived variables' corresponds to the total number of response variables measured in the study, *i.e.* p = number of 'derived variables'.

In sorting and grouping procedures, an important aspect of the analysis rests with the concept of *distance* of either single observations, or samples of observations, from the centre, or mean, of distinct groups of data. Group means are usually referred to as the *group centroid*. Distance provides a measure of the similarity or dissimilarity of samples, or groups of samples, to one another if they could be plotted in multi-dimensional space. In other words, it measures the level of commonality of characteristics shared by groups of samples. Generally, small distance is indicative of similarity, while large distance suggests dissimilarity. For example, a distance measure could be determined, on the basis of inherent characteristics, to distinguish between two chemical compounds. A distance near zero would imply that the two compounds share many common characteristics, *i.e.* compounds are similar, while a larger distance would suggest little commonality of characteristics, *i.e.* compounds have different chemical properties.

Several distance measures can be used within sorting and grouping techniques. Three commonly applied measures are the *Euclidean distance*, the *Mahalanobis D^2 distance*, and the *Minkowski metric*, each derived from the information about the objects to be classified, *i.e.* the collected experimental data. The Euclidean distance between two samples i and j, based on measurements from p variables, is expressed as

$$d_{ij} = \left[\sum_{k=1}^{p} (x_{ik} - x_{jk})^2 \right]^{1/2} \tag{8.4}$$

while the Mahalanobis distance between the centroids $\bar{d}$ of two sample groupings i and j is presented as

$$d_{ij}^2 = (\bar{d}_i - \bar{d}_j)^2 \tag{8.5}$$

The Mahalanobis distance measure accounts for correlations between the response variables while the Euclidean measure does not, in general. This results in the former being more appropriate if there is a suggestion of significant correlations between the measured variables. When dealing with more than two groups, the set of group distance measures is generally summarised in matrix form using a *distance matrix*.

2 PRINCIPAL COMPONENT ANALYSIS

The technique of *principal component analysis* (*PCA*), an unsupervised learning method, was first described by Karl Pearson at the turn of the century as a means of fitting planes by orthogonal least squares. It is essentially an exploratory tool with its modern application due to Harold Hotelling who, in the 1930s, used it to analyse correlational structures between a series of measured responses, such as could occur in chemical identification. In chemistry, PCA was first introduced by Malinowski around 1960 under the name principal factor analysis and since 1970, a large number of chemical applications of PCA have been published in such areas as mixture analysis, pattern recognition, and multivariate calibration. PCA is based on the linear transformation of correlated response variables (related response measurements) to pair-wise uncorrelated components (unrelated functions of the responses) to help explain the patterns of variation inherent in the set of measured responses. It is basically an ordination method that attempts to extract and interpret information from multivariate data conforming to the structure illustrated in Table 8.1.

In PCA, interest lies in assessing the variables as a set, the inter-relationships (correlations) between them, and the information these relationships contain jointly about the samples on which measurement has taken place. The description of PCA in this section will be based on using the *original measured data*. A simple introduction to the application of PCA in chemistry is provided by Smith.[6]

[6] G.L. Smith, *Anal. Proc.*, 1991, **28**, 150.

2.1 Objective of PCA

The general objective of PCA is simplification, or dimensionality reduction, whereby we try to summarise a multivariate data set by relatively few components with minimal loss of information. PCA can

- be exploratory (examining data by simplification),
- help understand the structures of the variables (the inter-relationships between the variables),
- help determine the dimensionality of the data (how 'multivariate' they are),
- help derive a low dimensional representation of the data (graphical representation of the relationships between variables and between samples).

It, therefore, offers a means of reducing the dimensionality of multi-variate data through the replacement of response variables by inter-pretable components, expressed as combinations of the response variables, which explain the structure and patterning inherent in the collected data.

PCA is based on the derivation of linear combinations of the p measured variables X_1, X_2,...., X_p to produce *indices*, or 'derived variables', that are uncorrelated and are such that each explains a different 'dimension' within the data. Such indices are referred to as *principal components* (PCs). The PCs are derived in such a way that they are all mutually orthogonal, *i.e.* independent and at right angles to one another in multi-dimensional space. These components are also arranged in order such that the first accounts for the largest portion of explainable variability in the measured data [largest eigenvalue solution of equation (8.1)], the second accounts for the second largest portion of explainable variability subject to being uncorrelated with the first [second largest eigenvalue solution of equation (8.1)], and so on.

As there are p response variables within the data set, p principal components can be derived, *i.e.* number of components = number of response variables. It is hoped, however, that most of the explainable variability in the data can be accounted for by k *components* where k is very much less than p. In other words, PCA attempts to condense the measured data but in such a way that enables the important features to appear in a small number of interpretable components. The derived PCs 'replace' the response variables with minimal loss of information and provide the means of extracting information on the structure of the data.

The first PC, denoted PC_1, is expressed in the form

$$PC_1 = \alpha_{11}X_1 + \alpha_{12}X_2 + \ldots + \alpha_{1p}X_p \qquad (8.6)$$

where the α terms refer to the *weights*, or *loadings*, for each variable within this principal component. These terms are unique to each PC as they are functions of the angles between the variables and the component in p dimensional space. The second PC, denoted PC_2, can be similarly expressed as

$$PC_2 = \alpha_{21}X_1 + \alpha_{22}X_2 + \ldots + \alpha_{2p}X_p \qquad (8.7)$$

and provides the linear combination, orthogonal to PC_1, which explains the next highest level of response variation. Orthogonality, in this context, means independent of PC_1 and lying at right angles to PC_1 in two-dimensional space. The remaining PCs, PC_3 to PC_p, can be similarly defined with each orthogonal to their immediate predecessors.

In PCA applications, for the eigenvalue equation (8.1), S refers to the variance–covariance matrix for the p variables and the elements of the eigenvector w are scaled to satisfy $w'w = 1$. This scaling means that the calculated eigenvalue is equivalent to the variance of the associated PC. Hence, use of eigenvalues to measure the importance of each PC in respect of the amount of variation it explains. The weights α of each PC represent the eigenvector solutions of equation (8.1) which maximise the variance of each PC.

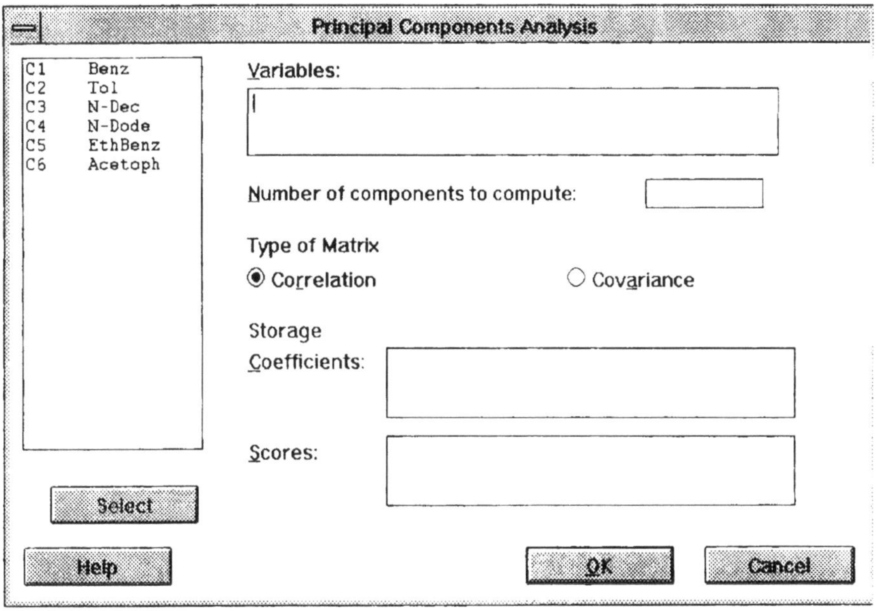

Figure 8.1 *Principal components analysis dialog window in Minitab*

Implementation of PCA within Minitab is uses the menu commands **Stat ▷ Multivariate ▷ Principal Components** which produce the PCA dialog window shown in Figure 8.1. As previously, data require to be selected and boxes filled in or checked before invoking the in-built PCA routine.

2.2 Number of Components

Where there are correlations between the response variables, it can be expected that PCA will result in some degree of data condensation enabling the essential features of the data to be explained by a small number of uncorrelated PCs. How do we decide how many components are necessary to provide an acceptable and efficient explanation of the data collected?

The importance of each PC, in terms of level of data variation explained, is specified by its eigenvalue, the λ term, with $\Sigma\lambda$ representing the total of the p eigenvalues. A measure of the proportion of data variation accounted for by each PC, based on the equivalence of eigenvalue and PC variance, is provided by the expression

$$\lambda/(\Sigma\lambda) \tag{8.8}$$

Expression (8.8) will always be large for the first PC, PC_1, less so for the second, PC_2, less again for the third, PC_3, and so on.

The general guideline in PCA applications is to select those PCs which account, cumulatively, for *at least 80% to 90%* of the data variation, *i.e.* the addition of equation (8.8) for each important PC must generally exceed 0.8. In practice, it is hoped that little variability is accounted for by successive PCs beyond the third or fourth suggesting that most variation may be adequately described by the first two or three PCs. If this is possible, we will have achieved dimensionality reduction in that we can replace the p correlated response variables by two or three uncorrelated components which explain most of the data structure and patterning. This reduction of multivariate data is also called *feature reduction*, or *ordination*.

Example 8.1

An investigation was conducted into the concentration of organic air pollutants at an air quality monitoring station. Air samples were taken weekly over a study period of 12 weeks and the concentrations of the chemicals benzene, toluene, *n*-decane, *n*-dodecane, ethylbenzene, and acetophenone measured by capillary GC. The collected data, in arbitrary units, are presented in Table B.6. Are there subsets of chemicals with similar characteristics? Are certain chemical air pollution patterns prevalent?

This is a data set with $p = 6$ variables and $n = 12$ samples providing $6 \times 12 = 72$ observations in total. We are interested in assessing whether the data are sufficiently structured to be able to explain the construction of the chemical air pollution and any similarities between the weekly air samples collected.

Data entry in Minitab
To perform PCA in Minitab, we require to enter the chemical data into separate columns for each measured variable using C1 for benzene, C2 for toluene, C3 for *n*-decane, and so on. We also create two further columns of labels, Vars (C7) and Weeks (C8), corresponding to abbreviations for the variables measured and the weekly samples collected. The label columns will be used for PCA plotting (see Examples 8.2 and 8.3).

Minitab output
Output 8.1 represents the PCA output obtained from the Principal Components procedure within Minitab where the variable labels used are self-explanatory. The Storage options in the Menu procedure enable the PC weights and PC scores to be stored for later analysis (see

Output 8.1 *PCA information for Example 8.1*

Select **Stat** ▷ **Multivariate** ▷ **Principal Components** ▷ for *Variables*, highlight **Benz to Acetoph** and click **Select** ▷ for *Storage*, select the **Coefficients** box and enter **C9–C14** ▷ for *Storage*, select the **Scores** box and enter **C15–C20** ▷ click **OK**.

Principal Component Analysis
Eigenanalysis of the Correlation Matrix

Eigenvalue	4.2990	1.0260	0.4188	0.2111	0.0292	0.0160
Proportion	0.716	0.171	0.070	0.035	0.005	0.003
Cumulative	0.716	0.887	0.957	0.992	0.997	1.000

Variable	PC1	PC2	PC3	PC4	PC5	PC6
Benz	−0.477	−0.005	−0.104	−0.089	0.521	0.694
Tol	−0.462	−0.144	−0.143	−0.443	0.345	−0.656
N-Dec	0.432	−0.001	0.408	−0.777	0.099	0.184
N-Dode	−0.035	0.978	−0.153	−0.115	0.018	−0.069
EthBenz	−0.470	−0.045	−0.092	−0.361	−0.774	0.198
Acetoph	0.388	−0.142	−0.878	−0.220	0.006	0.101

Examples 8.2 and 8.3). The output presented provides details of the eigenvalues and form of each of the six PCs which describe the chemicals data set.

Number of components
The first part of Output 8.1 indicates that the first PC (PC_1) has the largest eigenvalue at 4.2990 ('Eigenvalue') and accounts for 71.6% of the variation in the collected data ('Proportion 0.716'). The second PC (PC_2) has next largest eigenvalue at 1.0260 and accounts for 17.1% of the data variation ('Proportion 0.171'). The variation accounted for by the remaining PCs are third PC 7% ('Proportion 0.070'), fourth PC 3.5% ('Proportion 0.035'), fifth PC 0.5% ('Proportion 0.005'), and sixth PC 0.3% ('Proportion 0.003').

Cumulatively, the first two PCs, PC_1 and PC_2, account for 88.7% of the total variation ('Cumulative 0.887') indicating that two PCs appear sufficient to be able to explain the structures and patterns in the data. Including the third PC raises this accountability to 95.7% ('Cumulative 0.957') so it could be argued that the first three PCs could provide a better platform for data interpretation.

Specification of the first two PCs
From the second part of Output 8.1, we are able to obtain the format of each PC in terms of the measured chemicals.

PC_1 The first PC ('PC1') is specified as

PC_1 = −0.477 Benzene − 0.462 Toluene + 0.432 n-Decane − 0.035 n-Dodecane − 0.47 Ethylbenzene + 0.388 Acetophenone

Only *n*-dodecane has no influence on PC_1 while the other chemicals split into two distinct groups (benzene, toluene, ethylbenzene) and (*n*-decane, acetophenone). The weights, except that for n-dodecane, are all middling suggesting all have moderate influence on whatever PC_1 explains of the pollution data.

PC_2 The second PC ('PC2') is

PC_2 = −0.005 Benzene − 0.144 Toluene − 0.001 n-Decane + 0.978 n-Dodecane − 0.045 Ethylbenzene − 0.142 Acetophenone.

This component is dominated by *n*-dodecane (highest weight). All other variables appear to make no contribution to this PC (low weights).

PC_3 The third PC ('PC3') is

PC_3 = −0.104 Benzene − 0.143 Toluene + 0.408 n-Decane − 0.153 n-Dodecane − 0.092 Ethylbenzene − 0.878 Acetophenone

Acetophenone and *n*-decane dominate PC_3 but with opposite signs and different magnitudes of their coefficients. This suggests they have opposite effects on the aspect of the pollution data explained by this PC.

For the purposes of PCA illustration, only the first two PCs from the pollution data of Example 8.1 will be assessed. They account for an acceptable level of variation in the collected data at 88.7%, in excess of the accepted target of 80% accountability for PCA applications. Another reason for only choosing PC_1 and PC_2 for analysis lies in the fact that each variable has a weighting markedly different from zero within these two PCs, five in PC_1 and one in PC_2. In other words, structural explanation of data through the first two PCs picks up all six variables.

2.3 Analysis of PC Weights

Interpretation of the results of PCA can be difficult dependent as it is on an investigator's ability to translate the PC information into a chemically interpretable solution. The first PC is generally an overall measure of the chemical information contained in the variables. The PC weights refer to the correlations between the variables and the PC with their sign and magnitude reflecting the directions and relative importance of each response to the information explained by the PC. Weights markedly different from zero, either positive or negative, signify strong correlation while weights near zero, either positive or negative, are indicative of weak correlation. Thus, the higher the weighting the more the variable has in common with the PC and the more it contributes to what the PC explains of the data structure. If several variables load together on one or more PCs, the patterns in the PC weights may reflect important relationships among the variables and so provide an explanation of the data structure. It is this effect that must be explained on a chemical basis.

To help in this aspect of PCA, it is advisable to plot the PC weights for pairings of the k principal components necessary to describe the data adequately. This helps to provide simple visual representations of the data structure which will hopefully highlight trends or variable clusters from which a chemical interpretation can be formulated. If the number of PCs required to provide an adequate description is small, this aspect can be simple to implement.

Example 8.2

Having established that the chemicals data of Example 8.1 can be
adequately described by means of the first two PCs, we need to now
interpret these components to assess how they explain the data structure. We can do this by examining the PC weights and by plotting them
to assess for evidence of clustering or patterning which may help
explain the chemical make-up of air pollution in the area studied.

Interpretation of the PC weights
A Minitab plot of the weights for the first two PCs is presented in
Output 8.2. PC_1 has similar weights (*numerical* value of coefficient) for

Output 8.2 *PC weights plot for Example 8.2*

Select **Graph** ▷ **Plot** ▷ for *Graph 1 Y*, select **PC2W** (C10) and click **Select** ▷
for *Graph 1 X*, select **PC1W** (C9) and click **Select**.
  Select **Frame** ▷ **Axis** ▷ select the **Label 1** box and enter **PC1 Weights** ▷
select the **Label 2** box and enter **PC2 Weights** ▷ click **OK**.
  Select **Annotation** ▷ **Title** ▷ for *Title 1*, enter **PC Weights Plot, PC1 and
PC2** ▷ click **OK**.
  Select **Edit Attributes** ▷ for *Graph 1 Type*, enter **None** in the box ▷ click
OK.
  Select **Annotation** ▷ **Data Labels** ▷ select **Show data labels** ▷ select **Use
labels**, select the empty box, select **Vars** and click **Select** ▷ click **OK**.
  Select **Frame** ▷ **Reference** ▷ for *Direction 1*, enter **Y** ▷ select the **Position 1**
box and enter **0** ▷ select the **Direction 2** box and enter **X** ▷ select the **Position
2** box and enter **0** ▷ click **OK** ▷ click **OK**.

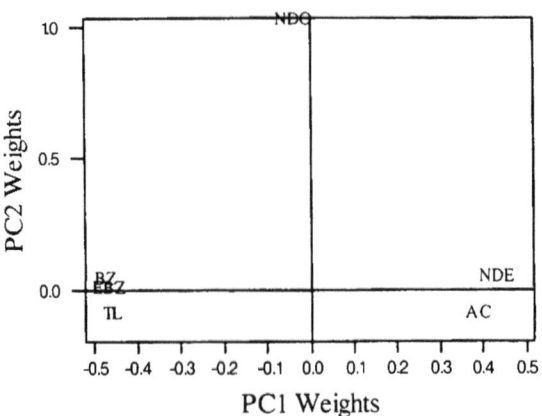

PC Weights Plot, PC1 and PC2

BZ Benz, EBZ EthylBenz, TL Toluene, NDO N-Dodec, NDE N-Dec, AC Acetoph benzene, toluene, *n*-decane, ethylbenzene, and acetophenone suggesting that these substances contribute most to the air pollution. n-Decane and acetophenone have positive weights while benzene, toluene, and ethylbenzene have negative weights. PC_2 appears to be dominated by n-dodecane and so may be measuring the effect of this chemical within the air pollution data.

Examining the plot with respect to the PC_1 weights axis shows two distinct groupings, one of benzene, ethylbenzene, and toluene at the negative end (BZ, EBZ, TOL) and one of *n*-decane and acetophenone at the positive end (NDE, AC). The former corresponds to aromatic hydrocarbons while the latter comprises aliphatic hydrocarbons. PC_1 may therefore be a measure of the difference between these two types of pollution whereby when one is high, the other may well be low due to the different signs of the respective group coefficients (negative correlation).

Examining the plot with respect to the PC_2 weights axis, we see that *n*-dodecane (NDO) is the dominant chemical (positive coefficient) suggesting that this PC is explaining the effect of this aliphatic hydrocarbon within air pollution such that when it is high, this type of chemical pollution is high due to the positive value of the associated coefficient in the PC.

For the chemicals as a whole, there is distinct evidence of clustering with groupings of (benzene, ethylbenzene, toluene), (*n*-decane, aceto-phenone), and *n*-dodecane very much apparent. This indicates that the chemicals data appear to break down into aromatic versus aliphatic hydrocarbons suggesting pollution causes may differ with source of the two types of chemical pollution.

2.4 Analysis of PC Scores

The value for a principal component corresponding to a specific sample of response measurements is called the *PC score* for that sample. PC scores are evaluated by either substituting the standardised values of each response measurement into the PC expression and calculating the resultant numerical value, or by substituting the original response measurements and standardising the resultant scores. Such data are used to assess for similarity across the samples with similar PC scores being indicative of samples with comparable characteristics. Plots of PC scores for each pair of selected components are therefore an integral part of the data interpretation enabling samples to be visually compared for similarities in characteristics. Exact sample comparisons for those deemed similar can then be carried out by referring back to the original data measurements.

Example 8.3

In Example 8.2, analysis of the PC weights for the chemicals data of Example 8.1 was carried out. The final stage of PCA concerns analysis of the PC scores for the air samples to discern if there are groupings of samples which have similar chemical characteristics through the comparability of their respective scores. As with PC weights analysis, graphical approaches are best for this stage of PCA as the Minitab generated plot in Output 8.3 demonstrates.

Interpretation of PC scores
The plot in Output 8.3 has a wide scatter of points with no substantive evidence of sample clustering. Some minor grouping of samples is occurring with samples from weeks (5, 6, 10, 12) and (7, 8, 9) producing reasonably similar score estimates in respect of both PCs. This suggests that these weeks may well have produced reasonably similar measurement responses for all variables.

We must first consider an examination of the scores plot with respect to the PC_1 axis. This entails assessing sample similarity in a vertical direction. From this check, we can see that weeks with similar PC scores include (8, 9), (5, 10, 12), (6, 11), and possibly (3, 4). Assessing the original data in Table B.6, we can see that samples 3 and 4, for instance, show similarities in the measurement of the aromatic hydrocarbons but differences in aliphatic hydrocarbon measurements though the way the differences occur and the PC coefficient signs means such an effect cancels itself out. In all other cases, all measurements show some difference but not sufficiently for any real, chemically interpretable pattern to be forthcoming.

Examining the plot with respect to the PC_2 axis, *i.e.* assessing sample similarity in a horizontal direction, we see that weeks with similar scores include (7, 9, 12) and (4, 5, 6, 10). On examination of the recorded data in Table B.6, it can be seen that these weeks have similar *n*-dodecane measurements, the dominant response of PC_2. Weeks 1, 2, and 11 have very different *n*-dodecane measurements and hence the obvious separation of these weeks in the plot in Output 8.3.

In Examples 8.1 to 8.3, we have illustrated the principals underpinning PCA and have been able to show how it can be used to extract chemical information on data structure. Such interpretation, however, relies heavily on the data conforming to a chemically interpretable structure. Inclusion of information on meteorological characteristics, time of sampling, location of industry, and location of local roads network would help further the understanding of the chemicals data of

Output 8.3 *PC scores and plot for Example 8.3*

Select **File** ▷ **Display Data** ▷ for *Columns, constants, and matrices to display,*
enter **C15-C20** ▷ click **OK**.

Data Display
Row	PC1S	PC2S	PC3S	PC4S	PC5S	PC6S
1	−3.34416	1.04027	0.00087	−0.306186	0.208668	0.055648
2	−2.37647	0.64446	−0.08432	0.498906	0.182255	0.060158
3	2.68123	−0.93539	0.73933	−0.321579	0.200991	0.076056
4	3.08385	−0.10418	−0.88865	0.233854	0.218873	−0.211860
5	0.77526	−0.12788	0.62667	0.710973	−0.067527	0.047763
6	1.40037	0.02417	0.32252	−0.018282	−0.056007	0.139298
7	−0.98790	−0.77743	−1.01916	0.083844	−0.285044	0.011175
8	−1.82368	−1.16036	0.73204	−0.733050	−0.081024	−0.176987
9	−1.96642	−0.64598	−0.05831	0.361885	0.020118	−0.134486
10	0.42370	0.06533	0.50688	0.433715	−0.177808	0.028927
11	1.60136	2.49825	0.08395	−0.417343	−0.167984	−0.091380
12	0.53286	−0.52125	−0.96182	−0.526736	0.004490	0.195688

Select **Graph** ▷ **Plot** ▷ for *Graph 1 Y*, select **PC2S** (C16) and click **Select** ▷
for *Graph 1 X*, select **PC1S** (C15) and click **Select**.

Select **Frame** ▷ **Axis** ▷ select the **Label 1** box and enter **PC1 Scores** ▷
select the **Label 2** box and enter **PC2 Scores** ▷ click **OK**.

Select **Annotation** ▷ **Title** ▷ for *Title 1*, enter **PC Scores Plot, PC1 and PC2**
▷ click **OK**.

Select **Edit Attributes** ▷ for *Graph 1 Type*, enter **None** in the box ▷ click
OK.

Select **Annotation** ▷ **Data Labels** ▷ select **Show data labels** ▷ select **Use
labels**, select the empty box, select **Weeks** and click **Select** ▷ click **OK**.

Select **Frame** ▷ **Reference** ▷ for *Direction 1*, enter **Y** ▷ select the **Position 1**
box and enter **0** ▷ select the **Direction 2** box and enter **X** ▷ select the **Position
2** box and enter **0** ▷ click **OK** ▷ click **OK**.

PC Scores Plot, PC1 and PC2

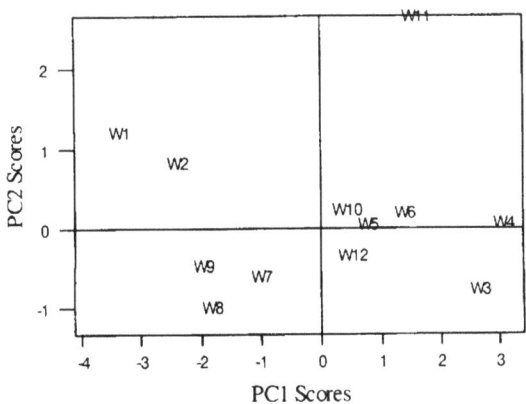

W1 - Week 1. W2 - Week 2. etc.

Example 8.1. In the illustration of PCA, we had $n = 12$ samples and $p = 6$ response variables. This combination of variables and samples is smaller than would be ideal for proper application of PCA when we should attempt to have the number of samples much larger than the number of response variables measured on the samples.

Examples 8.1 to 8.3 also illustrated that the chemicals data could be adequately explained by only two PCs providing suitable dimensionality reduction and an interpretable picture of the inherent structure of the data with minimal loss of information. It is hoped that, in most applications of PCA, only a few PCs will be necessary to describe the data. However, there may be cases where four, five, or more PCs are necessary with analysis requiring consideration of all possible combinations of selected PCs from both weights and scores perspective. Simple illustrations of the use of PCA in practical chemical studies are provided in Scott[7] and Shattuck *et al.*[8]

PCA can also be implemented from two other starting points, the *correlation matrix* and *covariance matrix*, each of which represent summary forms of the recorded data. The former corresponds to the correlation coefficients for each pair of variables and the latter to the 'variances' for each pair of variables. The resultant PCA solution is the same irrespective of chosen starting point. If there is a need to compensate for experimental error, the principle of *weighted PCA* (WPCA) can be considered. One method of weighting that can be used is based on weighting each observation by dividing it by the standard deviation of the associated response measurements, a technique analogous to WLS (see Chapter 5, Section 10). PCA is then performed on the transformed data matrix.

Exercise 8.1

The data presented in Table B.7 represent measurements, in arbitrary units, of air pollution variables recorded at 12 noon in a particular urban area over a 20 day period. The variables measured in each air sample were solar radiation levels (SolRad), wind levels (Wind), and the concentrations of carbon monoxide (CO), nitric oxide (NO), hydrocarbons (HC), nitrogen dioxide (NO_2), and ozone (O_3). Assess these data using PCA. Review the results obtained, particularly considering the effectiveness of the first few PCs and whether or not they have any plausible interpretation. Is it possible to summarise these data in less than seven dimensions? Can the data be summarised in less than three dimensions?

[7] C.J.P. Scott, *Anal. Proc.*, 1991, **28**, 245.
[8] T.W. Shattuck, M.S. Germani and P.R. Buseck, *Anal. Chem.*, 1991, **63**, 2646.

3 PRINCIPAL COMPONENTS REGRESSION

In multivariate data handling, we can often come across a situation where we wish to model a response Y, *e.g.* concentration of a compound, as function of many Xs which may refer to absorbance of different substances within the compound. This type of problem occurs in multivariate calibration and in UV/VIS spectrophotometry when analysing mixtures consisting of four or five compounds monitored over 20 to 30 wavelengths. We could consider multiple regression to model Y as a function of the Xs, or we could carry out PCA on the X variables (component measurements) and use the simplification obtained as the basis of fitting a multiple regression model. The latter approach is called *principal components regression* (PCR) and represents a 'supervised' MVA technique which has found application in areas such as structure–activity relationships (SAR) and modelling chemical properties of molecules.

PCR combines PCA and regression techniques and is particularly appropriate in the presence of multi-collinearity among the Xs. To use PCR, we would first carry out PCA on the recorded X data to, hopefully, derive a low dimension representation of the data through a few principal components. This enables the X measurements to be replaced by a set of independent components, the PCs, which show no evidence of collinearity. We then model the response Y as a function of the scores of these PCs with suitability of modelling Y in this way dependent on response prediction being acceptable. In other words, the PC scores become the independent variables in the modelling process.

Linear models based on the multiple model

$$Y = a + b_1(PC_1 \text{ scores}) + b_2(PC_2 \text{ scores}) + \ldots + b_p(PC_p \text{ scores})$$

are always used. A sequential building approach is adopted, similar to forward selection in stepwise multiple regression (see Chapter 5, Section 8). We first fit a one-component model

$$Y = a + b(PC_1 \text{ scores})$$

then a two-component model

$$Y = a + b_1(PC_1 \text{ scores}) + b_2(PC_2 \text{ scores})$$

then a three-component model,

$$Y = a + b_1(PC_1 \text{ scores}) + b_2(PC_2 \text{ scores}) + b_3(PC_3 \text{ scores})$$

and so on. At each stage, adequacy of model fit is monitored to ensure that the model is providing acceptable predictive accuracy. The most suitable model will be the one which provides best predictions based on least number of PCs, either the number specified by PCA for data explanation or fewer. It must be noted, however, that PCR can never improve on the variation accounted for by a simple multiple linear regression model based on all the controlled variables. *Partial least squares* (PLS) regression, representing a generalisation of the methods of PCA and multiple regression, is often used as an alternative to PCR in chemical studies.

4 FACTOR ANALYSIS

The technique of *factor analysis*,[9] which can be considered an extension of PCA, has been applied primarily in psychology to investigate animal and human processes for underlying observable traits. With the advent of powerful computers and statistical software, FA has come to be applied in chemistry in a variety of forms to enable multivariate chemical data to be investigated. Areas where factor analysis has been applied include pyrolysis mass spectra (PyMS) experiments and GC–MS where several mass spectra may be acquired across the chromatographic profile. Factor analysis is based on specification of a mathematical model of the observations in the form of factors plus error where we assume the common factors and error are uncorrelated, and the error conforms to a specified covariance structure.

The main aim of factor analysis is the orderly simplification of the variables into groups whereby within-group correlations are high and inter-group correlations low. As with PCA, it is hoped that each group of measured variables represents a conceptually meaningful independent factor such that the inherent data structures become more apparent in the 'factor space' as opposed to the 'data space'. Factor analysis attempts to reduce the number of variables by grouping them into basic factors which more directly explain the chemical information contained within the data.

By way of an example, suppose data have been collected on several characteristics associated with molecules. Using factor analysis, a molecule score matrix, which depends solely upon the characteristics of the molecules, and a measurement loading matrix, which depends solely upon the nature of the measurements, may be obtained. This separation of the features of the molecules from the features of the measurements

[9] E.R. Malinowski, 'Factor Analysis in Chemistry', 2nd Edn., Wiley, New York, 1991.

may help to provide a better insight into the true nature of the chemical phenomena present in the data.

5 STATISTICAL DISCRIMINANT ANALYSIS

Statistical discriminant analysis (SDA) differs from data reduction methods such as PCA in that it is concerned with deriving a mechanism for classifying k known groups of samples based on a common set of p variable measurements. SDA is a 'supervised' learning tool which requires the data to be structured in such a way that sample clusters can be easily distinguished between according to measured characteristics. Through these differences, it may be possible to build a mechanism for sample classification. SDA, therefore, can be used for discrimination (recognition, classification) and prediction, the latter concerned with assigning unknown to known classifications based on their measured characteristics.

This MVA technique requires a group structure for the original data, such as shown in Table 8.2, and tries to assess if such a structure can form the basis of a discrimination procedure. Using multiple observations to discriminate between groups of samples was first discussed by Fisher (1936)[10] for the analysis of flower measurements for three species of iris. In chemistry, for example, developing classification systems for spectral, chromatographic, or compositional data from HPLC methods may be desirable in order to produce mechanisms for source identification, food quality testing, and chemical taxonomy.

For example, suppose data were collected on the various fatty acid profiles of three related bacterial strains. Using the inherent strain grouping of the data, we could use SDA to build a mechanism for distinguishing bacterial strains through their fatty acid profiles. Structuring the approach in this way would provide an objective classification system though such could be readily revised and improved if more appropriate discriminatory characteristics were forthcoming.

5.1 Objective of SDA

The main objective of SDA is to determine *discriminant functions* ('derived variables') of the p measured variables that separate the k groupings as distinctly as possible. Discriminant functions represent linear combinations of the response variables, such as

$$Z = a_1 X_1 + a_2 X_2 + \ldots + a_k X_p + c,$$

[10] R.A. Fisher, *Ann. of Eugenics*, 1936, **7**, 179.

where the parameters a_1, a_2, ..., a_p, and c depend on the data collected. In SDA, this linear combination is generally referred to as *Fisher's linear discriminant function* with the numerical value of this function for each sample of measurements called the *discriminant score*. These scores are such that they should have the same sign or similar magnitude for all samples within each distinct grouping. Plotting such scores, for each pair of variables, can be used to help assess the discrimination power of response variables and to indicate how group discrimination is operating.

Derivation of an SDA solution is based on the same mechanisms as PCA through solving the eigenvalue equation (8.1). Measures of the group differences are provided by the eigenvalues while the eigenvectors provide the estimates of the parameters of the associated discriminant functions. The values of the first discriminant function (associated with the largest eigenvalue) best separate the groups and reflect most of the group differences. The second discriminant function (associated with the second largest eigenvalue) captures as much as possible of the group differences not explained by the first function. The other discriminant functions, if appropriate, can be similarly interpreted.

For multiple group discrimination, based on k groups and p measured variables, the number of discriminant functions that can be derived is the minimum of p and $(k-1)$. It is generally hoped that few, *i.e.* one or two, discriminant functions will be sufficient to explain group differences though this depends on the number of groupings involved and number of response variables measured. For two groups ($k = 2$), there can only be one discriminant function (minimum of p and 1) irrespective of number of response variables measured. For three groups ($k = 3$), it is likely that two discriminant functions will exist (minimum of p and 2) because the number of response variables will probably exceed 2.

5.2 Analysis Concepts Associated with SDA

SDA is a more statistically based multivariate method than PCA. Analysis can be split into distinct parts with each coming together to provide an overall assessment of the feasibility of discrimination. The outlined analysis elements of Box 8.1 are based on using Minitab to provide the SDA results. Minitab's SDA procedure can be accessed through the menu commands **Stat ▷ Multivariate ▷ Discriminant Analysis**. The resulting dialog window is shown in Figure 8.2 with the usual selection, checking, and filling in required before invoking the procedure. Other software packages may provide SDA results in a different format requiring the adoption of other procedures for assessment of the discrimination information.

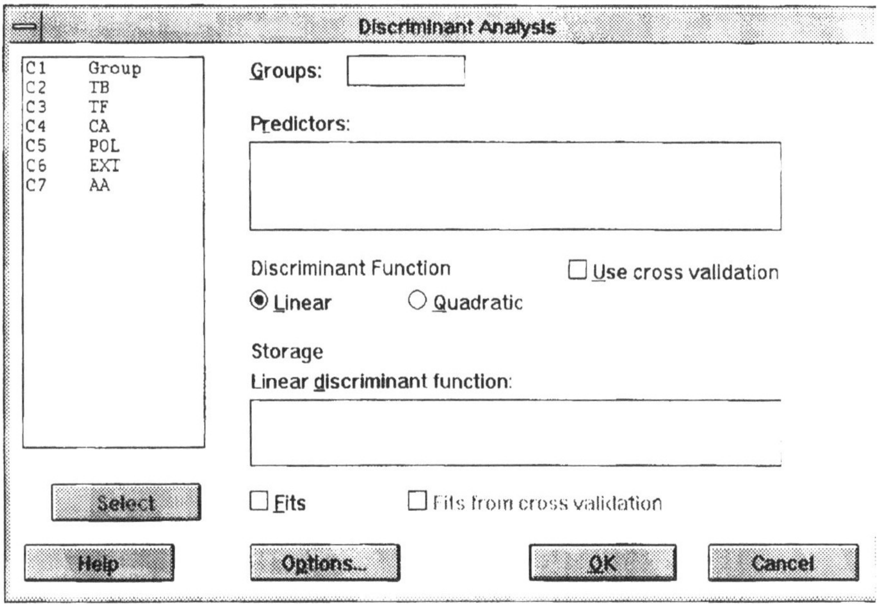

Figure 8.2 *Discriminant analysis dialog window in Minitab*

The *assumptions* underpinning SDA are comparable to those for the parametric inference procedures illustrated in earlier chapters. SDA is, in general, robust to departures from these assumptions though, in such cases, the statistical validity of the derived SDA procedure may be more difficult to establish.

Ideally, all experimental measurements will be correctly grouped. However, it is often the case that measurements may be incorrectly grouped and when building an SDA mechanism, it is important to know if any observations fall into this category. Testing for mis-classifications provides this data check. It is based on the principle that correctly grouped observations will produce a small distance measure for their specified group whereas a mis-classified observation is likely to produce a small distance measure for an alternative group. In other words, the mis-classified observation appears to have more character-istics in common with the alternative group than the group in which it was originally recorded. Developing a discriminant routine on the basis of incorrectly classified observations could introduce bias into the routine which may affect the structure of the routine and its ability to discriminate between the groupings correctly.

Test statistic (8.9) helps to check only the statistical validity of discrimination between each pair of groups. It does not provide a statistical check on the validity of the full discrimination routine

Box 8.1 *Assessment and interpretation concepts for SDA*

Assumptions. Within-group covariance matrix is the same for all groups (equal response variability for all groups) and response data follow the multivariate normal (each response variable is normally distributed).

Distance measure. The SDA procedure in Minitab utilises the Mahalanobis distance measure (8.5) for group discrimination where small values are indicative of similarity and large dissimilarity.

Mis-classifications. Once the discriminant functions have been derived, it is appropriate to check the effectiveness of the original data groupings, *i.e.* to check all the original samples have been classified correctly. This is known as *data exploration* and is based on determining each sample's distance measure with respect to each grouping with small values indicative of similarity.

Statistical validity. In any application of SDA, it is necessary to assess the statistical validity of the developed mechanism. One such test, based on the Mahalanobis distance measure, is described as follows.

The hypotheses for this test are specified as H_0: no discrimination between the compared groups against H_1: discrimination is possible between the compared groups. The test statistic is given by

$$F = \frac{n_1 n_2 (n_1 + n_2 - p - 1)}{(n_1 + n_2)(n_1 + n_2 - 2)p} D^2 \qquad (8.9)$$

based on degrees of freedom $df_1 = p$ and $df_2 = n_1 + n_2 - p - 1$ where n_1 is the number of samples in the first group being compared, n_2 the number in the second group being compared, p the number of response variables measured, and D^2 the Mahalanobis distance measure associated with the two groupings. The decision rule is such that small F values, generally corresponding to a small distance measure, implying no statistical discrimination between the groups are feasible, *i.e.* $F < F_{\alpha, df_1, df_2} \Rightarrow$ accept H_0.

Prediction. Provided the statistical validity of the classification mechanism is justified, we can use the developed procedure for data classification, whereby we try to predict group membership for test samples of known origin or samples of unknown origin. Classification, in Minitab, is based on calculation of distance measures and posterior probabilities for each test sample of observations with minimum group distance and maximum group probability, which always coincide, used for group classification.

Data summaries and plots. The final part of SDA assessment, provided classification is feasible, should be the assessment of variable summaries and plots to investigate whether we can detect the important variables driving the classification system and determine how the discrimination routine is classifying the observations. For summaries, this involves simple comparison of the mean and variability of each variable in respect of each group. For data plots, this checking entails plotting all pairs of variables and assessing each plot for any evidence of patterning or clustering which would explain the discrimination mechanism.

developed, except in the case of only two groups ($k = 2$). If statistical discrimination can be achieved for all group pairings, we can be reasonably confident that the full discrimination routine is valid and we have achieved a reduction in the dimensionality of the data.

If there are large numbers of variables, *i.e.* p is large, the data plotting approach can be cumbersome with many plots necessary to be produced. In addition, not all variable plots will necessarily provide a readily interpretable pattern.

Example 8.4

The data presented in Table B.8 correspond to chemical descriptors for two classes of tea, black and green. The latter is produced by drying and roasting the leaves while black tea requires that the leaves are additionally fermented. The chemical descriptors measured were theobromine (TB), theophylline (TP), caffeine (CA), total polyphenols (POL), aqueous extract (EXT), and total free amino acids (AA). Separate samples of each class were also experimented on to provide test data for validation of the predictive ability of any developed discrimination routine. The data from these additional samples are presented in Table 8.3.

Table 8.3 *Test chemical composition data for Example 8.4 (contents in % w/w, drybase)*

Sample number	TB	TP	CA	POL	EXT	AA
1 (Black)	0.11	0.04	3.6	18.0	33.8	0.86
2 (Black)	0.65	0.29	3.9	15.1	32.9	0.77
3 (Black)	0.34	0.14	3.2	20.5	32.0	1.43
4 (Green)	0.16	0.03	3.8	22.4	26.9	0.98
5 (Green)	0.22	0.04	4.3	24.8	27.6	1.37

We have data on two classes ($k = 2$) across six response variables ($p = 6$) for 12 samples of the Black class ($n_1 = 12$) and seven samples of the Green class ($n_2 = 7$). The data are presented according to class of tea and we want to apply SDA to assess whether such data can be separated into classes according to the measured chemical descriptors. We could attempt to look at the data for distinguishing features but it is clearly better to apply a more structured approach for which SDA is well suited.

Data entry
Entry of the chemical composition data into Minitab requires the first column C1 (Group) to be codes (1 and 2) for the tea classes checked.

Given the data structure in Table B.8, this would correspond to twelve
1s then seven 2s. The second to seventh columns, C2 (TB) to C7 (AA),
are used for the composition data. As with PCA, we create a column of
labels, Class (C8), corresponding to abbreviations for the tea classes
tested for use in any data plots. The measurements for the test samples
of Table 8.3 are entered as columns C9 (UnkTB), C10 (UnkTP), C11
(UnkCA), C12 (UnkPOL), C13 (UnkEXT), and C14 (UnkAA) where
each row would refer to a set of sample measurements.

Minitab output
Output 8.4 contains the initial results produced from the Discriminant
Analysis procedure in Minitab. The 'Predict group membership' aspect
of the Options sub-menu produces the test sample prediction information of Output 8.6 while choice of 'Above plus complete classification summary' in the Options sub-menu produces the summary of
classified observations in Output 8.5 and the summary statistics of
Output 8.7.

Summary
From 'Summary of Classification' presented in Output 8.4, we can see
that all the samples in group 2 (Green) were correctly placed. Unfortunately, it appears that one sample in group 1 (Black) may have been
incorrectly classified in the original data as its chemical descriptor
characteristics appear more similar to the Green class than the Black
class. The proportion of correctly grouped observations is given as
'Prop. Correct = 0.947' (94.7%) which is acceptable despite the
presence of a potentially mis-classified observation. Using this incorrectly classified observation could affect, though only marginally, the
ability of the developed routine to discriminate correctly between the
classes of tea.

Distance measure
The Minitab derived Mahalanobis distance measure for class separation
is presented in Output 8.4 in the section 'Squared Distance Between
Groups' with 1 referring to Black and 2 to Green. The estimate of
6.08478 is sufficient to suggest that there appears to be a reasonable
degree of separation between the two classes on the basis of their
chemical composition, indicating that discrimination between the
classes may be feasible.

Statistical validity
The next step is to use the distance measure and the F test statistic (8.9)
to assess the statistical validity of the classification system. For this

Output 8.4 *SDA information for Example 8.4*

Select **Stat** ▷ **Multivariate** ▷ **Discriminant Analysis** ▷ for *Groups*, select **Group** and click **Select** ▷ for *Predictors*, highlight **TB to AA** and click **Select**.

Select **Options** ▷ select the **Predict group membership for** box, highlight **UnkTB to UnkAA** ▷ for *Display of Results*, select **Above plus complete classification summary** ▷ click **OK** ▷ click **OK**.

Discriminant Analysis

Linear Method for Response: Group
Predictors: TB   TP   CA   POL   EXT   AA

Group	1	2
Count	12	7

Summary of Classification
Put into   . . . .True Group. . . .

Group	1	2
1	11	0
2	1	7
Total N	12	7
N Correct	11	7
Proport.	0.917	1.000

N = 19   N Correct = 18   Prop. Correct = 0.947

Squared Distance Between Groups

	1	2
1	0.00000	6.08478
2	6.08478	0.00000

Linear Discriminant Function for Group

	1	2
Constant	−87.669	−86.165
TB	−52.353	−49.257
TP	−35.715	−89.557
CA	14.764	11.794
POL	−0.629	0.145
EXT	4.666	4.098
AA	−11.405	0.075

illustration, we have $p = 6$ (number of variables measured), $n_1 = 12$ (number of Black samples), and $n_2 = 7$ (number of Green samples). We will test statistical validity of class discrimination at the 5% significance level. The degrees of freedom associated with the test statistic (8.9) are $df_1 = p = 6$ and $df_2 = n_1 + n_2 - p - 1 = 12 + 7 - 6 - 1 = 12$. The critical value (5% significance level) from Table A.3 for this comparison is $F_{0.05,6,12} = 3.00$.

The null hypothesis to be tested is H_0: no discrimination between the tea classes. For the test statistic, we have $n_1 = 12$, $n_2 = 7$, $p = 6$, and $D^2 = 6.08478$ resulting in test statistic (8.9) being calculated as

$$F = \frac{12(7)(12 + 7 - 6 - 1)}{(12 + 7)(12 + 7 - 2)6}(6.08478) = (0.5201)(6.08478) = 3.165$$

As test statistic exceeds the 5% critical value of 3.00, we can reject H_0, though only just, and conclude that there appears sufficient evidence to indicate that discrimination on the basis of the chemical descriptors presented appears possible between the tea classes ($p < 0.05$). As there are only two classes of tea, we can use this result as confirmation of the statistical validity of the developed discrimination routine. Given the closeness of test statistic and critical value, this result has to be treated with some caution as it suggests, even though statistical validity is technically shown, that the developed SDA routine may not be fulfilling its purpose as well as would be hoped.

Assessment of mis-classification
The information presented in Output 8.5 under 'Summary of Classified Observations' provides a detailed summary of how the discrimination routine has classified each of the original tea samples. In this table, we examine the results for 'True Group' (original class), 'Pred Group' (predicted class), 'Sqrd Distnc' (distance of observation from each class centroid), and 'Probability' (probability of sample being a member of a class) to check the classification procedure. 'Pred Group' specifies the group with which the tea class sample shows greatest similarity, while the latter two measures are used to provide numerical back-up to explain how this specification was arrived at.

Sample classification is based on assigning a sample to a group which has minimum distance and maximum probability in respect of that particular sample. Checking these measures for each observation, based on observation 1 to 11 as Black tea and 12 to 19 as Green tea, we can see that original group classification (True Group = Pred Group) is acceptable for all the samples bar two. These two results correspond to observations 1 and 11.

Output 8.5 *Summary of observation classifications for Example 8.4*

Summary of Classified Observations

Observation	True Group	Pred Group	Group	Sqrd Distnc	Probability
1	1	1	1	3.440	0.577
			2	4.064	0.423
2	1	1	1	5.098	0.910
			2	9.734	0.090
3	1	1	1	3.892	0.988
			2	12.781	0.012
4	1	1	1	7.997	0.870
			2	11.804	0.130
5	1	1	1	2.929	0.957
			2	9.148	0.043
6	1	1	1	3.415	0.949
			2	9.251	0.051
7	1	1	1	9.981	1.000
			2	26.850	0.000
8	1	1	1	5.400	0.941
			2	10.944	0.059
9	1	1	1	4.814	0.957
			2	11.009	0.043
10	1	1	1	8.680	0.926
			2	13.721	0.074
11 **	1	2	1	7.372	0.092
			2	2.792	0.908
12	1	1	1	12.41	0.999
			2	26.34	0.001
13	2	2	1	7.361	0.171
			2	4.200	0.829
14	2	2	1	12.216	0.065
			2	6.890	0.935
15	2	2	1	4.406	0.288
			2	2.597	0.712
16	2	2	1	8.711	0.042
			2	2.464	0.958
17	2	2	1	15.799	0.002
			2	3.283	0.998
18	2	2	1	9.238	0.074
			2	4.188	0.926
19	2	2	1	11.440	0.014
			2	2.956	0.986

Observation 11, clearly marked by '**', is the mis-classified Black tea sample highlighted in the summary earlier. We see that this observation is classified as Pred Group 2 and so appears to exhibit chemical composition characteristics more similar to group 2 (class Green) than its original class (Black). The reasons why this is arrived at can be seen in the group 2 distance measure and group 2 probability for this observation, both of which differ markedly from the comparable measures for the sample's original grouping (group 1, Black). From this assessment, we can see how and why this observation is defined as a mis-classification.

Observation 1, on the other hand, has not been mis-classified but the similarities in the distance measures and group probabilities associated with it give cause for concern in respect of the acceptability of its classification. The information presented suggests that, though criteria of minimum distance and maximum probability are satisfied for group 1 prediction (class Black), this sample's classification may not be on a firm foundation.

Predictions for the test samples
Additional data on samples of known class, presented in Table 8.3, were withheld to be used to test any developed discrimination routine. Output 8.6 contains the summary classifications for these test samples

Output 8.6 *Prediction results for test observations of Example 8.4*

Prediction for Test Observations

Observation	Pred. Group	From Group	Sqrd Distnc	Probability
1	1			
		1	4.522	0.998
		2	17.321	0.002
2	1			
		1	110.337	1.000
		2	154.030	0.000
3	1			
		1	19.339	0.594
		2	20.103	0.406
4	2			
		1	8.704	0.083
		2	3.896	0.917
5	2			
		1	18.755	0.001
		2	5.748	0.999

and corresponds to the information generated by the Predict option of Minitab's Discriminant Analysis routine. Interest, as in assessment of mis-classification, lies in examining the 'Pred. Group' column using the distances ('Sqrd Distnc') and probabilities ('Probability') to check class prediction.

All the prediction results correctly predict sample class using the minimum distance and maximum probability criteria. Predictions for test samples 1 to 3 (Observations 1, 2, and 3) are specified as Pred. Group 1 (class Black) agreeing with their original designation. Samples 4 and 5 (Observation 4 and 5) are predicted to be Pred. Group 2 (class Green) also in agreement with their original designation. For sample 2, however, the associated distance measures are very large, indicative of dissimilarity. Though prediction correctly specifies the sample's tea class, we should be cautious of accepting such a prediction based on a distance measure of such magnitude. For sample 3, the predicted class is correctly specified as Pred. Group 1 (class Black) but the large distance measures and probabilities for the two possible tea classes are very similar. This difference within each prediction mechanism is not as marked as for the other test samples. We should, therefore, treat this prediction with caution and conclude that it may not be feasible to identify this sample adequately using the developed discrimination routine.

Overall, we can say that the discrimination routine appears to work reasonably well and appears capable of correctly predicting class membership though the criteria on which this is based may not be as valid as it appears. Such checking of the discrimination process and how it operates should be routine, irrespective of how good predictive capacity is.

Data summaries and plots
As a final assessment of group differences, we should consider looking at group summaries and variable plots of the composition measurements to investigate if we can detect which variables play the major roles in the routine. Through this, we may be able to highlight the important chemical measurements which may be driving the classification system. The mean composition measurements for each class are presented in the first part of Output 8.7.

Checking these, we can see that differences in average measurements, in a relative sense, are most important for theophylline (TP), total polyphenols (POL), and total free amino acids (AA). Examination of the summary standard deviations show the Green class to have low variability of measurements for the same three characteristics and theobromine (TB). Such results indicate that there are class differences

Output 8.7 *Summary statistics for Example 8.4*

Variable	Pooled Mean	Means for Group 1	Means for Group 2
TB	0.19316	0.19667	0.18714
TP	0.05053	0.05833	0.03714
CA	4.1053	4.0250	4.2429
POL	23.658	22.400	25.814
EXT	32.574	32.925	31.971
AA	1.0632	0.9850	1.1971

Variable	Pooled Stdev	Stdev for Group 1	Stdev for Group 2
TB	0.08925	0.10697	0.03988
TP	0.03524	0.04218	0.01604
CA	0.5347	0.5294	0.5442
POL	3.712	4.280	2.337
EXT	2.969	2.927	3.044
AA	0.2686	0.3047	0.1850

in the composition measurements with perhaps, TP, POL, and AA the important drivers of the classification system.

The variable plots in this case, of which 15 are possible, did not provide for good separation and so were of no use to this aspect of SDA assessment.

The mode of presentation of SDA for Example 8.4 has been based on the way in which the Minitab software presents its SDA information. Other ways of examining SDA results also exist and these may be the base of the information produced by other software packages, *e.g.* SAS. It may, therefore, be necessary to use different analysis approaches from those described here, such as discriminant function values and group centroids, to assess the validity of the discrimination routine produced.

An alternative method of group prediction based on evaluating each group's discriminant function, such as that shown in Output 8.4, is available. It is based on evaluating each discriminant function for each test sample and classifying the sample to the group with maximum function value. Further graphical approaches such as variable standard error plots for each group and discriminant score dotplots for each group could also be considered to help understand how group classification is occurring.

Exercise 8.2

A study was set up to investigate whether a discrimination scheme, based on measurements of trace metals by FAAS, could be developed to distinguish samples from three nearby river systems. The trace metals measured in samples from each system were chromium (Cr), copper (Cu), nickel (Ni), lead (Pb), and zinc (Zn). The collected data are presented in Table B.9 together with test samples from each system for testing the SDA scheme developed. Is discrimination between the water systems feasible based on the trace measurements? Is test sample prediction acceptable?

6 FURTHER SDA APPROACHES

Extensions to the SDA principles and assessment mechanisms outlined in Section 5 are also available. Several of these are described here.

6.1 Tests of the Overall Effectiveness of an SDA Routine

The *F* test (8.9) provides a means of statistically validating the discrimination between group pairings. It does not provide for assessing the overall significance of a *k* group SDA routine developed for *p* variables. Two tests which offer this facility are *Bartlett's V test statistic* and *Wilks's λ test statistic*. The null hypothesis assessed by each test is H_0: discrimination routine not viable, with alternative expressed as H_1: discrimination viable.

Bartlett's *V test statistic* is based on the eigenvalues of the derived discriminant functions and is expressed as

$$V = [N - (p + k)/2 - 1] \sum_{m=1}^{s} \log_e(1 + \lambda_m) = \sum_{m=1}^{s} V_m \qquad (8.10)$$

where N is the total number of observations ($\sum n_i$), s is the number of discriminant functions capable of being derived (minimum of p and $(k-1)$), λ_m is the eigenvalue of the *m*th discriminant function, and

$$V_m = [N - (p + k)/2 - 1] \log_e(1 + \lambda_m) \qquad (8.11)$$

V is assumed to be a χ^2 statistic with degrees of freedom $p(k-1)$ while V_m is also assumed to be a χ^2 statistic but with degrees of freedom $(p + k - 2m)$. Application of Bartlett's V statistic is made sequentially as Table 8.4 summarises.

Table 8.4 *Operation of Bartlett's V statistic*

Test	Test of	Approx χ^2 statistic	Degrees of freedom
1	Validity of SDA routine	V	$p(k-1)$
2	Are discriminant functions beyond the first necessary? ($s \geq 2$)	$V - V_1$	dfTest1 $- (p + k - 2)$
3	Are discriminant functions beyond the first two necessary? ($s \geq 3$)	$V - V_1 - V_2$	dfTest2 $- (p + k - 4)$
.	.	.	.
.	.	.	.

Test 1 is carried out first to check overall effectiveness of the developed SDA procedure. For $k = 2$ (s = 1), this test will provide the same result as test statistic (8.9). Tests 2, 3, *etc.* are based on cumulatively subtracting V_1, V_2, and so on from V to provide a remainder which enables the residual discrimination to be tested in order to assess how many discriminant functions are necessary to classify the grouped data adequately.

Testing stops once an insignificant result is achieved. This often occurs before s (minimum of p and $(k-1)$) is reached. For $k = 3$ groups with p very much greater than k, *i.e.* more response variables than groups, we have $s = 2$ and Test 1 will provide the test of overall SDA effectiveness, while Test 2 would provide a test of the validity of the second discriminant function.

ANOVA principles form the base of Wilks's λ measure, expressed as

$$\lambda = SS(\text{within group})/SSTotal \qquad (8.12)$$

This term provides a measure of the proportion of variation in discriminant scores not accounted for by the derived discriminant functions. As this represents a measure of unaccounted for variation, *i.e.* error, values near 0 will be indicative of good discrimination. The approximate χ^2 test statistic, based on Wilks's λ, is expressed as

$$\chi^2 = -[N - (p + k)/2 - 1] \log(\lambda) \qquad (8.13)$$

with of freedom $p(k-1)$ where N is the total number of observations ($\sum n_i$). Small values of λ will provide large χ^2 values associated with rejection of the null hypothesis of unviable discrimination. Such values will, therefore, be indicative of the derived SDA routine being a potentially effective tool for group discrimination.

6.2 Training/Test Set

In applications of SDA, it is sometimes useful to split the data set into two parts, particularly if it is large. One part is treated as the *training/calibrating* set with the second as the *test/validation* set. We would use the training set to develop and assess the SDA procedure. The test set is then used to monitor the predictive ability of the discrimination procedure developed. Splitting the data set in this way can ensure that the validity of a discrimination procedure is rigorously checked with independent data not used to develop the discrimination. The use of test data in Example 8.4 and again in Exercise 8.2 represents a simple illustration of this principle.

6.3 'Best' Discriminating Variables

Choice of response variables to form the basis of an SDA routine can be important as some may be good at discriminating between groups and others poor. In such applications, it may be appropriate to base discrimination on a 'best' set of variables which provide the best discrimination power. Adoption of this approach requires preliminary examination of the data to judge which variables may be 'best' and which may be discarded.

A related mechanism for selecting 'best' variables is that of s*tepwise discriminant analysis*. Similar to stepwise multiple regression (see Chapter 5, Section 8), this procedure builds up the discriminant functions one variable at a time until the necessary discriminant functions, based on a set of 'best' variables (acceptable group discrimination), are arrived at.

6.4 *A Priori* Probabilities

The practical illustration of SDA provided in Example 8.4 was based on the assumption that each class of tea occurs with equal probability in general circulation. This equality of occurrence may not always be the case, necessitating building the discrimination routine around a different base structure. Inequality of occurrence reflects imbalance between groups which must be reflected in the SDA routine developed. For example, if trying to classify oil types by their chemical composition, we may need to build the development process around the non-uniform rates of occurrence of different oil types in general use. Accounting for group imbalance corresponds to the assignment of *a priori probabilities* reflecting the nature of the imbalance, large probabilities associated with greater occurrence and small with less occurrence. Such assignment then becomes an integral part of the developed SDA routine and helps to ensure that any developed discrimination routine properly reflects natural group imbalance.

7 CLUSTER ANALYSIS

One possible objective of MVA may be to try to identify, if possible, groups of 'similar' samples sharing common characteristics for data mirroring the layout shown in Table 8.1, *i.e.* data which are not pre-grouped. Such groupings are called *clusters* with the techniques used to find them constituting *cluster analysis*.[11] This grouping technique is based on no prior knowledge of how many groups there are likely to be, nor how to distinguish between 'good' and 'bad' groupings. It is a primitive, unsupervised learning procedure.

Cluster analysis uses clustering algorithms, either *hierarchical* or *non-hierarchical*, to search the data for likely groupings which can be displayed on a dendrogram, similar to a family tree. Such algorithms are searching tools and so usage of software for algorithm implementation is essential. The clusters produced by such a process may have no real meaning so caution must be applied to the analysis of the produced 'clusters' to ensure that they can be interpreted appropriately from a chemical standpoint.

8 CORRESPONDENCE ANALYSIS

This MVA method[12] is a member of the family of factor analysis methods and has some resemblance to PCA. It treats the data matrix symmetrically by weighting the rows and columns of the matrix (see Table 8.1) equally, *i.e.* simultaneously represents rows and columns of the matrix. One advantage of correspondence analysis is that it can be applied to most data types (discrete, continuous, logical, nominal, ordinal) in order to gain an insight into the structures and patterns prevalent in the data set through the construction of new factors. It is an ordination method which is often applied to sample data on widely differing scales of measurement as no transformation of data is necessary for its implementation, unlike PCA.

[11] B.S. Everitt, 'Cluster Analysis', Arnold, London, 1993.
[12] M. Mellinger, *Chemom. Intell. Lab. Syst.*, 1987, **2**, 61.

Statistical Tables

Table A.1 *Critical values of the Student's t distribution*

df = degrees of freedom

df	$t_{0.1}$	$t_{0.05}$	$t_{0.025}$	$t_{0.01}$	$t_{0.005}$	df
1	3.078	6.314	12.710	31.820	63.660	1
2	1.886	2.920	4.303	6.965	9.925	2
3	1.638	2.353	3.182	4.541	5.841	3
4	1.533	2.132	2.776	3.747	4.604	4
5	1.476	2.015	2.571	3.365	4.032	5
6	1.440	1.943	2.447	3.143	3.707	6
7	1.415	1.895	2.365	2.998	3.499	7
8	1.397	1.860	2.306	2.896	3.355	8
9	1.383	1.833	2.262	2.821	3.250	9
10	1.372	1.812	2.228	2.764	3.169	10
11	1.363	1.796	2.201	2.718	3.106	11
12	1.356	1.782	2.179	2.681	3.055	12
13	1.350	1.771	2.160	2.650	3.012	13
14	1.345	1.761	2.145	2.624	2.977	14
15	1.341	1.753	2.131	2.602	2.947	15
16	1.337	1.746	2.120	2.583	2.921	16
17	1.333	1.740	2.110	2.567	2.898	17
18	1.330	1.734	2.101	2.552	2.878	18
19	1.328	1.729	2.093	2.539	2.861	19
20	1.325	1.725	2.086	2.528	2.845	20
21	1.323	1.721	2.080	2.518	2.831	21
22	1.321	1.717	2.074	2.508	2.819	22
23	1.319	1.714	2.069	2.500	2.807	23
24	1.318	1.711	2.064	2.492	2.797	24
25	1.316	1.708	2.060	2.485	2.787	25
26	1.315	1.706	2.056	2.479	2.779	26
27	1.314	1.703	2.052	2.473	2.771	27
28	1.313	1.701	2.048	2.467	2.763	28
29	1.311	1.699	2.045	2.462	2.756	29
30	1.310	1.697	2.042	2.457	2.750	30
32	1.309	1.694	2.037	2.449	2.738	32
34	1.307	1.691	2.032	2.441	2.728	34
36	1.306	1.688	2.028	2.434	2.719	36
38	1.304	1.686	2.024	2.429	2.712	38
40	1.303	1.684	2.021	2.423	2.704	40
50	1.299	1.676	2.009	2.403	2.678	50
60	1.296	1.671	2.000	2.390	2.660	60
90	1.291	1.662	1.987	2.369	2.632	90

The critical values presented are appropriate for both two- and one-tailed testing.

For a *two-tailed test* at the 5% ($\alpha = 0.05$) significance level, $\alpha/2 = 0.025$ and the value is read from column $t_{0.025}$.

For a *one-tailed test* at the 5% significance level, $\alpha = 0.05$ and the value is read from column $t_{0.05}$.

Computed with Minitab® by W.P. Gardiner.

Table A.2 *Critical values of the χ^2 distribution*

df = degrees of freedom

df	$\chi^2_{0.995}$	$\chi^2_{0.99}$	$\chi^2_{0.975}$	$\chi^2_{0.95}$	$\chi^2_{0.05}$	$\chi^2_{0.025}$	$\chi^2_{0.01}$	$\chi^2_{0.005}$	df
1	0.00	0.00	0.00	0.00	3.84	5.02	6.63	7.88	1
2	0.01	0.02	0.05	0.10	5.99	7.38	9.21	10.60	2
3	0.07	0.11	0.22	0.35	7.81	9.35	11.34	12.84	3
4	0.21	0.30	0.48	0.71	9.49	11.14	13.28	14.86	4
5	0.41	0.55	0.83	1.15	11.07	12.83	15.09	16.75	5
6	0.68	0.87	1.24	1.64	12.59	14.45	16.81	18.55	6
7	0.99	1.24	1.69	2.17	14.07	16.01	18.48	20.28	7
8	1.34	1.65	2.18	2.73	15.51	17.53	20.09	21.95	8
9	1.73	2.09	2.70	3.33	16.92	19.02	21.67	23.59	9
10	2.16	2.56	3.25	3.94	18.31	20.48	23.21	25.19	10
11	2.60	3.05	3.82	4.57	19.68	21.92	24.72	26.76	11
12	3.07	3.57	4.40	5.23	21.03	23.34	26.22	28.30	12
13	3.57	4.11	5.01	5.89	22.36	24.74	27.69	29.82	13
14	4.07	4.66	5.63	6.57	23.68	26.12	29.14	31.32	14
15	4.60	5.23	6.26	7.26	25.00	27.49	30.58	32.80	15
16	5.14	5.81	6.91	7.96	26.30	28.85	32.00	34.27	16
17	5.70	6.41	7.56	8.67	27.59	30.19	33.41	35.72	17
18	6.26	7.01	8.23	9.39	28.87	31.53	34.81	37.16	18
19	6.84	7.63	8.91	10.12	30.14	32.85	36.19	38.58	19
20	7.43	8.26	9.59	10.85	31.41	34.17	37.57	40.00	20
21	8.03	8.90	10.28	11.59	32.67	35.48	38.93	41.40	21
22	8.64	9.54	10.98	12.34	33.92	36.78	40.29	42.80	22
23	9.26	10.20	11.69	13.09	35.17	38.08	41.64	44.18	23
24	9.89	10.86	12.40	13.85	36.42	39.36	42.98	45.56	24
25	10.52	11.52	13.12	14.61	37.65	40.65	44.31	46.93	25
26	11.16	12.20	13.84	15.38	38.89	41.92	45.64	48.29	26
27	11.81	12.88	14.57	16.15	40.11	43.19	46.96	49.65	27
28	12.46	13.56	15.31	16.93	41.34	44.46	48.28	50.99	28
29	13.12	14.26	16.05	17.71	42.56	45.72	49.59	52.34	29
30	13.79	14.95	16.79	18.49	43.77	46.98	50.89	53.67	30

The critical values presented are appropriate for both two- and one-tailed testing.

For a *two-tailed test* at the 5% ($\alpha = 0.05$) significance level, $\alpha/2 = 0.025$, $1 - (\alpha/2) = 0.975$ and the necessary values are read from columns $\chi^2_{0.025}$ and $\chi^2_{0.975}$.

For a *one-tailed testing*, critical values depend on the form of the alternative hypothesis. For H_1 with the inequality < at the 5% significance level, $\alpha = 0.05$, $1 - \alpha = 0.95$, and the value is read from column $\chi^2_{0.95}$. For H_1 with the inequality > at the 5% significance level, $\alpha = 0.05$ and the value is read from column $\chi^2_{0.05}$.

Computed with Minitab® by W.P. Gardiner.

Table A.3 *Critical values of the F distribution*

Values of $F_{0.1}$ ($\alpha = 0.1$)

df_1 = degrees of freedom of numerator df_2 = degrees of freedom of denominator

df_2	1	2	3	4	5	6	7	8	df_2
2	8.53	9.00	9.16	9.24	9.29	9.33	9.35	9.37	2
3	5.54	5.46	5.39	5.34	5.31	5.28	5.27	5.25	3
4	4.54	4.32	4.19	4.11	4.05	4.01	3.98	3.95	4
5	4.06	3.78	3.62	3.52	3.45	3.40	3.37	3.34	5
6	3.78	3.46	3.29	3.18	3.11	3.05	3.01	2.98	6
7	3.59	3.26	3.07	2.96	2.88	2.83	2.78	2.75	7
8	3.46	3.11	2.92	2.81	2.73	2.67	2.62	2.59	8
9	3.36	3.01	2.81	2.69	2.61	2.55	2.51	2.47	9
10	3.29	2.92	2.73	2.61	2.52	2.46	2.41	2.38	10
11	3.23	2.86	2.66	2.54	2.45	2.39	2.34	2.30	11
12	3.18	2.81	2.61	2.48	2.39	2.33	2.28	2.24	12
14	3.10	2.73	2.52	2.39	2.31	2.24	2.19	2.15	14
16	3.05	2.67	2.46	2.33	2.24	2.18	2.13	2.09	16
18	3.01	2.62	2.42	2.29	2.20	2.13	2.08	2.04	18
20	2.97	2.59	2.38	2.25	2.16	2.09	2.04	2.00	20
22	2.95	2.56	2.35	2.22	2.13	2.06	2.01	1.97	22
26	2.91	2.52	2.31	2.17	2.08	2.01	1.96	1.92	26
30	2.88	2.49	2.28	2.14	2.05	1.98	1.93	1.88	30
40	2.84	2.44	2.23	2.09	2.00	1.93	1.87	1.83	40
50	2.81	2.41	2.20	2.06	1.97	1.90	1.84	1.80	50
60	2.79	2.39	2.18	2.04	1.95	1.87	1.82	1.77	60

df_2	9	10	12	15	20	24	30	df_2
2	9.38	9.39	9.41	9.42	9.44	9.45	9.46	2
3	5.24	5.23	5.22	5.20	5.18	5.18	5.17	3
4	3.94	3.92	3.90	3.87	3.84	3.83	3.82	4
5	3.32	3.30	3.27	3.24	3.21	3.19	3.17	5
6	2.96	2.94	2.90	2.87	2.84	2.82	2.80	6
7	2.72	2.70	2.67	2.63	2.59	2.58	2.56	7
8	2.56	2.54	2.50	2.46	2.42	2.40	2.38	7
9	2.44	2.42	2.38	2.34	2.30	2.28	2.25	8
10	2.35	2.32	2.28	2.24	2.20	2.18	2.16	9
11	2.27	2.25	2.21	2.17	2.12	2.10	2.08	10
12	2.21	2.19	2.15	2.10	2.06	2.04	2.01	11
14	2.12	2.10	2.05	2.01	1.96	1.94	1.91	12
16	2.06	2.03	1.99	1.94	1.89	1.87	1.84	16
18	2.00	1.98	1.93	1.89	1.84	1.81	1.78	18
20	1.96	1.94	1.89	1.84	1.79	1.77	1.74	20
22	1.93	1.90	1.86	1.81	1.76	1.73	1.70	22
26	1.88	1.86	1.81	1.76	1.71	1.68	1.65	26
30	1.85	1.82	1.77	1.72	1.67	1.64	1.61	30
40	1.79	1.76	1.71	1.66	1.61	1.57	1.54	40
50	1.76	1.73	1.68	1.63	1.57	1.54	1.50	50
60	1.74	1.71	1.66	1.60	1.54	1.51	1.48	60

Table A.3 (*continued*)

Values of $F_{0.05}$ ($\alpha = 0.05$)

df_1 = degrees of freedom of numerator df_2 = degrees of freedom of denominator

df_2	1	2	3	4	df_1 5	6	7	8	df_2
2	18.51	19.00	19.16	19.25	19.30	19.33	19.35	19.37	2
3	10.13	9.55	9.28	9.12	9.01	8.94	8.89	8.85	3
4	7.71	6.94	6.59	6.39	6.26	6.16	6.09	6.04	4
5	6.61	5.79	5.41	5.19	5.05	4.95	4.88	4.82	5
6	5.99	5.14	4.76	4.53	4.39	4.28	4.21	4.15	6
7	5.59	4.74	4.35	4.12	3.97	3.87	3.79	3.73	7
8	5.32	4.46	4.07	3.84	3.69	3.58	3.50	3.44	8
9	5.12	4.26	3.86	3.63	3.48	3.37	3.29	3.23	9
10	4.96	4.10	3.71	3.48	3.33	3.22	3.14	3.07	10
11	4.84	3.98	3.59	3.36	3.20	3.09	3.01	2.95	11
12	4.75	3.89	3.49	3.26	3.11	3.00	2.91	2.85	12
14	4.60	3.74	3.34	3.11	2.96	2.85	2.76	2.70	14
16	4.49	3.63	3.24	3.01	2.85	2.74	2.66	2.59	16
18	4.41	3.55	3.16	2.93	2.77	2.66	2.58	2.51	18
20	4.35	3.49	3.10	2.87	2.7f	2.60	2.51	2.45	20
22	4.30	3.44	3.05	2.82	2.66	2.55	2.46	2.40	22
26	4.23	3.37	2.98	2.74	2.59	2.47	2.39	2.32	26
30	4.17	3.32	2.92	2.69	2.53	2.42	2.33	2.27	30
40	4.08	3.23	2.84	2.61	2.45	2.34	2.25	2.18	40
50	4.03	3.18	2.79	2.56	2.40	2.29	2.20	2.13	50
60	4.00	3.15	2.76	2.53	2.37	2.25	2.17	2.10	60

df_2	9	10	12	15	df_1 20	24	30	df_2
2	19.38	19.40	19.41	19.43	19.45	19.45	19.46	2
3	8.81	8.79	8.74	8.70	8.66	8.64	8.62	3
4	6.00	5.96	5.91	5.86	5.80	5.77	5.75	4
5	4.77	4.74	4.68	4.62	4.56	4.53	4.50	5
6	4.10	4.06	4.00	3.94	3.87	3.84	3.81	6
7	3.68	3.64	3.57	3.51	3.44	3.41	3.38	7
8	3.39	3.35	3.28	3.22	3.15	3.12	3.08	8
9	3.18	3.14	3.07	3.01	2.94	2.90	2.86	9
10	3.02	2.98	2.91	2.85	2.77	2.74	2.70	10
11	2.90	2.85	2.79	2.72	2.65	2.61	2.57	11
12	2.80	2.75	2.69	2.62	2.54	2.51	2.47	12
14	2.65	2.60	2.53	2.46	2.39	2.35	2.31	14
16	2.54	2.49	2.42	2.35	2.28	2.24	2.19	16
18	2.46	2.41	2.34	2.27	2.19	2.15	2.11	18
20	2.39	2.35	2.28	2.20	2.12	2.08	2.04	20
22	2.34	2.30	2.23	2.15	2.07	2.03	1.98	22
26	2.27	2.22	2.15	2.07	1.99	1.95	1.90	26
30	2.21	2.16	2.09	2.01	1.93	1.89	1.84	30
40	2.12	2.08	2.00	1.92	1.84	1.79	1.74	40
50	2.07	2.03	1.95	1.87	1.78	1.74	1.69	50
60	2.04	1.99	1.92	1.84	1.75	1.70	1.65	60

Table A.3 (*continued*)

Values of $F_{0.025}$ ($\alpha = 0.025$)

df_1 = degrees of freedom of numerator df_2 = degrees of freedom of denominator

df_2	1	2	3	4	df_1 5	6	7	8	df_2
2	38.51	39.00	39.17	39.25	39.30	39.33	39.36	39.37	2
3	17.44	16.04	15.44	15.10	14.88	14.73	14.62	14.54	3
4	12.22	10.65	9.98	9.60	9.36	9.20	9.07	8.98	4
5	10.01	8.43	7.76	7.39	7.15	6.98	6.85	6.76	5
6	8.81	7.26	6.60	6.23	5.99	5.82	5.70	5.60	6
7	8.07	6.54	5.89	5.52	5.29	5.12	4.99	4.90	7
8	7.57	6.06	5.42	5.05	4.82	4.65	4.53	4.43	8
9	7.21	5.71	5.08	4.72	4.48	4.32	4.20	4.10	9
10	6.94	5.46	4.83	4.47	4.24	4.07	3.95	3.85	10
11	6.72	5.26	4.63	4.28	4.04	3.88	3.76	3.66	11
12	6.55	5.10	4.47	4.12	3.89	3.73	3.61	3.51	12
14	6.30	4.86	4.24	3.89	3.66	3.50	3.38	3.29	14
16	6.12	4.69	4.08	3.73	3.50	3.34	3.22	3.12	16
18	5.98	4.56	3.95	3.61	3.38	3.22	3.10	3.01	18
20	5.87	4.46	3.86	3.51	3.29	3.13	3.01	2.91	20
22	5.79	4.38	3.78	3.44	3.22	3.05	2.93	2.84	22
26	5.66	4.27	3.67	3.33	3.10	2.94	2.82	2.73	26
30	5.57	4.18	3.59	3.25	3.03	2.87	2.75	2.65	30
40	5.42	4.05	3.46	3.13	2.90	2.74	2.62	2.53	40
50	5.34	3.97	3.39	3.05	2.83	2.67	2.55	2.46	50
60	5.29	3.93	3.34	3.01	2.79	2.63	2.51	2.41	60

df_2	9	10	12	15	df_1 20	24	30	df_2
2	39.39	39.40	39.41	39.43	39.45	39.46	39.46	2
3	14.47	14.42	14.34	14.25	14.17	14.12	14.08	3
4	8.90	8.84	8.75	8.66	8.56	8.51	8.46	4
5	6.68	6.62	6.52	6.43	6.33	6.28	6.23	5
6	5.52	5.46	5.37	5.27	5.17	5.12	5.07	6
7	4.82	4.76	4.67	4.57	4.47	4.42	4.36	7
8	4.36	4.30	4.20	4.10	4.00	3.95	3.89	8
9	4.03	3.96	3.87	3.77	3.67	3.61	3.56	9
10	3.78	3.72	3.62	3.52	3.42	3.37	3.31	10
11	3.59	3.53	3.43	3.33	3.23	3.17	3.12	11
12	3.44	3.37	3.28	3.18	3.07	3.02	2.96	12
14	3.21	3.15	3.05	2.95	2.84	2.79	2.73	14
16	3.05	2.99	2.89	2.79	2.68	2.63	2.57	16
18	2.93	2.87	2.77	2.67	2.56	2.50	2.44	18
20	2.84	2.77	2.68	2.57	2.46	2.41	2.35	20
22	2.76	2.70	2.60	2.50	2.39	2.33	2.27	22
26	2.65	2.59	2.49	2.39	2.28	2.22	2.16	26
30	2.57	2.51	2.41	2.31	2.20	2.14	2.07	30
40	2.45	2.39	2.29	2.18	2.07	2.01	1.94	40
50	2.38	2.32	2.22	2.11	1.99	1.93	1.87	50
60	2.33	2.27	2.17	2.06	1.94	1.88	1.82	60

Table A.3 *(continued)*

Values of $F_{0.01}$ ($\alpha = 0.01$)

df_1 = degrees of freedom of numerator df_2 = degree of freedom of denominator

df_2	1	2	3	4	df_1 5	6	7	8	df_2
2	98.50	99.00	99.17	99.25	99.30	99.33	99.36	99.37	2
3	34.12	30.82	29.46	28.71	28.24	27.91	27.67	27.49	3
4	21.20	18.00	16.69	15.98	15.52	15.21	14.98	14.80	4
5	16.26	13.27	12.06	11.39	10.97	10.67	10.46	10.29	5
6	13.75	10.92	9.78	9.15	8.75	8.47	8.26	8.10	6
7	12.25	9.55	8.45	7.85	7.46	7.19	6.99	6.84	7
8	11.26	8.65	7.59	7.01	6.63	6.37	6.18	6.03	8
9	10.56	8.02	6.99	6.42	6.06	5.80	5.61	5.47	9
10	10.04	7.56	6.55	5.99	5.64	5.39	5.20	5.06	10
11	9.65	7.21	6.22	5.67	5.32	5.07	4.89	4.74	11
12	9.33	6.93	5.95	5.41	5.06	4.82	4.64	4.50	12
14	8.86	6.51	5.56	5.04	4.69	4.46	4.28	4.14	14
16	8.53	6.23	5.29	4.77	4.44	4.20	4.03	3.89	16
18	8.29	6.01	5.09	4.58	4.25	4.01	3.84	3.71	18
20	8.10	5.85	4.94	4.43	4.10	3.87	3.70	3.56	20
22	7.95	5.72	4.82	4.31	3.99	3.76	3.59	3.45	22
26	7.72	5.53	4.64	4.14	3.82	3.59	3.42	3.29	26
30	7.56	5.39	4.51	4.02	3.70	3.47	3.30	3.17	30
40	7.31	5.18	4.31	3.83	3.51	3.29	3.12	2.99	40
50	7.17	5.06	4.20	3.72	3.41	3.19	3.02	2.89	50
60	7.08	4.98	4.13	3.65	3.34	3.12	2.95	2.82	60

df_2	9	10	12	15	df_1 20	24	30	df_2
2	99.39	99.40	99.42	99.43	99.45	99.46	99.46	2
3	27.35	27.23	27.05	26.87	26.69	26.60	26.50	3
4	14.66	14.55	14.37	14.20	14.02	13.93	13.84	4
5	10.16	10.05	9.89	9.72	9.55	9.47	9.38	5
6	7.98	7.87	7.72	7.56	7.40	7.31	7.23	6
7	6.72	6.62	6.47	6.31	6.16	6.07	5.99	7
8	5.91	5.81	5.67	5.52	5.36	5.28	5.20	8
9	5.35	5.26	5.11	4.96	4.81	4.73	4.65	9
10	4.94	4.85	4.71	4.56	4.41	4.33	4.25	10
11	4.63	4.54	4.40	4.25	4.10	4.02	3.94	11
12	4.39	4.30	4.16	4.01	3.86	3.78	3.70	12
14	4.03	3.94	3.80	3.66	3.51	3.43	3.35	14
16	3.78	3.69	3.55	3.41	3.26	3.18	3.10	16
18	3.60	3.51	3.37	3.23	3.08	3.00	2.92	18
20	3.46	3.37	3.23	3.09	2.94	2.86	2.78	20
22	3.35	3.26	3.12	2.98	2.83	2.75	2.67	22
26	3.18	3.09	2.96	2.81	2.66	2.58	2.50	26
30	3.07	2.98	2.84	2.70	2.55	2.47	2.39	30
40	2.89	2.80	2.66	2.52	2.37	2.29	2.20	40
50	2.78	2.70	2.56	2.42	2.27	2.18	2.10	50
60	2.72	2.63	2.50	2.35	2.20	2.12	2.03	60

Computed with Minitab® by W.P. Gardiner.

Table A.4 *Table of z scores for the standard normal distribution*

α	z_α	α	z_α	α	z_α	α	z_α
0.20	0.8416	0.10	1.2816	0.0333	1.8344	0.0036	2.6875
0.19	0.8779	0.09	1.3408	0.025	1.9600	0.0033	2.7164
0.18	0.9154	0.08	1.4051	0.0167	2.1272	0.0025	2.8070
0.17	0.9542	0.07	1.4758	0.0125	2.2414	0.0024	2.8202
0.16	0.9945	0.06	1.5548	0.0083	2.3954	0.0017	3.9290
0.15	1.0364	0.05	1.6449	0.0075	2.4324	0.0012	3.0357
0.14	1.0803	0.04	1.7507	0.0067	2.4730	0.001	3.0902
0.13	1.1264	0.03	1.8808	0.005	2.5758		
0.12	1.1750	0.02	2.0537	0.0048	2.5899		
0.11	1.2265	0.01	2.3263	0.0042	2.6356		

Values refer to $P(z > z_\alpha) = \alpha$ where z corresponds to the standard, or unit, normal distribution, mean $\mu = 0$ and variance $\sigma^2 = 1$. They represent the $100\alpha\%$ critical values of the standard normal distribution.

Computed with Minitab® by W.P. Gardiner.

Table A.5 *5% Critical values for the studentised range statistic*

r = number of steps between ordered means   resdf = residual degrees of freedom

resdf	2	3	4	5	6	7	8	*r* 9	10	11	12	13	14	15	resdf
1	18.00	27.00	32.80	37.10	40.40	43.10	45.40	47.40	49.10	50.60	52.00	53.20	54.30	55.40	1
2	6.09	8.30	9.80	10.90	11.70	12.04	13.00	13.50	14.00	14.40	14.70	15.01	15.40	15.70	2
3	4.50	5.91	6.82	7.50	8.04	8.48	8.85	9.18	9.46	9.72	9.95	10.20	10.40	10.50	3
4	3.93	5.04	5.76	6.29	6.71	7.05	7.35	7.60	7.83	8.03	8.21	8.37	8.52	8.66	4
5	3.64	4.60	5.22	5.67	6.03	6.33	6.58	6.80	6.99	7.17	7.32	7.47	7.60	7.72	5
6	3.46	4.34	4.90	5.31	5.63	5.89	6.12	6.32	6.49	6.65	6.79	6.92	7.03	7.14	6
7	3.34	4.16	4.69	5.06	5.36	5.61	5.82	6.00	6.16	6.30	6.43	6.55	6.66	6.76	7
8	3.26	4.04	4.53	4.89	5.17	5.40	5.60	5.77	5.92	6.05	6.18	6.29	6.39	6.48	8
9	3.20	3.95	4.42	4.76	5.02	5.24	5.43	5.60	5.74	5.87	5.98	6.09	6.19	6.28	9
10	3.15	3.88	4.33	4.65	4.91	5.12	5.30	5.46	5.60	5.72	5.83	5.93	6.03	6.11	10
11	3.11	3.82	4.26	4.57	4.82	5.03	5.20	5.35	5.49	5.61	5.71	5.81	5.90	5.99	11
12	3.08	3.77	4.20	4.51	4.75	4.95	5.12	5.27	5.40	5.51	5.62	5.71	5.80	5.88	12
13	3.06	3.73	4.15	4.45	4.69	4.88	5.05	5.19	5.32	5.43	5.53	5.63	5.71	5.79	13
14	3.03	3.70	4.11	4.41	4.64	4.83	4.99	5.13	5.25	5.36	5.46	5.55	5.64	5.72	14
16	3.00	3.65	4.05	4.33	4.56	4.74	4.90	5.03	5.15	5.26	5.35	5.44	5.52	5.59	16
18	2.97	3.61	4.00	4.28	4.49	4.67	4.82	4.96	5.07	5.17	5.27	5.35	5.43	5.50	18
20	2.95	3.58	3.96	4.23	4.45	4.62	4.77	4.90	5.01	5.11	5.20	5.28	5.36	5.43	20
24	2.92	3.53	3.90	4.17	4.37	4.54	4.68	4.81	4.92	5.01	5.10	5.18	5.25	5.32	24
30	2.89	3.49	3.84	4.10	4.30	4.46	4.60	4.72	4.83	4.92	5.00	5.08	5.15	5.21	30
40	2.86	3.44	3.79	4.04	4.23	4.39	4.52	4.63	4.74	4.82	4.91	4.98	5.05	5.11	40
60	2.83	3.40	3.74	3.98	4.16	4.31	4.44	4.55	4.65	4.73	4.81	4.88	4.94	5.00	60

$r = j + 1 - i$ is the number of steps the treatments in positions i and j are apart in the ordered list of treatment mean responses.

Table A.6 *Power values for treatment F test based on the non-central F distribution for 5% significance testing*

f_1 = degrees of freedom of numerator MS f_2 = degrees of freedom of denominator MS

$f_1 = 2$ f_2	0.5	1.0	1.2	1.4	1.6	ϕ 1.8	2.0	2.2	2.6	3.0	f_2
6	8.7	21.1	28.7	37.4	47.0	56.7	66.0	74.5	87.4	94.8	6
8	9.2	23.3	31.9	41.8	52.3	62.7	72.3	80.5	91.8	97.3	8
10	9.5	24.8	34.1	44.6	55.7	66.4	75.9	83.8	93.9	98.3	10
12	9.7	25.8	35.6	46.6	58.0	68.8	78.3	85.8	95.1	98.7	12
14	9.9	26.6	36.8	48.1	59.7	70.6	79.9	87.2	95.8	99.0	14
16	10.0	27.3	37.7	49.2	61.0	71.9	81.1	88.2	96.3	99.2	16
18	10.1	27.8	38.4	50.1	62.0	72.9	82.0	88.9	96.7	99.3	18
20	10.2	28.2	38.9	50.8	62.8	73.7	82.7	89.5	96.9	99.4	20
22	10.3	28.5	39.4	51.4	63.4	74.3	83.2	89.9	97.2	99.4	22
24	10.3	28.8	39.8	51.9	64.0	74.8	83.7	90.3	97.3	99.5	24
26	10.4	29.0	40.2	52.3	64.4	75.3	84.1	90.6	97.4	99.5	26
28	10.4	29.2	40.4	52.7	64.8	75.7	84.4	90.8	97.5	99.5	28
30	10.5	29.4	40.7	53.0	65.2	76.0	84.7	91.0	97.6	99.6	30
32	10.5	29.6	40.9	53.3	65.5	76.3	84.9	91.2	97.7	99.6	32
34	10.5	29.7	41.1	53.5	65.7	76.5	85.2	91.4	97.8	99.6	34
36	10.6	29.9	41.3	53.7	66.0	76.7	85.3	91.5	97.8	99.6	36
38	10.6	30.0	41.5	53.9	66.2	76.9	85.5	91.7	97.9	99.6	38
40	10.6	30.1	41.6	54.1	66.3	77.1	85.7	91.8	97.9	99.6	40

$f_1 = 3$ f_2	0.5	1.0	1.2	1.4	1.6	ϕ 1.8	2.0	2.2	2.6	3.0	f_2
6	8.6	20.9	28.7	37.7	47.6	57.7	67.4	75.9	88.7	95.6	6
8	9.1	23.6	32.7	43.2	54.3	65.2	75.0	83.1	93.6	98.2	8
10	9.4	25.5	35.5	46.9	58.7	69.8	79.4	86.9	95.8	99.0	10
12	9.7	26.9	37.5	49.5	61.7	72.9	82.2	89.3	97.0	99.4	12
14	9.9	27.9	39.1	51.5	63.9	75.1	84.2	90.8	97.6	99.6	14
16	10.1	28.8	40.3	53.0	65.6	76.7	85.6	91.8	98.0	99.7	16
18	10.2	29.5	41.3	54.2	66.9	78.0	86.6	92.6	98.3	99.8	18
20	10.4	30.0	42.1	55.2	67.9	78.9	87.4	93.2	98.5	99.8	20
22	10.4	30.5	42.7	56.0	68.8	79.7	88.0	93.6	98.7	99.8	22
24	10.5	30.9	43.3	56.7	69.5	80.4	88.6	94.0	98.8	99.8	24
26	10.6	31.3	43.8	57.2	70.1	80.9	89.0	94.3	98.9	99.9	26
28	10.7	31.6	44.2	57.7	70.6	81.4	89.4	94.5	99.0	99.9	28
30	10.7	31.8	44.6	58.2	71.1	81.8	89.7	94.7	99.0	99.9	30
32	10.8	32.1	44.9	58.6	71.5	82.1	89.9	94.9	99.1	99.9	32
34	10.8	32.3	45.2	58.9	71.8	82.4	90.2	95.1	99.1	99.9	34
36	10.8	32.4	45.4	59.2	72.1	82.7	90.4	95.2	99.2	99.9	36
38	10.9	32.6	45.7	59.5	72.4	83.0	90.5	95.3	99.2	99.9	38
40	10.9	32.8	45.9	59.7	72.6	83.2	90.7	95.4	99.2	99.9	40

Table A.6 (*continued*)

f_1 = degrees of freedom of numerator MS f_2 = degrees of freedom of denominator MS

$f_1 = 4$

f_2	0.5	1.0	1.2	1.4	1.6	ϕ 1.8	2.0	2.2	2.6	3.0	f_2
6	8.5	20.9	28.9	38.2	48.4	58.8	68.6	77.2	89.7	96.2	6
8	9.1	24.0	33.5	44.5	56.1	67.3	77.1	85.1	94.9	98.7	8
10	9.5	26.2	36.8	48.9	61.2	72.6	82.1	89.2	97.0	99.4	10
12	9.8	27.9	39.3	52.1	64.8	76.2	85.2	91.7	98.0	99.7	12
14	10.0	29.2	41.3	54.5	67.5	78.7	87.3	93.2	98.6	99.8	14
16	10.2	30.3	42.8	56.4	69.5	80.6	88.8	94.2	98.9	99.9	16
18	10.4	31.2	44.0	57.9	71.0	82.0	89.9	95.0	99.1	99.9	18
20	10.6	31.9	45.0	59.1	72.3	83.1	90.7	95.5	99.3	99.9	20
22	10.7	32.5	45.9	60.1	73.3	83.9	91.4	95.9	99.4	99.9	22
24	10.8	33.0	46.6	60.9	74.1	84.7	91.9	96.2	99.5	100	24
26	10.9	33.5	47.2	61.7	74.8	85.3	92.3	96.5	99.5	100	26
28	10.9	33.9	47.7	62.3	75.5	85.8	92.7	96.7	99.6	100	28
30	11.0	34.2	48.2	62.8	76.0	86.2	93.0	96.9	99.6	100	30
32	11.1	34.5	48.6	63.3	76.4	86.6	93.3	97.0	99.6	100	32
34	11.1	34.8	49.0	63.7	76.8	86.9	93.5	97.2	99.7	100	34
36	11.2	35.0	49.3	64.1	77.2	87.2	93.7	97.3	99.7	100	36
38	11.2	35.3	49.6	64.4	77.5	87.4	93.8	97.4	99.7	100	38
40	11.3	35.5	49.9	64.7	77.8	87.7	94.0	97.5	99.7	100	40

$f_1 = 5$

f_2	0.5	1.0	1.2	1.4	1.6	ϕ 1.8	2.0	2.2	2.6	3.0	f_2
6	8.5	21.0	29.1	38.7	49.1	59.7	69.6	78.2	90.5	96.7	6
8	9.1	24.4	34.3	45.7	57.6	69.0	78.9	86.6	95.7	99.0	8
10	9.5	26.9	38.1	50.7	63.4	74.9	84.2	90.9	97.7	99.6	10
12	9.9	28.9	41.0	54.3	67.5	78.8	87.5	93.4	98.6	99.8	12
14	10.2	30.4	43.2	57.1	70.4	81.6	89.7	94.8	99.1	99.9	14
16	10.4	31.7	45.0	59.3	72.7	83.5	91.2	95.8	99.4	99.9	16
18	10.6	32.7	46.5	61.0	74.4	85.0	92.3	96.5	99.5	100	18
20	10.8	33.6	47.7	62.5	75.8	86.2	93.1	97.0	99.6	100	20
22	10.9	34.4	48.7	63.6	77.0	87.1	93.7	97.3	99.7	100	22
24	11.0	35.0	49.6	64.6	77.9	87.9	94.2	97.6	99.7	100	24
26	11.1	35.5	50.3	65.5	78.7	88.5	94.6	97.8	99.8	100	26
28	11.2	36.0	51.0	66.2	79.4	89.0	94.9	98.0	99.8	100	28
30	11.3	36.5	51.6	66.9	80.0	89.4	95.2	98.1	99.8	100	30
32	11.4	36.8	52.1	67.4	80.5	89.8	95.4	98.3	99.8	100	32
34	11.4	37.2	52.5	67.9	80.9	90.2	95.6	98.4	99.9	100	34
36	11.5	37.5	52.9	68.3	81.3	90.4	95.8	98.4	99.9	100	36
38	11.6	37.8	53.3	68.7	81.7	90.7	96.0	98.5	99.9	100	38
40	11.6	38.0	53.6	69.1	82.0	90.9	96.1	98.6	99.9	100	40

Table A.6 (*continued*)

f_1 = degrees of freedom of numerator MS f_2 = degrees of freedom of denominator MS

$f_1 = 6$						ϕ					
f_2	0.5	1.0	1.2	1.4	1.6	1.8	2.0	2.2	2.6	3.0	f_2
6	8.4	21.1	29.4	39.1	49.7	60.4	70.4	79.0	91.1	97.0	6
8	9.1	24.8	34.9	46.7	58.9	70.4	80.2	87.8	96.3	99.2	8
10	9.6	27.6	39.1	52.2	65.2	76.7	85.8	92.2	98.2	99.7	10
12	10.0	29.8	42.4	56.3	69.6	80.9	89.2	94.6	99.0	99.9	12
14	10.3	31.5	44.9	59.4	72.9	83.8	91.4	96.0	99.4	99.9	14
16	10.5	33.0	47.0	61.8	75.3	85.9	92.9	96.9	99.6	100	16
18	10.8	34.2	48.7	63.8	77.2	87.4	93.9	97.5	99.7	100	18
20	10.9	35.2	50.1	65.4	78.8	88.6	94.7	97.9	99.8	100	20
22	11.1	36.1	51.3	66.7	80.0	89.5	95.3	98.2	99.8	100	22
24	11.2	36.8	52.3	67.8	81.0	90.3	95.8	98.4	99.9	100	24
26	11.4	37.5	53.1	68.8	81.8	90.9	96.1	98.6	99.9	100	26
28	11.5	38.0	53.9	69.6	82.6	91.4	96.4	98.7	99.9	100	28
30	11.6	38.5	54.6	70.3	83.2	91.9	96.7	98.9	99.9	100	30
32	11.7	39.0	55.2	70.9	83.7	92.2	96.9	99.0	99.9	100	32
34	11.7	39.4	55.7	71.5	84.2	92.5	97.1	99.0	99.9	100	34
36	11.8	39.8	56.2	72.0	84.6	92.8	97.2	99.1	99.9	100	36
38	11.9	40.1	56.6	72.4	85.0	93.1	97.3	99.1	100	100	38
40	11.9	40.4	57.0	72.8	85.3	93.3	97.4	99.2	100	100	40

Entries refer to estimated power for the planned treatment F test based on (f_1, f_2) degrees of freedom and parameter $\phi = \sqrt{[(n\delta^2)/(2k\sigma^2)]}$. The values presented refer to percentages.

Computed with SAS® by W.P. Gardiner.

Table A.7 *Lower critical values for the Mann–Whitney test*

m = number in sample 1 n = number in sample 2

m	Lower critical values	2	3	4	5	6	7	8	9	10	11	12	13	14	15	16	17	18	19	20
2	$T_{0.005}$	0	0	0	0	0	0	0	0	0	0	0	0	0	0	0	0	0	1	1
	$T_{0.01}$	0	0	0	0	0	0	0	0	0	0	0	1	1	1	1	1	1	2	2
	$T_{0.025}$	0	0	0	0	0	0	1	1	1	1	2	2	2	2	2	3	3	3	3
	$T_{0.05}$	0	0	0	1	1	1	2	2	2	2	3	3	4	4	4	4	5	5	5
3	$T_{0.005}$	0	0	0	0	0	0	0	1	1	1	2	2	2	3	3	3	3	4	4
	$T_{0.01}$	0	0	0	0	0	1	1	2	2	2	3	3	3	4	4	5	5	5	6
	$T_{0.025}$	0	0	0	1	2	2	3	3	4	4	5	5	6	6	7	7	8	8	9
	$T_{0.05}$	0	1	1	2	3	3	4	5	5	6	6	7	8	8	9	10	10	11	12
4	$T_{0.005}$	0	0	0	0	1	1	2	2	3	3	4	4	5	6	6	7	7	8	9
	$T_{0.01}$	0	0	0	1	2	2	3	4	4	5	6	6	7	8	8	9	10	10	11
	$T_{0.025}$	0	0	1	2	3	4	5	5	6	7	8	9	10	11	12	12	13	14	15
	$T_{0.05}$	0	1	2	3	4	5	6	7	8	9	10	11	12	13	15	16	17	18	19
5	$T_{0.005}$	0	0	0	1	2	2	3	4	5	6	7	8	8	9	10	11	12	13	14
	$T_{0.01}$	0	0	1	2	3	4	5	6	7	8	9	10	11	12	13	14	15	16	17
	$T_{0.025}$	0	1	2	3	4	6	7	8	9	10	12	13	14	15	16	18	19	20	21
	$T_{0.05}$	1	2	3	5	6	7	9	10	12	13	14	16	17	19	20	21	23	24	26
6	$T_{0.005}$	0	0	1	2	3	4	5	6	7	8	10	11	12	13	14	16	17	18	19
	$T_{0.01}$	0	0	2	3	4	5	7	8	9	10	12	13	14	16	17	19	20	21	23
	$T_{0.025}$	0	2	3	4	6	7	9	11	12	14	15	17	18	20	22	23	25	26	28
	$T_{0.05}$	1	3	4	6	8	9	11	13	15	17	18	20	22	24	26	27	29	31	33
7	$T_{0.005}$	0	0	1	2	4	5	7	8	10	11	13	14	16	17	19	20	22	23	25
	$T_{0.01}$	0	1	2	4	5	7	8	10	12	13	15	17	18	20	22	24	25	27	29
	$T_{0.025}$	0	2	4	6	7	9	11	13	15	17	19	21	23	25	27	29	31	33	35
	$T_{0.05}$	1	3	5	7	9	12	14	16	18	20	22	25	27	29	31	34	36	38	40
8	$T_{0.005}$	0	0	2	3	5	7	8	10	12	14	16	18	19	21	23	25	27	29	31
	$T_{0.01}$	0	1	3	5	7	8	10	12	14	16	18	21	23	25	27	29	31	33	35
	$T_{0.025}$	1	3	5	7	9	11	14	16	18	20	23	25	27	30	32	35	37	39	42
	$T_{0.05}$	2	4	6	9	11	14	16	19	21	24	27	29	32	34	37	40	42	45	48
9	$T_{0.005}$	0	1	2	4	6	8	10	12	14	17	19	21	23	25	28	30	32	34	37
	$T_{0.01}$	0	2	4	6	8	10	12	15	17	19	22	24	27	29	32	34	37	39	41
	$T_{0.025}$	1	3	5	8	11	13	16	18	21	24	27	29	32	35	38	40	43	46	49
	$T_{0.05}$	2	5	7	10	13	16	19	22	25	28	31	34	37	40	43	46	49	52	55
10	$T_{0.005}$	0	1	3	5	7	10	12	14	17	19	22	25	27	30	32	35	38	40	43
	$T_{0.01}$	0	2	4	7	9	12	14	17	20	23	25	28	31	34	37	39	42	45	48
	$T_{0.025}$	1	4	6	9	12	15	18	21	24	27	30	34	37	40	43	46	49	53	56
	$T_{0.05}$	2	5	8	12	15	18	21	25	28	32	35	38	42	45	49	52	56	59	63

Table A.7 (*continued*)

m	Lower critical values	2	3	4	5	6	7	8	9	n 10	11	12	13	14	15	16	17	18	19	20
11	$T_{0.005}$	0	1	3	6	8	11	14	17	19	22	25	28	31	34	37	40	43	46	49
	$T_{0.01}$	0	2	5	8	10	13	16	19	23	26	29	32	35	38	42	45	48	51	54
	$T_{0.025}$	1	4	7	10	14	17	20	24	27	31	34	38	41	45	48	52	56	59	63
	$T_{0.05}$	2	6	9	13	17	20	24	28	32	35	39	43	47	51	55	58	62	66	70
12	$T_{0.005}$	0	2	4	7	10	13	16	19	22	25	28	32	35	38	42	45	48	52	55
	$T_{0.01}$	0	3	6	9	12	15	18	22	25	29	32	36	39	43	47	50	54	57	61
	$T_{0.025}$	2	5	8	12	15	19	23	27	30	34	38	42	46	50	54	58	62	66	70
	$T_{0.05}$	3	6	10	14	18	22	27	31	35	39	43	48	52	56	61	65	69	73	79
13	$T_{0.005}$	0	2	4	8	11	14	18	21	25	28	32	35	39	43	46	50	54	58	61
	$T_{0.01}$	1	3	6	10	13	17	21	24	28	32	36	40	44	48	52	56	60	64	68
	$T_{0.025}$	2	5	9	13	17	21	25	29	34	38	42	46	51	55	60	64	68	73	77
	$T_{0.05}$	3	7	11	16	20	25	29	34	38	43	48	52	57	62	66	71	76	81	85
14	$T_{0.005}$	0	2	5	8	12	16	19	23	27	31	35	39	43	47	51	55	59	64	68
	$T_{0.01}$	1	3	7	11	14	18	23	27	31	35	39	44	48	52	57	61	66	70	74
	$T_{0.025}$	2	6	10	14	18	23	27	32	37	41	46	51	56	60	65	70	75	79	84
	$T_{0.05}$	4	8	12	17	22	27	32	37	42	47	52	57	62	67	72	78	83	88	93
15	$T_{0.005}$	0	3	6	9	13	17	21	25	30	34	38	43	47	52	56	61	65	70	74
	$T_{0.01}$	1	4	8	12	16	20	25	29	34	38	43	48	52	57	62	67	71	76	81
	$T_{0.025}$	2	6	11	15	20	25	30	35	40	45	50	55	60	65	71	76	81	86	91
	$T_{0.05}$	4	8	13	19	24	29	34	40	45	51	56	62	67	73	78	84	89	95	101
16	$T_{0.005}$	0	3	6	10	14	19	23	28	32	37	42	46	51	56	61	66	71	75	80
	$T_{0.01}$	1	4	8	13	17	22	27	32	37	42	47	52	57	62	67	72	77	83	88
	$T_{0.025}$	2	7	12	16	22	27	32	38	43	48	54	60	65	71	76	82	87	93	99
	$T_{0.05}$	4	9	15	20	26	31	37	43	49	55	61	66	72	78	84	90	96	102	108
17	$T_{0.005}$	0	3	7	11	16	20	25	30	35	40	45	50	55	61	66	71	76	82	87
	$T_{0.01}$	1	5	9	14	19	24	29	34	39	45	50	56	61	67	72	78	83	89	94
	$T_{0.025}$	3	7	12	18	23	29	35	40	46	52	58	64	70	76	82	88	94	100	106
	$T_{0.05}$	4	10	16	21	27	34	40	46	52	58	65	71	78	84	90	97	103	110	116
18	$T_{0.005}$	0	3	7	12	17	22	27	32	38	43	48	54	59	65	71	76	82	88	93
	$T_{0.01}$	1	5	10	15	20	25	31	37	42	48	54	60	66	71	77	83	89	95	101
	$T_{0.025}$	3	8	13	19	25	31	37	43	49	56	62	68	75	81	87	94	100	107	113
	$T_{0.05}$	5	10	17	23	29	36	42	49	56	62	69	76	83	89	96	103	110	117	124
19	$T_{0.005}$	1	4	8	13	18	23	29	34	40	46	52	58	64	70	75	82	88	94	100
	$T_{0.01}$	2	5	10	16	21	27	33	39	45	51	57	64	70	76	83	89	95	102	108
	$T_{0.025}$	3	8	14	20	26	33	39	46	53	59	66	73	79	86	93	100	107	114	120
	$T_{0.05}$	5	11	18	24	31	38	45	52	59	66	73	81	88	95	102	110	117	124	131

Table A.7 (*continued*)

m	Lower critical values	2	3	4	5	6	7	8	9	10	11	12	13	14	15	16	17	18	19	20
20	$T_{0.005}$	1	4	9	14	19	25	31	37	43	49	55	61	68	74	80	87	93	100	106
	$T_{0.01}$	2	6	11	17	23	29	35	41	48	54	61	68	74	81	88	94	101	108	115
	$T_{0.025}$	3	9	15	21	28	35	42	49	56	63	70	77	84	91	99	106	113	120	128
	$T_{0.05}$	5	12	19	26	33	40	48	55	63	70	78	85	93	101	108	116	124	131	139

The values presented refer to the lower critical values for the Mann–Whitney procedure. Upper critical values are determined as $(mn - T_\alpha)$.

For a *two-tailed test* at the 5% ($\alpha = 0.05$) significance level, $\alpha/2 = 0.025$ with the lower critical value read from the $T_{0.025}$ entry and the upper critical value calculated as $(mn - T_{0.025})$.

For *one-tailed testing*, critical values depend on the form of the alternative hypothesis. For H_1 with the inequality < at the 5% significance level, $\alpha = 0.05$ and the critical value is read from the $T_{0.05}$ entry. For H_1 with the inequality > at the 5% significance level, the critical value is calculated as $(mn - T_{0.05})$.

Table A.8 *Lower critical values for the Wilcoxon signed-ranks test*

n = number of non-zero differences

n	$T_{0.005}$	$T_{0.01}$	$T_{0.025}$	$T_{0.05}$	$T_{0.1}$	n
		Lower critical values				
4	0	0	0	0	1	4
5	0	0	0	1	3	5
6	0	0	1	3	4	6
7	0	1	3	4	6	7
8	1	2	4	6	9	8
9	2	4	6	9	11	9
10	4	6	9	11	15	10
11	6	8	11	14	18	11
12	8	10	14	18	22	12
13	10	13	18	22	27	13
14	13	16	22	26	32	14
15	16	20	26	31	37	15
16	20	24	30	36	43	16
17	24	28	35	42	49	17
18	28	33	41	48	56	18
19	33	38	47	54	63	19
20	38	44	53	61	70	20
21	43	50	59	68	78	21
22	49	56	66	76	87	22
23	55	63	74	84	95	23
24	62	70	82	92	105	24
25	69	77	90	101	114	25
26	76	85	99	111	125	26
27	84	93	108	120	135	27
28	92	102	117	131	146	28
29	101	111	127	141	158	29
30	110	121	138	152	170	30

The values presented refer to the lower critical values for the Wilcoxon Signed-ranks procedure. Upper critical values are given by $[n(n + 1)/2 - T_\alpha]$ where n is the number of non-zero differences.

For a *two-tailed test* at the 5% ($\alpha = 0.05$) significance level, $\alpha/2 = 0.025$ with the lower critical value read from the $T_{0.025}$ column and the upper critical value calculated as $[n(n + 1)/2 - T_{0.025}]$.

For *one-tailed testing*, critical values depend on the form of the alternative hypothesis. For H_1 with the inequality < at the 5% significance level, $\alpha = 0.05$ and the critical value is read from column $T_{0.05}$. For H_1 with the inequality > at the 5% significance level, the critical value is calculated as $[n(n + 1)/2 - T_{0.05}]$.

APPENDIX B

Tables of Large Data Sets

Table B.1 *CRF recordings for Example 4.6*

Acetic acid (mol dm^{-3})	Methanol:water (%)	Citric acid (g dm^{-3})	CRF measurements	
0.004	70	2	9.8	10.2
0.004	70	4	9.6	9.7
0.004	70	6	9.1	9.5
0.004	80	2	10.8	11.2
0.004	80	4	11.4	11.3
0.004	80	6	11.9	11.9
0.007	70	2	9.6	9.5
0.007	70	4	9.3	9.4
0.007	70	6	9.0	9.7
0.007	80	2	10.7	10.9
0.007	80	4	11.1	11.2
0.007	80	6	11.4	11.4
0.010	70	2	9.3	9.7
0.010	70	4	8.7	9.8
0.010	70	6	9.5	10.1
0.010	80	2	10.6	10.8
0.010	80	4	11.2	11.3
0.010	80	6	11.6	11.8

Source: Adapted from 'Chemometrics: Experimental Design' (E. Morgan) with the permission of the University of Greenwich.

Table B.2 *Atrazine experimental data for Example 5.8*

pH	N	CM (% methanol)	CR (mg l^{-1})	F (ml min^{-1} for 15 min)	Atrazine retention time (min)
5	9	45	2	1.5	16.34
6.5	7	40	6	1.25	29.00
5	5	35	2	1.5	57.50
6.5	7	42.5	6	1.25	23.86
8	5	35	10	1.5	39.49
6.5	7	40	3.5	1.25	31.28
8	9	45	2	1	22.84
6.5	7	40	8.5	1.25	27.53
8	9	35	10	1	27.53
5	7	40	6	1.25	29.41
6.5	9	40	6	1.25	25.67
5	5	45	10	1.5	16.81
6.5	7	40	6	1.6	24.69
8	9	35	2	1.5	32.85
8	5	35	2	1	65.33
6.5	7	40	1	1.25	34.57
5	9	35	2	1	67.69
6.5	5	40	6	1.25	35.08
5	5	45	2	1	22.12
6.5	7	37.5	6	1.25	34.17
8	5	45	2	1.5	16.67
5	5	35	10	1	71.12
6.5	7	40	6	0.9	44.30
5	9	35	10	1.5	31.73
8	5	45	10	1	23.69
8	7	40	6	1.25	25.72
5	9	45	10	1	21.42
6.5	7	40	11	1.25	27.98

Source: Reprinted from E. Marengo, M.C. Gennaro and C. Abrigo, *Analytica Chimica Acta*, 1996, **321**, 225–236 with kind permission of Elsevier Science-NL, Sara Burger-hartstraat 25, 1055 KV Amsterdam, The Netherlands.

Table B.3 *Fluorescence measurements for Exercise 5.4*

SVRS	pH	Heating Time	Delay Time	Cool Time	Fluorescence
5.00	5.00	10	120	240	91
2.30	5.00	10	120	240	267
3.38	8.00	8	72	288	196
2.30	5.00	4	120	240	345
2.89	5.30	9	94	266	197
5.00	5.00	10	120	240	91
2.30	2.00	10	120	240	245
2.68	5.05	9	94	266	271
1.86	4.72	8	88	272	332
1.34	4.43	8	85	275	320
2.27	4.91	6	49	311	178
2.30	5.00	10	0	240	205
1.74	4.80	7	57	303	323
1.92	4.90	9	105	255	306
2.30	5.00	10	120	360	243
1.75	4.77	5	196	332	337
1.47	4.65	3	294	378	366

Table B.4 *Design matrix and fluorescence measurements for Example 7.5*

Run	SVRS	pH	Heating Time	Delay Time	Fluorescence
1	1	2	15	120	60
2	1	6	5	240	162
3	1	6	5	120	219
4	1	2	15	240	127
5	1	2	5	120	159
6	5	6	15	240	49
7	5	6	15	120	45
8	5	2	15	120	86
9	5	2	15	240	90
10	5	6	5	120	71
11	1	2	5	240	182
12	5	2	5	120	117
13	1	6	15	240	211
14	1	6	15	120	172
15	5	6	5	240	71
16	5	2	5	240	123

Table B.5 *Enzyme activity data for Exercise 7.2*

Run	A	B	C	D	E	Enzyme activity (IU)
			Factors			
1	−	−	−	+	−	119
2	+	−	−	+	+	128
3	+	+	+	+	−	148
4	+	+	+	−	−	123
5	+	+	+	+	+	132
6	−	−	−	+	+	96
7	−	−	−	−	+	106
8	−	+	−	−	+	113
9	+	−	−	+	−	146
10	+	+	−	−	−	113
11	−	+	−	+	+	95
12	−	−	+	−	−	103
13	+	−	+	+	−	145
14	+	+	+	−	+	115
15	−	+	−	+	−	111
16	+	+	−	+	−	143
17	−	+	+	−	+	105
18	+	−	−	−	−	113
19	+	−	−	−	+	120
20	+	−	+	+	+	131
21	+	+	−	+	+	127
22	−	−	+	−	+	109
23	+	−	+	−	+	117
24	−	−	−	−	−	109
25	+	+	−	−	+	115
26	+	−	+	−	−	104
27	−	−	+	+	+	99
28	−	+	+	−	−	106
29	−	+	+	+	+	92
30	−	+	−	−	−	103
31	−	+	+	+	−	110
32	−	−	+	+	−	116

Source: Reproduced from 'Chemometrics: Experimental Design' (E. Morgan) with the permission of the University of Greenwich.

Table B.6 *Chemicals data for Example 8.1*

Week	Benzene	Toluene	n-Decane	n-Dodecane	Ethylbenzene	Acetophenone
1	48	26	2	11	11	1
2	44	20	1	9	9	2
3	26	5	14	2	3	7
4	24	4	10	6	2	11
5	31	7	7	5	5	4
6	30	7	10	6	5	6
7	38	18	3	4	9	7
8	40	24	7	2	10	2
9	41	21	2	4	9	3
10	32	9	7	6	6	4
11	28	6	11	16	5	6
12	35	14	8	5	7	9

Source: D.L. Massart, B.G.M. Vandeginste, S.N. Deming, Y. Michotte and L. Kaufman, 'Chemomterics: A Textbook', Elsevier, Amsterdam, 1988: reproduced with the permission of Elsevier Science.

Table B.7 *Air pollution data for Exercise 8.1*

Day No	SolRad	Wind	CO	NO	HC	NO_2	O_3
1	89	8	7	2	2	12	8
2	103	7	4	3	3	5	6
3	91	6	4	2	3	8	10
4	84	9	7	4	5	12	15
5	82	7	5	1	3	11	11
6	71	6	5	4	3	10	3
7	72	7	7	4	3	18	10
8	72	10	4	1	3	8	10
9	76	8	4	1	3	7	7
10	67	9	4	2	3	13	2
11	62	10	5	3	4	14	4
12	80	8	4	2	4	13	11
13	83	6	5	1	4	10	23
14	78	6	4	2	3	11	11
15	62	6	4	3	3	9	8
16	71	8	4	1	3	10	7
17	48	5	6	5	3	8	4
18	35	10	4	1	2	6	9
19	86	5	3	1	2	6	12
20	79	7	7	4	3	9	25

Source: R.A. Johnson and D.W. Wichern, 'Applied Multivariate Statistical Analysis', 3rd Edn., Prentice Hall, Englewood Cliffs, New York, 1992: reproduced with the permission of Prentice Hall, Inc.

Table B.8 *Chemical composition data for Example 8.4 (contents in % w/w, drybase)*

Class	TB	TP	CA	POL	EXT	AA
Black	0.19	0.08	4.0	23.4	35.1	1.36
	0.16	0.09	3.4	18.1	30.0	1.19
	0.13	0.07	3.5	17.7	33.8	1.16
	0.26	0.05	3.3	28.0	35.9	0.61
	0.12	0.03	3.5	23.0	31.7	0.63
	0.05	0.01	3.9	19.3	31.2	0.90
	0.26	0.03	4.2	18.7	30.7	0.55
	0.34	0.06	4.4	29.7	35.3	0.70
	0.04	0.02	4.0	16.8	28.2	0.98
	0.39	0.10	4.8	23.1	34.6	1.41
	0.18	0.01	4.8	26.0	30.6	1.07
	0.24	0.15	4.5	25.0	38.0	1.26
Green	0.16	0.03	3.4	22.8	32.4	1.05
	0.23	0.07	3.6	25.2	28.3	1.00
	0.20	0.03	4.7	26.0	34.2	1.19
	0.16	0.02	4.2	23.2	27.6	1.08
	0.25	0.04	4.8	27.8	32.7	1.52
	0.15	0.03	4.5	26.4	36.1	1.36
	0.16	0.04	4.5	29.3	32.5	1.18

Source: Reprinted from P. Valera, F. Pablos and A.G. Gonzalez, *Talanta*, 1996, **43**, 415–419 with kind permission of Elsevier Science-NL, Sara Burgerhartstraat 25, 1055 KV Amsterdam, The Netherlands.

Table B.9 *Trace metal concentrations for Exercise 8.2 (all measurements* mg kg^{-1} *dryweight)*

Water System	Cr	Cu	Ni	Pb	Zn
A	94.34	39.62	39.62	96.23	299.06
	84.70	36.30	48.40	78.12	264.76
	68.87	29.84	36.73	84.94	246.33
	57.60	29.62	46.08	82.29	235.68
	85.91	32.93	51.55	80.18	253.44
	61.68	28.78	49.34	75.38	228.89
	78.48	29.15	49.33	73.99	226.91
B	63.92	45.65	20.09	109.57	193.75
	32.70	31.07	31.07	83.39	211.90
	44.57	24.20	28.02	88.77	194.87
	36.02	27.02	21.61	99.06	197.41
	65.69	33.78	26.28	98.76	231.47
	46.13	27.68	19.99	81.49	209.87
	44.80	24.64	17.92	87.37	175.63
C	58.32	33.05	36.94	108.86	208.20
	48.92	22.83	32.62	84.80	183.63
	60.10	22.32	30.91	70.40	208.45
	44.04	20.55	32.30	80.74	151.79
	46.67	28.00	14.00	52.89	201.62
	82.69	74.42	30.32	84.07	465.13
	71.17	40.93	23.13	99.72	225.80

Test Observations

	Cr	Cu	Ni	Pb	Zn
Water System A	67.20	18.73	53.76	78.41	245.97
Water System B	52.85	38.76	12.33	95.14	187.10
Water System C	60.39	21.14	36.61	108.70	205.01

Answers to Exercises

Exercise 2.1

The data plot shows a distinct difference between the signal measurements and those obtained after 10 minutes standing higher and more variable. For 10 minutes, the mean is 26.189 g, standard deviation 4.371 g, and *RSD* 16.7%. For 24 hours, the mean is 11.291 g, standard deviation 0.285 g, and *RSD* 2.5%. The summaries highlight the wide differences in the signal measurements with the results for 10 minutes being much higher and considerably more variable than those measured after 24 hours. The results suggest that standing time of water samples before filtration can affect measured analytical signal markedly.

Exercise 2.2

For both mean and precision (variability), interest lies in ascertaining whether there is a difference between the two procedures, so both sets of hypotheses will be similarly specified. The null hypothesis for both cases would be expressed as H_0: no difference between the procedures while the alternative for both would be two-sided and expressed as H_1: difference between the procedures.

Exercise 2.3

Data plotting highlights that all the measurements are reasonably similar with one exception, the value 27.03 which appears lower than the rest and may be a possible outlier (z score $= -2.51$). The mean is 27.69 titre cm^{-3} near to nominal level. The standard deviation is 0.265 titre cm^{-3} which is above target though the *RSD* of 0.96% suggests very consistent measurements. The test for the mean (2.9) is not significant ($t = -0.67$, $p = 0.5209$) while the variability test (2.11) indicates that variability is statistically different from the target ($\chi^2 = 20.64$, $p = 0.0143$). The 95% confidence interval for the mean is (7.504, 27.884) titre cm^{-3} which straddles the target, backing-up the conclusion on the test of the mean though with a hint of more values lying towards the lower end. Removal of 27.03 changes the variability conclusion ($p > 0.05$), as would be expected, and produces a more uniform confidence interval for the mean of (27.624, 27.886) titre cm^{-3}.

Exercise 2.4

For both comparison aspects, only difference needs to be tested so both forms of analysis would be based on specification of a two-sided alternative. Graphically, XRF results show up as lower and more variable with the summaries providing a back-up to this conclusion (XRF: mean 120.3, *RSD* 8.9%; FAAS: mean 129.9, *RSD* 5.2%). Testing the means, using test statistic (2.13), results in rejection of the null hypothesis suggesting evidence exists of a difference between the procedures ($t = -2.391, p = 0.0279$). The 95% confidence interval (2.17) is given as $(-18.975, -0.225)$ µg l^{-1} Zn agreeing with the test conclusions and suggesting that XRF is producing lower results in the main. The variability test statistic (2.18) is $F = 2.51$ and $p = 0.1858$ (doubled Excel entry) leading to the conclusion that no variability differences can be detected.

Exercise 2.5

Since assessment is concerned with checking for differences only, both aspects of analysis are based on specification of two-sided alternatives. Graphically, the 35 °C results differ from those for 55 °C with some differences substantially in favour of 55 °C (higher results). The differences (35–55) are mixed, but the negative differences are numerically larger than the positive differences suggesting a possible difference in the steroid solubility with temperature. Testing the mean difference, using test statistic (2.20), results in acceptance of no difference ($t = -1.88$, $p = 0.097$). The 95% confidence interval (2.21) for the mean difference of $(-8.89, 0.91)$ shows that, though no significant difference is detected, there is a suggestion that steroid solubility is generally higher at higher temperature. The variability test (2.22) results in rejection of no difference in variability ($t = 2.8$, $p = 0.0265$).

Exercise 2.6

From the information provided, $d = 0.4$, $\sigma = 0.3$, and ES = 1.3333. The test is two-tailed (difference between soils) at 5% significance level ($\alpha = 0.05$) with power at least 90%. These test constraints provide $\beta = 0.1$,

$z_\beta = 1.2816$, and $z_{\alpha/2} = z_{0.025} = 1.96$. Equation (2.24) provides an estimate of n of 12. The advice would therefore be to base the experiment on at least 12 samples of each soil type.

Exercise 2.7

The experimenter has suggested that values of $d = 0.15$ and $\sigma_D = 0.19$ (ES = 0.7895) are appropriate for the planned experiment. The associated inference is based on a two-tailed test (method difference) at the 5% significance level ($z_{0.025} = 1.96$) with power at least 80% ($\beta = 0.2$, $z_\beta = 0.8416$). Using equation (2.26), an estimate of 13 is produced, meaning that at least 13 beer samples would need to be used if the planned experiment is to be capable of detecting the level of method difference specified.

Exercise 3.1

The response model of percent impurities removed is μ + temperature + error. The removal results cluster into three distinct temperature groupings of (I, II), (III, IV), and V with an indication of increase in percent impurity removed as temperature increases. Variability is similar for all temperature settings except I which exhibits largest variability. The summaries of mean and *RSD* provide back-up to these interpretations. Statistically, the temperature settings differ ($F = 32.64$, $p = 0.000$) and use of the SNK multiple comparison provides the result II I IV III V indicating significant differences between temperatures with only settings III and IV unable to be distinguished statistically. Diagnostic checking hints at variability and possible normality problems with the percent impurity response as may be expected given nature of the measured response.

Exercise 3.2

The response model for selenium content is μ + subject + procedure + error. All oxidation procedures appear to provide measurements of similar magnitude but with different trends. Mean selenium measurements show difference and hint at groupings of D, (B, C), and (A, E) emerging. All procedures result in high RSDs, greater than 20%, suggesting inconsistent measurements for all procedures. On a statistical basis, the procedures appear to differ ($F = 33.86$, $p = 0.000$) as does the subject blocking factor ($p = 0.000$). Application of the SNK multiple comparison provides the result D B C E A showing that the procedures split into three statistically different groupings. The contrast of procedure A against the rest also provides evidence of statistical difference, interval (2.193, 3.793) mg Se per 100 ml. Diagnostic checking hints at no major violations of the ANOVA assumptions.

Exercise 3.3

The proposed structure provides the following information: $k = 5$, $n = 4$, $\delta = 7.85$ ppm, and $\sigma^2 = 8.65$. The treatment F test is planned for the 5% significance level with degrees of freedom $f_1 = 4$ and $f_2 = 15$. From equation (3.12), ϕ is estimated to be 1.69 with Table A.6 providing a power estimate of around 75%. This is below the ideal so a change to design structure may be necessary. For example, changing the number of depths sampled to five increases power to a more acceptable level of around 86% ($f_2 = 20$, $\phi = 1.89$).

Exercise 3.4

The background information on the proposed structure provides the following information: $k = 3$, $f_1 = 2$, $\delta = 0.6$ mg ml^{-1}, and $\sigma^2 = 0.21$ with 5% being the significance level for the planned F test. Starting with $n = 4$, power is estimated to be approximately 28% ($f_2 = 9$, $\phi = 1.07$). Increasing n to 6 changes the estimated power to approximately 66% ($f_2 = 15 \approx 16$, $\phi = 1.71$) while $n = 8$ suggests a power of approximately 93% ($f_2 = 21 \approx 22$, $\phi = 2.29$). From these results, the best result appears to be $n = 8$ so the advice would be to use eight samples of each type.

Exercise 4.1

The model for percent yield response is μ + temperature + duration + temperature × duration + error. Yield results are similar for temperatures 30 °C and 50 °C, but higher and less variable for a temperature of 40 °C. Yield increases as duration lengthens but no marked difference in variability of yield is apparent. The interaction test result is significant ($F = 20.62$, $p = 0.000 < 0.001$) providing evidence that interaction of temperature and duration affects process yield. The interaction plot shows three distinct trends with respect to duration and suggests (40 °C, 40 minutes) and (50 °C, 40 minutes) are the optimal combinations. The contrast of temperature 40 °C against the other temperature levels produces a 95% confidence interval (2.58%, 5.86%) indicative of significantly higher yield for 40 °C. Diagnostic checking highlights unequal variability for both temperature and duration factors and non-normality for the yield response.

Exercise 4.2

Based on the planned experiment, the following information is available: $a = 3$, $b = 3$, $\delta = 105.2$, $\sigma^2 = 464$, $f_1 = 4$, $f_2 = 9(n-1)$, $n' = n$, and $k' = ab = 9$. For the statistical test of interaction, 5% is the proposed significance level. A starting point of $n = 2$ provides a power estimate of 58% ($f_2 = 9 \approx 10$, $\phi = 1.63$) which is too low to consider. Changing n to 3 increases power to 90% ($f_2 = 18$, $\phi = 1.99$) while $n = 4$ provides a power estimate of at least 96% ($f_2 = 27 \approx 28$, $\phi = 2.30$). These calculations suggest at least three replications would be advised to satisfy the power constraint of at least 85%.

Exercise 4.3

Based on the response of percent recovery of Amaranth, the response model is μ + E-123 + E-124 + E-123 × E-124 + E-120 + E-123 × E-120 + E-124 × E-120 + E-123 × E-124 × E-120 + error. Recovery differs with level of E-123 tested as does consistency of results (24 mg 1^{-1} lowest). No obvious difference is apparent for E-124 in respect of mean and variability in recovery. The two levels of E-120 tested show some difference in recovery with 12 mg 1^{-1} being marginally higher. Statistically, the three factor interaction is highly significant ($F = 59.43$, $p = 0.000 < 0.001$). Follow-up through the interaction plot of the E-123 and E-124 interaction at each level of E-120 highlights different trends in the recovery data, especially at 12 mg 1^{-1} of E-123. Diagnostic checking highlights unequal variability for all factors and non-normality for the percent recovery response as might be expected considering nature of the measured response.

Exercise 5.1

Based on testing linearity, it is thought the model solubility $= \alpha + \beta$ pressure $+ \varepsilon$ may fit the presented data. The data, when plotted, are indicative of a positive linear trend but the points show a hint of a curved effect. Fitting a linear model produces the equation solubility $= -175.104 + 2.791$ pressure which is highly significant ($t = 24.94$, $F = 622.1$, $p < 0.001$, $R^2 = 98.7\%$). The predicted solubility is not good with most predictions differing markedly from the measured results. The predictions pattern appears to be over-, under-, and finally over-prediction suggesting presence of a trend in the results which the linear model is not explaining. The pressure residual plot hints at a quadratic/cubic trend, while the normality plot looks reasonably linear. The unexplained trend confirms the curved effect shown in the initial data plot.

Exercise 5.2

The data conform almost exactly to a perfect straight line. The fitted linear equation is intensity = 13.537 + 22.383 calcium which is statistically acceptable ($t = 106.3$, $F = 11306.8$, $p < 0.001$, $R^2 = 99.9\%$). The average intensity measurement from the two readings is 75.9225 mV resulting in an estimated calcium value of 2.787 ppm. The 95% confidence interval (5.16) is calculated to be (2.602, 2.972) ppm indicative of acceptable accuracy in the calcium concentration prediction.

Exercise 5.3

The plot shows an obvious non-parallel effect with the filter photometer readings rising very much more rapidly compared with the spectro-photometer readings. The lines appear to cross near the origin. Both fitted equations are highly significant on a statistical basis based on the following information: spectrophotometer-absorbance = 0.0125 + 0.0022 concentration, $t = 24.2$, $F = 584.4$, $p = 0.0002$, $R^2 = 99.5\%$; filter photometer-absorbance = -0.0056 + 0.0071 concentration, $t = 67.9$, $F = 4607.9$, $p < 0.0001$, $R^2 = 99.9\%$. The two equations differ in both slope and intercept. Use of Test 1 produces a test statistic (5.17) of 1277.6 ($p < 0.01$) indicative that the two data sets must be treated separately. Test 2 produces a test statistic (5.20) of 1292.1 ($p < 0.01$) providing evidence of the non-parallel nature of the two lines. The point of intersection is estimated to be (3.665 μg ml^{-1}, 0.0204).

Exercise 5.4

The MLR model fitted to the fluorescence data is fluorescence = $\mu + \beta_1$ SVRS + β_2 pH + β_3 Heating time + β_4 Cooling time + β_5 Delay time + ε. Statistically, the MLR model of fluorescence = f(SVRS, pH, HT, CT, DT) is highly significant ($F = 19.42$, $p = 0.000 < 0.001$, $R^2_{adj} = 85.2\%$, $s = 32.51$) with no evidence of collinearity between the regressor variables (no correlations exceed 0.6 numerically). Both best subsets and stepwise indicate two possible 'best' models: the original five variable one ($C_p = 6$) and a four variable model fluorescence = f(SVRS, pH, CT, DT) dropping HT. The latter provides $R^2_{adj} = 85.5\%$, $C_p = 4.7$, and $s = 32.16$ corresponding to a small increase in R^2_{adj} and small decrease in s. Comparing the predictive ability of each model shows that the full model is possibly better though predictions are reasonable on only a few occasions. Diagnostic checking reveals no distinct trends. Perhaps inclusion of interaction or power terms involving the factors could improve validity of model fit.

Exercise 6.1

The data are to be tested for method difference, a general directional difference, so the alternative hypothesis will be difference in methods (two-sided hypothesis). There is a strong overlap between the two sets of results though the Ibuprofen figures are more variable. These interpretations are reflected in the similarity of medians (188.63 for Internal and 189.15 for Ibuprofen) and in the differences in the *RSD*s (1.7% for Internal and 3.4% for Ibuprofen) though the latter are low, indicative of consistent measurements. Testing for difference in medians using the Mann–Whitney test statistic (6.2) results in acceptance of the null hypothesis indicating that there appears insufficient evidence in favour of a difference ($S = 64$, $m = 8$, $T = 28$, $p > 0.05$).

Exercise 6.2

Since difference in accuracy, determined as $D = $ Butylphenol $-$ AFNor, is the basis of the assessment, the associated alternative hypothesis will refer to method difference (two-sided). The data plot indicates very strong similarity in the measurements of the two procedures in respect of both accuracy and precision. The differences (Butylphenol $-$ AFNor) are roughly evenly split with comparable numbers of negative and positive differences occurring with a median difference of 0.3 reported. The Wilcoxon signed-ranks test statistic (6.4) is equal to 35 with $n = 10$ as there are no non-zero differences. Comparing this with the appropriate 5% critical values leads to acceptance of the null hypothesis and the conclusion that there appears insufficient evidence of a difference in procedure accuracy ($p > 0.05$).

Exercise 6.3

Exploratory analysis highlights that 'pH > 8' appears to have most effect on recovery of propazine, with the measurements collected lower, though variability appears similar for each pH level tested. The summaries show different mean recoveries (89.45, 87.58, 82.63) and minor differences in *RSD*s (3%, 2.5%, 2.7%), the latter not sufficient for precision differences to be indicated. The test statistic (6.7) is 10.85 which is significant at the 5% level. Application of Dunn's procedure using a 10% experimentwise error rate ($\alpha = 0.1$, $z_{0.0167} = 2.1272$) results in the difference between 'pH > 8' adjustment and the others being statistically significant. Only 'no adjustment' and 'pH = 5' show no evidence of statistical difference.

Exercise 6.4

The results obtained show no real difference between analysts with all providing similar numerical measurements though analyst C is perhaps the most consistent. The median potash measurements are identical for analysts B and C and almost the same for A (15.2, 15.25, 15.25). The *RSD*s differ more (1.8%, 2.6%, 1.1%) indicative of different precision in the analyst's measurements. The test statistic (6.11) is 2.58 which is less than the corresponding 5% critical of $\chi^2_{0.05,2} = 5.99$ indicating no evidence to suggest a statistical difference between the analysts.

Exercise 6.5

The data presented have mean 48.87%, median 50%, and standard deviation 7.94%. The marginal difference between mean and median suggests normality could be in doubt. The normal plot provides a wave-shape which has a positive trend but which is marginally different from linearity. In conclusion, it would appear there is some doubt about normality for the impurity data as might be expected considering the nature of the measured response.

Exercise 7.1

The normal plot of effect estimates does not conform to the hoped for trend with only the E-123 effect looking to be unimportant. All other effects look important and worthy of investigation. Assessing the main effects plot shows that increasing the amount of both E-124 and E-120 causes recovery of Ponceau 4R to decrease markedly. Both the interactions E-123 × E-124 and E-123 × E-120 show a crossover effect indicative of differing influence on recovery of factor-level combinations. Best recovery results appear to occur when all factors are set at 12 mg l^{-1} (low level).

Exercise 7.2

From the normal plot of effect estimates, effects A, A × D, D, E, and D × E stand out as very different from the unimportant effects. The dotplot of effect estimates specifies the same conclusion. The main effects plot highlights the positive effect of A and D and the negative effect of E on enzyme activity when increasing the factor levels. For the A × D interaction, when changing levels of D, the low level of A has no effect, while the high level of A causes enzyme activity to rise. A similar comparison for the D × E interaction, based on changing levels of D, shows only the low level of E causing a substantive increase in enzyme activity. All elements of a proposed model enzyme activity $= A + A \times D + D + D \times E + E +$ error are highly significant ($p < 0.01$) providing evidence that the effects arrived at in the analysis are statistically important in their effect on level of enzyme activity measured. Diagnostic checking revealed column length differences for A, C, D, and E suggesting important influence of these factors on variability of enzyme activity. Fits and normal plots hint at two outliers, but suggest acceptable fit and response normality.

Exercise 8.1

The data set comprises $n = 20$ samples measured over $p = 7$ variables. PC_1 accounts for 32.1% of variation, PC_2 22%, PC_3 17.5%, PC_4 11.2%, PC_5 9.2%, PC_6 6.6%, and PC_7 1.4%. Cumulatively, at least the first four PCs would be required to explain the data adequately (82.7% contribution). For PC_1, CO, NO, HC, and NO_2 dominate with similar positive weights. All represent pollutants generated by human activity so PC_1 may be a measure of human generated air pollution. PC_2 is dominated by SolRad, Wind, and O_3 though Wind is of opposite sign to the other two variables, suggesting that the PC may be a weather/atmosphere component explaining how pollutants are broken down and dispersed. The PC scores plot for PC_2 against PC_1 hints at some clustering of samples 2 and 3, 8 and 16, 1 and 12, and 5 and 14 though these are few in number.

Exercise 8.2

All samples, bar one (system B sample), appear correctly classified giving a proportion correct of 95.2%. The distance measures show definite difference between systems A and B ($D^2 = 29.0983$), and A and C ($D^2 = 13.7607$) but not B and C ($D^2 = 3.6344$). Statistical validity of the SDA routine developed is acceptable for discrimination between A and B ($F = 13.58$, $p < 0.01$) and between A and C ($F = 6.42$, $p < 0.05$). However, validity of discrimination between B and C cannot be accepted ($F = 1.70$, $p > 0.1$). Checking mis-classifications shows that observation 12 (sample 5 from system B) is the mis-classified observation though the difference in distance measures is small (3.335 for B and 3.083 for C) and probabilities similar (0.408 for B and 0.531 for C). One other observation (sample 1 from system C) is a dubious classification as B and C distance measures (5.995 and 5.808) and probabilities (0.476 and 0.522) do not differ substantially. Predictions for the A and B specimens agree with original groupings with distance measures and probabilities very different. The prediction for the C specimen does predict system C but the distance measures (15.03 for B and 14.82 for C) and probabilities (0.459 for B and 0.509 for C) do not differ sufficiently to be sure that the prediction is valid.

Subject Index